D1674565

Edited by
Roland Kontermann

Therapeutic Proteins

Related Titles

Wang, W., Roberts, C. J. (eds.)

Aggregation of Therapeutic Proteins

2010
ISBN: 978-0-470-41196-4

Dübel, S. (ed.)

Handbook of Therapeutic Antibodies

Technologies, Emerging Developments and Approved Therapeutics

2010
ISBN: 978-3-527-32902-1

Jorgenson, L., Nielson, H. M. (eds.)

Delivery Technologies for Biopharmaceuticals

Peptides, Proteins, Nucleic Acids and Vaccines

2010
ISBN: 978-0-470-72338-8

An, Z. (ed.)

Therapeutic Monoclonal Antibodies

From Bench to Clinic

2009
ISBN: 978-0-470-11791-0

Jensen, K. (ed.)

Peptide and Protein Design for Biopharmaceutical Applications

2009
ISBN: 978-0-470-31961-1

Walsh, G. (ed.)

Post-translational Modification of Protein Biopharmaceuticals

2009
ISBN: 978-3-527-32074-5

Behme, S.

Manufacturing of Pharmaceutical Proteins

From Technology to Economy

2009
ISBN: 978-3-527-32444-6

Edited by Roland Kontermann

Therapeutic Proteins

Strategies to Modulate Their Plasma Half-Lives

WILEY-BLACKWELL

The Editor

Prof. Dr. Roland Kontermann
University of Stuttgart
Institute of Cell Biology and Immunology
Allmandring 31
70569 Stuttgart
Germany

Cover
Syringe: Corbis Images

■ **Limit of Liability/Disclaimer of Warranty:** While the publisher and author have used their best efforts in preparing this book, they make no representations or warranties with respect to the accuracy or completeness of the contents of this book and specifically disclaim any implied warranties of merchantability or fitness for a particular purpose. No warranty can be created or extended by sales representatives or written sales materials. The Advice and strategies contained herein may not be suitable for your situation. You should consult with a professional where appropriate. Neither the publisher nor authors shall be liable for any loss of profit or any other commercial damages, including but not limited to special, incidental, consequential, or other damages.

Library of Congress Card No.: applied for

British Library Cataloguing-in-Publication Data
A catalogue record for this book is available from the British Library.

Bibliographic information published by the Deutsche Nationalbibliothek
The Deutsche Nationalbibliothek lists this publication in the Deutsche Nationalbibliografie; detailed bibliographic data are available on the Internet at <http://dnb.d-nb.de>.

© 2012 Wiley-VCH Verlag & Co. KGaA, Boschstr. 12, 69469 Weinheim, Germany

Wiley-Blackwell is an imprint of John Wiley & Sons, formed by the merger of Wiley's global Scientific, Technical, and Medical business with Blackwell Publishing.

All rights reserved (including those of translation into other languages). No part of this book may be reproduced in any form – by photoprinting, microfilm, or any other means – nor transmitted or translated into a machine language without written permission from the publishers. Registered names, trademarks, etc. used in this book, even when not specifically marked as such, are not to be considered unprotected by law.

Cover Design   Formgeber, Eppelheim
Typesetting   Toppan Best-set Premedia Limited, Hong Kong
Printing and Binding   Markono Print Media Pte Ltd, Singapore
Printed in Singapore
Printed on acid-free paper

Print ISBN: 978-3-527-32849-9
ePDF ISBN: 978-3-527-64479-7
oBook ISBN: 978-3-527-64482-7
ePub ISBN: 978-3-527-64478-0
Mobi ISBN: 978-3-527-64480-3

Contents

Preface

At the end of 2011, roughly 200 biologics were approved for therapeutic applications and more than 600 were under clinical development. Many of these protein drugs, such as hormones, growth factors, cytokines, coagulation factors, and enzymes, are small in size and are rapidly cleared from circulation. Half-life extension strategies have therefore become increasingly important to improve the pharmacokinetic and pharmacodynamic properties of protein therapeutics, but also for reasons of compliance. Several half-life extension strategies are already utilized in approved drugs, including PEGylation, hyperglycosylation, binding to human serum albumin, and fusion to an immunoglobulin G (IgG) Fc region. However, there is a strong need for new strategies not only to further improve the pharmacokinetic properties but also to facilitate production and application of these half-life extended drugs. These strategies include those that increase the hydrodynamic radius of the drug, thus aiming at reducing the renal clearance, but also strategies that implement recycling by the neonatal Fc receptor (FcRn), which is responsible for the extraordinary long half-life of IgG molecules and serum albumin. In the past 5 to 10 years the field has experienced a rapid growth in the establishment of novel half-life extension strategies, including the application of novel hydrophilic polymers, the generation of recombinant PEG mimic polypeptide chains, and the development of various albumin-binding molecules. Furthermore, the half-life of IgG molecules was altered by engineering of the Fc region, which opens new opportunities for the development of next-generation antibody drugs.

This book is written by renowned experts in the field and is intended to provide a comprehensive overview of the various established but also emerging half-life extension strategies. It can be expected that in the near future many of these technologies will be evaluated in clinical trials and become established strategies to prolong the half-life and thus to improve the pharmacokinetic and pharmacodynamic properties of therapeutic proteins.

Stuttgart, October 2011 *Roland Kontermann*

Preface

List of Contributors

Gerd Bendas
University of Bonn
Pharmaceutical Department
Pharmaceutical Chemistry
An der Immenburg 4
53121 Bonn
Germany

Uli Binder
XL-protein GmbH
Lise-Meitner-Strasse 30
85354 Freising
Germany

Alan J. Bitonti
Biogen Idec Hemophilia
9 Fourth Avenue
Waltham, MA 02451
USA

Graeme J. Carroll
University of Notre Dame
Australia (Fremantle)
Department of Rheumatology
Fremantle Hospital and University of
Western Australia
Perth, WA 6959
Australia

Chen Chen
Imperial College London
Department of Life Sciences
Exhibition Road
London SW7 2AZ
UK

Antony Constantinou
Imperial College London
Department of Life Sciences
Exhibition Road
London SW7 2AZ
UK

Mahendra P. Deonarain
Imperial College London
Department of Life Sciences
Exhibition Road
London SW7 2AZ
UK

Jennifer A. Dumont
Biogen Idec Hemophilia
9 Fourth Avenue
Waltham, MA 02451
USA

Fuad Fares
Department of Biology and
Human Biology
Faculty of Natural Sciences
University of Haifa
Mount Carmel, Haifa 31095
Israel

Fredrik Y. Frejd
Affibody AB
Gunnar Asplunds Allé 24
SE-17163 Solna
Sweden
and
Uppsala University
Unit for Biomedical Radiation
Sciences
Department of Oncology
Radiology and Clinical Immunology
Rudbeck Laboratory
75185 Uppsala
Sweden

Astrid Hartung
University of Bonn
Pharmaceutical Department
Pharmaceutical Chemistry
An der Immenburg 4
53121 Bonn
Germany

Christopher Herring
Biopharm Research and
Development
GlaxoSmithKline

Thomas Hey
Fresenius Kabi Deutschland GmbH
Pfingstweide 53
61169 Friedberg
Germany

Jalal A. Jazayeri
School of Biomedical Sciences
Charles Sturt University
Wagga Wagga, NSW 2678
Australia

Simona Jevševar
Sandoz Biopharmaceuticals
Lek Pharmaceuticals d.d.
Kolodvorska 27
SI-1234 Menges
Slovenia

Xiaomei Jin
Biogen Idec Hemophilia
9 Fourth Avenue
Waltham, MA 02451
USA

Jonghan Kim
Harvard School of Public Health
Department of Genetics and
Complex Diseases
665 Huntington Avenue
Boston, MA 02115
USA

Helmut Knoller
Fresenius Kabi Deutschland GmbH
Pfingstweide 53
61169 Friedberg
Germany

Roland E. Kontermann
University of Stuttgart
Institute of Cell Biology and
Immunology
Allmandring 31
70569 Stuttgart
Germany

Menči Kunstelj
Sandoz Biopharmaceuticals
Lek Pharmaceuticals d.d.
Kolodvorska 27
SI-1234 Menges
Slovenia

Katharina Landfester
Max Planck Institute for Polymer
Research
Ackermannweg 10
55128 Mainz
Germany

Alvin Luk
Biogen Idec Hemophilia
9 Fourth Avenue
Waltham, MA 02451
USA

Volker Mailänder
Max Planck Institute for Polymer
Research
Ackermannweg 10
55128 Mainz
Germany

Bernd Meibohm
University of Tennessee Health
Science Center
College of Pharmacy
Departments of Pharmaceutical
Sciences
874 Union Avenue
Suite 5p
Memphis, TN 38163
USA

Hubert J. Metzner
CSL Behring GmbH
Preclinical Research and Development
Department
Emil-von-Behring-Str. 76
35041 Marburg
Germany

Anna Musyanovych
Max Planck Institute for Polymer
Research
Ackermannweg 10
55128 Mainz
Germany

Dario Neri
Swiss Federal Institute of Technology
(ETH)
Institute of Pharmaceutical Sciences
Department of Chemistry and Applied
Biosciences
Wolfgang-Pauli-Strasse 10
8093 Zurich
Switzerland

Raimund J. Ober
University of Texas Southwestern
Medical Center
Department of Immunology
6000 Harry Hines Blvd.
Dallas, TX 75390
USA
University of Texas at Dallas
Department of Electrical Engineering
Richardson, TX 75083
USA

Robert T. Peters
Biogen Idec Hemophilia
9 Fourth Avenue
Waltham, MA 02451
USA

Glenn F. Pierce
Biogen Idec Hemophilia
9 Fourth Avenue
Waltham, MA 02451
USA

Joerg Scheuermann
Swiss Federal Institute of Technology
(ETH)
Institute of Pharmaceutical Sciences
Department of Chemistry and Applied
Biosciences
Wolfgang-Pauli-Strasse 10
8093 Zurich
Switzerland

Oliver Schon
Biopharm Research and
Development
GlaxoSmithKline

Stefan Schulte
CSL Behring GmbH
Preclinical Research and Development
Department
Emil-von-Behring-Str. 76
35041 Marburg
Germany

Arne Skerra
XL-protein GmbH
Lise-Meitner-Strasse 30
85354 Freising
Germany
and
Technische Universität München
Lehrstuhl für Biologische Chemie
Emil-Erlenmeyer-Forum 5
85350 Freising-Weihenstephan
Germany

Sabrina Trüssel
Swiss Federal Institute of Technology
(ETH)
Institute of Pharmaceutical Sciences
Department of Chemistry and Applied
Biosciences
Wolfgang-Pauli-Strasse 10
8093 Zurich
Switzerland

Peter Vorstheim
Fresenius Kabi Deutschland GmbH
Pfingstweide 53
61169 Friedberg
Germany

E. Sally Ward
University of Texas Southwestern
Medical Center
Department of Immunology
6000 Harry Hines Blvd.
Dallas, TX 75390
USA

Thomas Weimer
CSL Behring GmbH
Preclinical Research and Development
Department
Emil-von-Behring-Str. 76
35041 Marburg
Germany

**Part One
General Information**

1
Half-Life Modulating Strategies – An Introduction

Roland E. Kontermann

1.1
Therapeutic Proteins

With roughly 200 biologics approved for therapeutic applications and more than 600 under clinical development [1], biotechnology products cover an increased proportion of all therapeutic drugs. Besides monoclonal antibodies and vaccines, which account for more than two-thirds of these produces, hormones, growth factors, cytokines, fusion proteins, coagulation factors, enzymes and other proteins are listed. An overview of the different classes of currently approved protein therapeutics is shown in Table 1.1. Except for antibodies and Fc fusion proteins, many of these proteins possess a molecular mass below 50 kDa and a rather short terminal half-life in the range of minutes to hours. In order to maintain a therapeutically effective concentration over a prolonged period of time, infusions or frequent administrations are performed, or the drug is applied loco-regional or subcutaneously utilizing a slow adsorption into the blood stream. These limitations of small size protein drugs has led to the development and implementation of half-life extension strategies to prolong circulation of these recombinant antibodies in the blood and thus improve administration and pharmacokinetic as well as pharmacodynamic properties.

1.2
Renal Clearance and FcRn-Mediated Recycling

The efficacy of protein therapeutics is strongly determined by their pharmacokinetic properties, including their plasma half-lives, which influence distribution and excretion. Although a small size facilitates tissue penetration, these molecules are often rapidly cleared from circulation. Thus, they have to be administered as infusion or repeated intravenous (i.v.) or subcutaneous (s.c.) bolus injections in order to maintain a therapeutically effective dose over a prolonged period of time,

Therapeutic Proteins: Strategies to Modulate Their Plasma Half-Lives, First Edition. Edited by Roland Kontermann.
© 2012 Wiley-VCH Verlag GmbH & Co. KGaA. Published 2012 by Wiley-VCH Verlag GmbH & Co. KGaA.

Table 1.1 Proteins used as therapeutics.

Protein class	Protein	Indication	Examples of approved drugs	M_r (kDa)	Terminal half-life
Hormones	Insulin	Diabetes	Humalog, Novolog	6	4–6 min
	hGH	Growth disturbance	Protropin, Humatrope,	22	2 h
	FSH	Infertility	Follistim, Fertavid	30	3–4 h
	Glucagon-like peptide 1	Type 2 diabetes	Victoza	4	2 min
	Parathyroid hormone	Osteoporosis	Preotach	10	4 min
	Calcitonin	Osteoporosis	Fortical	4	45–60 min
	Lutropin	Infertility	Luveris	23	20 min
	Glucagon	Hypoglycemia	Glucagon	4	3–6 min
Growth factors	Erythropoietin	Anemia	Epogen, Procrit	34	2–13 h
	G-CSF/GM-CSF	Neutropenia	Filgrastim,	20	4 h
	IGF-1	Growth failure	Increlex	8	10 min
Interferons	IFN-α	Hepatitis C (and B)	Roferon, Infergen	20	2–3 h
	IFN-β	Multiple sclerosis	Betaferon, Avonex	23	5–10 h
	IFN-γ	Chr. granulomatosis	Actimmune	25	30 min
Interleukins	IL-2	Renal cell carcinoma	Proleukin	16	5–7 min
	IL-11	Thrombocytopenia	Neumega	23	2 d
	IL-1Ra	Rheumatoid arthritis	Kineret	25	6 min
Coagulation factors	Factor VIII	Hemophilia A	Kogenate, ReFacto		12 h[a]
	Factor IX	Hemophilia B	Benefix	55	18–24 h
	Factor VIIa	Hemophilia	Novoseven	50	2–3 h
	Thrombin	Bleeding during surgery	Recothrom	36	2–3 d

Table 1.1 Continued

Protein class	Protein	Indication	Examples of approved drugs	M_r (kDa)	Terminal half-life
Thrombolytics and anti-coagulants	t-PA	Myocardial infarction	Tenecteplase	65	2–12 min
	Hirudin	Thrombozytopenia	Refludan	7	3 h
	Activated protein C	Severe sepsis	Xigris	62	1–2 h
Enzymes	α-glucosidase	Pompe disease	Myozyme, Lumizyme	109	2–3 h
	Glucocerebrosidase	Gaucher disease	Cerezyme	60	18 min
	Iduronate-2-sulfatase	Mucopolysaccharidose II	Elaprase	76	45 min
	Galactosidase	Fabry disease	Fabrazyme, Replagal	100	1–2 h
	Urate oxidase	Hyperuricemia	Fasturtec	140	17–19 h
	DNase	Cystic fibrosis	Pulmozyme	37	n.a.
Antibodies and antibody fragments	IgG	Cancer, inflammatory and infectious diseases, transplantation, etc.	Rituxan, Hereptin, Avastin, Remicade, Humira, Synagis, Zenapax, Xolair, etc.	150	days to weeks
	Fab	Prevention of blood clotting, AMD	ReoPro, Lucentis	50	30 min
Fusion proteins	TNFR2-Fc	Rheumatoid arthritis	Enbrel	150	3–6 d
	TMP-Fc	Thrombocytopenia	Nplate	60	1–34 d
	CTLA-4-Fc	Rheumatoid arthritis	Orenica	92	8–25 d
	IL-1R-Fc	CAPS	Arcalyst	251	9 d
	LFA-3-Fc	Plaque psoriasis	Amevive	92	11 d
	IL-2-DT	Cut. T-cell leukemia	Ontak	58	70–80 min

a) Bound to vWF; n.a., not available.

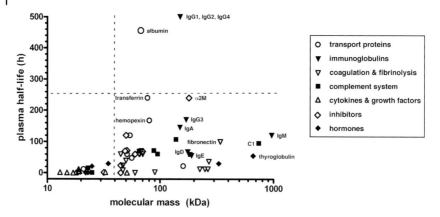

Figure 1.1 The molecular mass and half-life of plasma proteins. Proteins are allocated according to their function.

or are restricted to loco-regional treatment. The rapid elimination of these small molecules mainly occurs by renal filtration and degradation [2] (see also Chapter 2). A comparison of the half-lives of plasma proteins reveals the threshold for rapid excretion to be in the range of approximately 40–50 kDa, demonstrating that the size of the molecules is one of the determining factors (Figure 1.1). The glomerular filtration barrier is formed by the fenestrated endothelium, the glomerular basement membrane (GBM) and the slit diaphragm located between the podocyte foot processes [3]. The fenestrae between the glomerular endothelial cells have diameters between 50–100 nm, thus, allowing free diffusion of molecules. It was suggested that the slit diaphragm represents the ultimate macromolecular barrier, forming an isoporous, zipper-like filter structure with numerous small, 4–5 nm diameter pores and a lower number of 8–10 nm diameter pores [4–6]. In addition to size, the charge of a protein contributes to renal filtration. It has been suggested the proteoglycans of the endothelial cells and the GBM contribute to an anionic barrier, which partially prevents the passage of plasma macromolecules [3]. Consequently, the size of a protein therapeutic, that is, its hydrodynamic radius, and also its physiochemical properties, that is, charge, represent starting points in order to improve half-life. Interestingly, two kinds of molecules, serum albumin and IgGs, exhibit an extraordinary long half-life in humans. Thus, human serum albumin (HSA) has a half-life of 19 days and immunoglobulins (IgG1, IgG2 and IgG4) have half-lives in the range of 3 to 4 weeks [7, 8]. These long half-lives, which clearly set albumin and IgG apart from the other plasma proteins (Figure 1.1), are caused by a recycling process mediated by the neonatal Fc receptor (FcRn) [9–11] (see also Chapters 8 and 11). FcRn, expressed for example by endothelial cells, is capable of binding albumin and IgGs in a pH-dependent manner. Thus, after cellular uptake of plasma proteins through macropinocytosis, albumin and IgG will bind to FcRn in the acidic environment of the endosomes. This binding diverges albumin and IgG from degradation in the lysosomal compartment and redirects them to the plasma membrane, where they are released back into the blood plasma because of the neutral pH. This offers additional opportunities to extend or modulate the half-life of proteins, for example, through fusion to albumin or the Fc region of IgG [12].

1.3
Strategies to Modulate Plasma Half-Life

With an increasing number of protein therapeutics approved and developed (see Table 1.1) including various alternative antibody formats [13], many of them exhibiting a short plasma half-life, strategies to improve the pharmacokinetic properties are becoming increasingly important [14]. Based on the parameters described above which influence half-life, the strategies to extend half-lives of therapeutic proteins can be divided into those that (i) utilize an increased size and thus the hydrodynamic volume and (ii) in addition implement recycling by the neonatal Fc receptor (FcRn), the receptor that is responsible for the long half-lives of IgGs and albumin. These strategies comprise a variety of different approaches including chemical coupling of polymers and carbohydrates, post-translational modifications such as N-glycosylation, and fusion to recombinant polymer mimetics. Furthermore, conjugation, binding or fusion to an Fc region or serum albumin, respectively, results not only in an increased size but also incorporates FcRn-mediated recycling (Figure 1.2; Tables 1.2 and 1.3). More recently, nanoparticulate formulations have also been developed to improve the half-life and biodistribution of

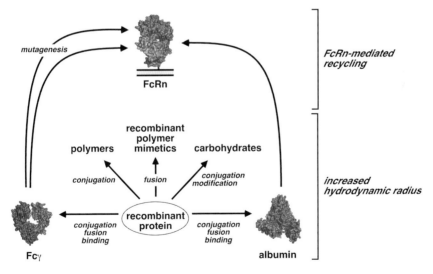

Figure 1.2 Overview of the different strategies to extend the plasma half-life of recombinant proteins, aiming at increasing the hydrodynamic radius as well as implementing FcRn-mediated recycling.

Table 1.2 Strategies to modulate half-life of therapeutic proteins.

Strategy	Modification	Effect on hydrodynamic radius	Effect on FcRn-mediated recycling	Effect on half-life
PEGylation	Chemical conjugation of methoxy polyethylene glycol (mPEG)	Increased	None	Prolonged, depending on PEG size and structure
Polysialylation	Chemical conjugation or attached by post-translational modification of polysialic acid	Increased (plus change of pI)	None	Prolonged, depending on extent of polysialylation
HESylation	Chemical conjugation of hydroxyethyl starch (HES)	Increased	None	Prolonged, depending on HES size and structure
Recombinant PEG mimetics	Genetic fusion of flexible, hydrophilic amino acid chains	Increased	None	Prolonged, depending on size and composition
N-glycosylation	Post-translational attachment of *N*-glycans	Increased	None	Moderately prolonged
O-glycosylation	Post-translational attachment of *O*-glycans	Increased	None	Moderately prolonged
Fc fusion	Genetic fusion of IgG Fc region	Increased	Utilized	Prolonged
Engineered Fc	Mutations introduced into the Fc region of IgGs or Fc fusion proteins	Unaltered	Increased or decreased (depending on mutations)	Prolonged or reduced
IgG binding	Genetic fusion or conjugation to IgG-binding moieties	Increased upon binding to IgG	Utilized or diminished (depending on binding site)	Prolonged
Albumin fusion	Genetic fusion to serum albumin	Increased	Utilized	Prolonged

Table 1.2 Continued

Strategy	Modification	Effect on hydrodynamic radius	Effect on FcRn-mediated recycling	Effect on half-life
Albumin binding	Genetic fusion or conjugation to albumin-binding moieties (peptides, protein domains, antibody fragments, antibody mimetic scaffolds, small chemicals, fatty acids, etc.)	Increased after binding to albumin	Utilized	Prolonged
Albumin coupling	Chemical conjugation to cysteine 34	Increased	Utilized	Prolonged
Nanoparticles	Encapsulation into or conjugation to nanoparticles (e.g., liposomes, polymeric capsules)	Strongly increased by the size of the nanoparticles (50–200 nm)	None	Prolonged (slow release)

therapeutic proteins, for example, through encapsulation into liposomes or polymeric capsules [15].

1.3.1
Strategies to Increase the Hydrodynamic Radius

The foremost approach in improving the half-life of a protein therapeutic is to reduce the renal clearance rate, for example, by increasing the size above the renal cut-off of 40–50 kDa. This can be achieved by several ways including chemical and post-translational modification as well as genetic engineering (Table 1.2).

PEGylation, that is, the chemical conjugation of polyethylene glycol (PEG), mainly of its methoxy derivative methoxy polyethylene glycol (mPEG), was established more then two decades ago [16] (see also Chapter 3). In 1990, the first PEGylated protein drug (pegadamase for the treatment of severe combined immunodeficiency [SCID]) was approved by the FDA [17, 18] and since then eight more PEGylated protein drugs have been approved, including enzymes, interferon-α2b, G-CSF, hGH, erythropoetin, and a Fab fragment [19, 20] (see Table 1.3). PEG is composed of ethylene oxide units connected in a linear or branched configuration

Table 1.3 Half-life extended protein therapeutics (approved or in clinical trials).

Modification/ Protein	Drug	Indication	Status
PEGylation			
Adenosine deaminase	Pegademase bovine (Adagen)	Severe combined immunodeficiency disease (SCID)	Approved 1990
L-asparaginase	Pegaspargase (Oncaspar)	Acute lymphoblastic leukemia	Approved 1994
Interferon α-2b	Peginterferon alfa-2b (Peg-Intron)	Hepatitis C	Approved 2000
Interferon α-2b	Peginterferon alfa-2b (Pegasys)	Hepatitis C	Approved 2002
G-CSF	Pegfilgrastim (Neulasta)	Chemotherapy-induced neutropenia	Approved 2002
Human growth hormone (hGH)	Pegvisomant (Somavert)	Acromegaly	Approved 2002
Erythropoietin	mPEG-epoetin beta (Mircera)	Anemia	Approved 2007
Anti-TNF Fab'	Certolizumab pegol (Cimzia)	Crohn's disease	Approved 2008
Uricase	Pegloticase (Krystexxa)	Chronic gout	Approved 2010
EPO mimetic peptide	Peginesatide (Hematide)	Anemia	Phase II
Phenylalanine ammonia lyase	rAvPAL-PEG	Phenylketonuria	Phase II
IL-29	PEG-rIL-29	Hepatitis C	Phase I
Coagulation factor IX	PEG-rFIX	Hemophilia B	Phase I
Arginase	PEG-rArgI	Liver cancer	Phase I
Hyaluronidase	PEGPH20	Advanced solid tumors	Phase I
N-glycosylation and polysialylation			
Erythropoietin	Darbepoetin alfa (Aranesp)	Anemia	Approved 2008
Erythropoietin	PSA-EPO (ErepoXen)	Anemia	Phase II
Insulin	PSA-insulin (SuliXen)	Diabetes mellitus	Phase I

Table 1.3 Continued

Modification/ Protein	Drug	Indication	Status
Albumin fusion			
Interferon α-2b	Albinterferon alfa-2b (Joulferon, Zalbin)	Hepatitis C	Phase III
Coagulation factor IX	rIX-FP	Hemophilia B	Phase I
HER2 + HER3 specific single-chain Fv	MM-111 (scFv-HSA-scFv)	Cancer	Phase I/II
Albumin binding through a conjugated fatty acid chain			
Insulin	Insulin detemir (Levemir)	Diabetes mellitus	Approved 2003
Glucagon-like peptide-1	Liraglutide (Victoza)	Diabetes mellitus type 2	Approved 2009
Fc fusion proteins			
TNF receptor 2	Etanercept (Enbrel)	Rheumatoid arthritis, ankylosing spondylitis, polyarticular juvenile idiopathic arthritis, psoriatic arthritis, plaque psoriasis	Approved 1998
LFA-3	Alefacept (Amevive)	Severe chronic plaque psoriasis	Approved 2003
CTLA-4	Abatacept (Orenica)	Rheumatoid arthritis, juvenile idiopathic arthritis	Approved 2005
IL-1R	Rilonacept (Arcalyst)	Cryopyrin-associated periodic syndromes	Approved 2008
TPO-mimetic peptide	Romiplostim (Nplate)	Chronic idiopathic thrombocytopenic purpura	Approved 2008
VEGF receptor	Aflibercept (VEGF trap)	Macular degeneration	Phase III
BR3	Briobacept (BR3-Fc)	Rheumatoid arthritis	Phase II
Coagulation factor IX	rFIXFc	Hemophilia B	Phase II/III
Coagulation factor VIII	rFVIII-Fc	Hemophilia A	Phase I

and of varying length. In the approved drugs, one or several PEG chains of 5 to 40 kDa are conjugated. Because PEG is highly hydrophilic, the molecular mass of PEGylated proteins is drastically increased through the binding of water molecules, reflected by a strong increase of the hydrodynamic radius. Different coupling methods have been established including random and site-directed approaches. The PEGylation strategy, that is, the site of PEGylation as well as the number and size of attached PEG chains, has to be carefully chosen in order to avoid a reduction or abrogation of the activity of the therapeutic protein [18]. Ideally, a single PEG chain is conjugated in a site-directed manner, for example, through the use of existing or genetically introduced cysteine residues. This is exemplified by Cimzia (certolizumab pegol), a bacterially produced anti-TNF Fab' fragment, where a 40 kDa PEG chain is attached to the free cysteine at the C-terminus of the heavy chain Fd chain, that is, opposite the antigen-binding site [21]. In general, PEGylation of proteins is considered to be safe and well tolerated [20], although in animals the occurrence of renal tubular vacuolization has been observed due to accumulation of the nondegradable PEG chains in the kidney.

Recently, alternative strategies to PEGylation have been established. For example, it was noted that the polypeptide backbone resembles, at least in part, the structure of PEG. This has led to the generation of long polypeptide chains of a hydrophilic and flexible structure, which can be genetically fused to recombinant proteins [22, 23] (see also Chapter 4). Thus, chemical conjugation and additional purification steps are avoided. Furthermore, the fused polypeptide chains are biodegradable and have also been shown to be immunologically inert. *In vivo* studies established that these recombinant PEG mimetics behave in a similar manner to PEG, that is, they result in a drastic increase of the hydrodynamic radius and extension of the half-life.

As an alternative, carbohydrate chains can be attached to therapeutic proteins. Because this process takes place naturally, for example, by post-translational modifications in mammalian cells, therapeutic proteins have been genetically modified to contain additional *N*- or *O*-glycosylation sites [24–26] (see Chapter 5). For example, *N*-glycosylation sites (Asn-X-Thr) can be introduced into the protein sequence of interest resulting in hyperglycosylated proteins (Table 1.3). A prominent example is darbepoetin alfa (Aranesp), a hyperglycosylated derivative of human erythropoietin containing two additional *N*-glycosylation sites [27]. Compared with recombinant human erythropoietin, darbepoetin alfa has a threefold longer terminal half-life and a strongly increased capacity to elicit an erythropoietic response [24]. Similarly, sequences obtained from *O*-glycosylated proteins, for example, from chorionic gonadotropin, have been fused to therapeutic proteins resulting in *O*-glycosylated proteins with prolonged half-life and improved bioactivity [28] (see Chapter 5).

Carbohydrates have also been chemically conjugated to therapeutic proteins in a similar way to the PEGylation approach (Table 1.3). For example, polysialic acid (PSA) has been investigated as an alternative to PEG [29, 30]. PSA is found on the surface of a variety of cells including mammalian cells, thus is a biocompatible and biodegradable natural polymer. Colominic acid, a linear polymer of α-(2,8)-

linked *N*-acetylneuraminic acid, was used for polysialylation of various proteins including asparaginase, insulin and antibody fragments and was shown to be capable of prolonging their half-lives [31] (see also Chapter 6).

Another carbohydrate structure, which has been established for half-life extension of therapeutic proteins, is hydroxyethyl starch (HES). HES is a modified, branched amylopectin, for example, isolated from waxy maize starch, composed of glucose units linked by α-1,4- and α-1,6-glycosidic bonds. HES is an approved plasma volume expander with a proven safety record. Because of its close similarity to glycogen, HES is not immunogenic [32, 33]. The size and structure of HES and thereby its stability can be adjusted by acidic hydrolysis and by chemical hydroxyethylation at positions 2, 3, and 6 of the glucose unit. The HESylation technology was pioneered by the company Fresenius Kabi to improve the pharmacokinetic and pharmacodynamic properties of therapeutic proteins. For example, it was successfully applied to produce improved derivatives of erythropoietin by chemical coupling of a 60 kDa HES (see also Chapter 7).

Another obvious approach to increase the size of a therapeutic protein is the fusion to another protein moiety, for example, a plasma protein. Mainly, fusion to immunoglobulin Fc regions or serum albumin has been utilized to extend the half-life of peptides and proteins [34, 35] (Table 1.3). Because albumin and Fc-containing proteins also utilize recycling by the FcRn, these strategies are summarized below (Section 1.3.2).

Pharmaceutical formulations of drugs are widely used to influence their pharmacokinetic properties, that is, administration, distribution, metabolism and excretion (ADME) [36, 37]. For example, small molecular weight drugs such as doxorobucin and amphothericin B have been encapsulated or incorporated, respectively, into liposomal carrier systems. In addition to liposomes, various other nanoparticulate carrier systems have been utilized to improve the pharmacokinetic and pharmacodynamic properties of proteins (see Chapters 16 and 17). The carrier systems combine various advantages. Because of their large size in the range of 50 to 200 nm, the half-life is strongly increased. Furthermore, nanoparticulate formulations influence biodistribution, for example, accumulation in tumors through an enhanced permeability and retention (EPR) in the tumor tissue [38]. In addition, they are also capable of protecting the protein drug from degradation, for example, through plasma proteases, and can act as slow release formulation, that is, as a drug depot in the body.

1.3.2
Strategies Implementing FcRn-Mediated Recycling

Fusion of a therapeutic protein (or peptide) to another protein results in a fusion protein with increased molecular mass. Mainly human plasma proteins, or fragments thereof, exhibiting per se a long half-life, are used for this purpose. This includes the Fc portion of IgG, especially of the γ1 subclass, and serum albumin. As described before, IgG and serum albumin possess a size above the renal filtration threshold and utilize recycling by the neonatal Fc receptor (FcRn).

Importantly, it has been shown that the binding sites for IgG and albumin are located at different regions of the FcRn, that is, the two molecules do not compete for the same binding site (see Chapter 8). In addition to direct fusion leading to a covalent linkage, binding to IgG or albumin has been employed for half-life extension strategies (Figure 1.2, Table 1.2). This noncovalent interaction allows for a reversible interaction, that is, dissociation from the plasma protein, which can be beneficial in respect to tissue penetration and bioactivity [39].

Therapeutic proteins and peptides have been mainly fused to the N-terminal of the Fc fragment, which often includes the hinge region in order to establish a covalent linkage between the heavy chain fragments [40] (see Chapters 9 and 10) (Table 1.3). Fusion to an Fc region increases the molecular mass by approximately 50 kDa but also results in a homodimeric molecule possessing two moieties of the therapeutic protein. This might further increase the therapeutic activity, for example, through stronger binding due to avidity effects and/or more efficient activation/neutralization of the targeted molecule [34]. This strategy has been realized for the generation of various soluble receptors (see Table 1.1), which are potent inhibitors of natural ligands such as TNF, a key mediator of inflammatory diseases, but also for ligands, for example, thrombopoietin mimetic peptides, leading to fusion proteins with strong receptor-activating properties. The binding site for the FcRn resides in the Fc region and is located between the CH2 and CH3 domains. Key residues involved in the interaction have been identified and mutants of the Fc region have been generated exhibiting either an increased or decreased affinity for FcRn at acidic pH. Incorporation of such mutations into whole IgG molecules results in antibodies with prolonged or reduced half-life [41–43] (see Chapter 11). Principally, these modifications can also be introduced in the Fc region of Fc fusion proteins in order to further improve the half-life of the therapeutic protein.

Using albumin, therapeutic proteins can be fused to either the N- or C-terminus or to both ends located at opposite sites of albumin (see Chapter 12). This strategy has been extensively applied to extend the half-life of a variety of different therapeutic proteins, including interferons, growth factors, hormones, cytokines, coagulation factors, and antibody fragments [39, 44]. Albinterferon alpha-2b (Zalbin, Joulferon) for the treatment of chronic hepatitis C is the most advanced albumin fusion proteins, which is currently in phase 3 clinical trials [45]. As an alternative to genetic fusion, therapeutic proteins can also be chemically conjugated to albumin. Because albumin exhibits a free and accessible cysteine residue at position 34, proteins, peptides but also other drugs can be chemically conjugated in a defined and site-directed manner, for example, with bifunctional crosslinkers. This strategy has, for example, been applied to extend the half-life of insulin and peptides with anti-HIV activity [46, 47].

A variety of albumin-binding strategies has been developed as an alternative for direct fusion [12]. Because serum albumin is a transport protein for different molecules, including fatty acids, such natural ligands with albumin-binding activity were used initially for half-life extension of small protein therapeutics. A prominent example is insulin determir (Levemir), where a myristyl chain is conjugated to a genetically modified insulin resulting in a long-acting insulin analog [48] (Table 1.3). Alternatively, albumin-binding synthetic peptides, for example, isolated from phage display libraries [49], or albumin-binding domains (ABD) from bacterial proteins, for example, streptococcal protein G [50], were employed for the generation of fusion proteins with albumin-binding activity (see Chapter 14). More recently, small chemicals with albumin-binding activity have been generated, which can be conjugated to therapeutic proteins and other molecules [51] (see Chapter 15). In addition, various recombinant antibody fragments (e.g., scFv, Fab, single domain antibodies, and nanobodies) or alternative scaffolds (e.g., Darpins) have been established as albumin-binding moieties [52–54] (see Chapter 13). All these strategies rely on complex formation with albumin after administration of the fusion protein into the blood stream. Several studies have shown that affinities in the micro- to nanomolar range are sufficient for half-life extension. For these strategies it is essential that binding of the fusion protein to albumin does not interfere with FcRn recycling and that binding is stable under acidic pH in order to avoid dissociation from albumin in the endosomal compartment.

1.4
Half-Life Extension Strategies Applied to a Bispecific Single-Chain Diabody – A Case Study

Bispecific antibodies were developed in the early 1980s, initially for the retargeting of immune effector cells to tumor cells for cellular cancer therapy [55]. For example, bispecific antibodies directed against a tumor-associated antigen (TAA) and CD3, which is part of the T-cell receptor complex, allow the retargeting and triggering of cytotoxic T lymphocytes leading to antibody-mediating killing of tumor cells. At that time, preparation methods included chemical crosslinking of two monoclonal antibodies or their fragments, or the use of somatic hybridization of two antibody-producing cell lines [56]. The high expectations were, however, not fulfilled in clinical trials, mainly because of low efficacy, the occurrence of severe side effects, such as cytokine storm syndrome, and the generation of a neutralizing human anti-mouse antibody response (HAMA) against the nonhuman bispecific antibodies [57]. A better understanding of effector cell biology and the implementation of antibody engineering resulted in the development of various recombinant bispecific antibody formats circumventing many of the problems associated with the original molecules. Among others, small bispecific antibodies such as tandem scFv, diabodies, and its single-chain version (single-chain diabodies) were developed and several of these molecules have meanwhile entered clinical trials, for example, for the treatment of hematologic malignancies [58]. These bispecific antibody molecules are composed only of the variable domains of two antibodies genetically connected by flexible linkers by a defined arrangement. Thus, in tandem scFvs two scFv moieties, which represent separate folding units, are connected by a middle linker of varying length in the order $(V_HA-V_LA)-(V_HB-V_LB)$. In contrast, in single-chain diabodies, which have a similar size as tandem scFv, the variable domains have the order $V_HA-V_LB-V_HB-V_LA$ [59]. The molecular mass of these antibody molecules is in the range of 50 to 60 kDa. Several studies have shown that these molecules are rapidly cleared from circulation with terminal half-lives of only a few hours. For example, a bispecific single-chain diabody (scDb) directed against carcinoembryonic antigen and CD3 (CEACD3) exhibits in mice a half-life of 1 to 2 hours [60]. Using this scDb, we have recently compared different half-life extension strategies, including PEGylation, *N*-glycosylation, fusion to HSA, and fusion to an albumin-binding domain from streptococcal protein G (Figure 1.3, Table 1.4).

PEGylation of the scDb was achieved by introducing an additional cysteine residue either in one of the flanking linkers or at the C-terminus. This allowed for a site-directed and defined conjugation of a branched 40 kDa PEG chain resulting in a PEGylated scDb with a calculated molecular mass of 100 kDa. Interestingly however, in size exclusion chromatography this molecule migrated with an apparent mass of around 800 kDa, that is, the hydrodynamic radius increased from 2.7 nm of the unmodified scDb to 7.9 nm of the PEGylated scDb (see also Table 1.2). This resulted in an increase of the terminal half-life to approximately 13 hours in CD1 mice receiving a single dose of 25 μg protein [61].

N-glycosylated scDb derivatives were generated by introducing *N*-glycosylation sites (Asn-X-Thr sequon) in the two flanking linkers as well as at a C-terminal extension of varying length [61]. Thus, three derivatives containing three, six or nine potential *N*-glycosylation sites (scDb-ABC$_1$, scDb-ABC$_4$, and scDb-ABC$_7$) were obtained and produced in 293 cells. *N*-glycosylation through post-translational modification resulted in molecules with an increased size, as revealed by SDS-PAGE and SEC analysis. However, this analysis in combination with MS

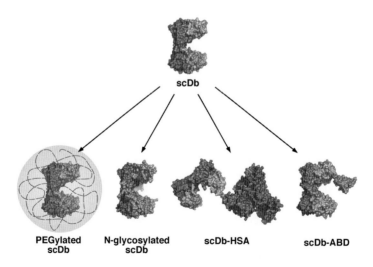

scDb

| PEGylated scDb | N-glycosylated scDb | scDb-HSA | scDb-ABD |

Figure 1.3 Half-life extension strategies applied to a bispecific single-chain diabody (scDb) including PEGylation, introduction of *N*-glycosylation sites, fusion to human serum albumin (HSA) or a 6 kDa albumin-binding domain (ABD) from streptococcal protein G.

Table 1.4 Biochemical and pharmacokinetic properties of an scDb and its half-life extended derivatives.

Construct	Length (aa)	calc. M_r[a] (kDa)	Apparent M_r (kDa)		Sr (nm)	$t_{1/2}\beta$ (h)	Tumor accumulation
			SDS-PAGE	SEC			AUC (%*h)
scDb	505	54.5	62	36	2.7	1.3	168 ± 20
scDb-ABC$_1$	497	53.6	62–68	8	2.9	8.9	n.d.
scDb-ABC$_4$	519	55.6	64–80	65	3.5	7.2	n.d.
scDb-ABC$_7$	544	57.8	68–87	78	3.8	6.2	150 ± 6
scDb-A'-PEG$_{40k}$	505	54.5	230	650	7.9	13.1	450 ± 14
chimeric IgG	1382	145.1	~280	280	5.7	163	n.d.
HSA	585	66.5	67	66	3.5	n.d.	
scDb-HSA	1080	119.6	121	90	3.9	25.0	n.d.
scDb-ABD	550	59.3	64	32	2.5	–	
scDb-ABD + HSA	–	125.8	–	150	4.8	27.6	753 ± 47

a) Calculated from the amino acid sequence without post-translational modifications. n.d., not determined.
Half-lives were determined in CD1 mice receiving a single dose of 25 µg. Half-lives were calculated for the first 3 days, except for IgG, which was calculated over a period of 7 days.

analysis also revealed substantial heterogeneity in respect to the number of glycosylated sites and carbohydrate composition. The moderate increase in size translated into a moderate extension of half-life, which compared well with the effects seen for example for the hyperglycosylated form of erythropoietin, darbepoetin alfa.

An scDb-albumin fusion protein was generated by fusing the scDb to the N-terminus of HSA. The purified scDb-HSA fusion protein had a M_r of 120 kDa and showed a strongly extended half-life (25 h) [60]. As an alternative, the scDb was fused to an albumin-binding domain (ABD) from streptococcal protein G. This 46 aa domain binds with a nanomolar affinity to human and mouse albumin. *In vitro* studies confirmed that the scDb-ABD fusion protein forms 1:1 complexes with HSA or MSA. *In vivo*, this resulted in a similar half-life extension as seen for scDb-HSA indicating that scDb-ABD binds also *in vivo* to the abundantly present

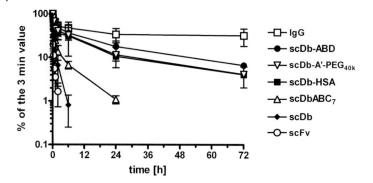

Figure 1.4 Clearance of scDb and its half-life extended derivatives after a single dose i.v. injection into mice. An scFv and a chimeric IgG were included for comparison (scDb-ABD, fusion of the scDb to an albumin-binding domain from protein G; scDb-A'-PEG40k, scDb chemically conjugated to a branched 40 kDa PEG; scDb-HSA, fusion of scDb with human serum albumin; scDb-ABC7, scDb with in total 9 N-glycosylation site in the linkers A and B as well as at a C-terminal extension).

serum albumin (Figure 1.4) [50]. In an attempt to further increase the half-life we compared an scDb-ABD derivative containing an ABD with increased affinity for albumin. However, *in vitro* studies showed that the increased affinity led only to a moderate increase of the terminal half-life [62]. In addition, we generated an scDb ABD fusion protein possessing two ABDs, one at the N-terminus and one at the C-terminus of the scDb (scDb-ABD$_2$). SEC studies showed that this derivative is indeed capable of binding two albumin molecules at the same time leading to a further increase of the apparent size. Interestingly, this did not translate in any improved half-life as compared with scDb-ABD [62].

All scDb derivatives were able to bind with unaltered strength to effector and target cells indicating that the modifications did not influence binding. However, *in vitro* recruitment assays, determining T-cell activation by measuring release of IL-2 and cytotoxicity towards CEA-positive tumor cells, revealed a reduce capacity of T-cell stimulation and activity. This finding demonstrates a negative effect of the various modifications in this specific setting, where two cells have to come into close contact. Importantly, however, further biodistribution studies clearly showed that a prolonged circulation time result in an increased accumulation of the antibody molecule in antigen-positive tumors, which was most pronounced for the scDb-ABD fusion protein [63]. Thus, these studies provide a rationale for modulating the half-life of recombinant proteins to various extent.

1.5
Conclusion

Proteins are established therapeutics for the treatment of a variety of diseases and the number of proteins under development, most of them produced in recombinant form, is steadily growing. Many of these therapeutic proteins exhibit a rather short half-life, which necessitates frequent injections or infusions. Improvement of the pharmacokinetic properties through half-life extension strategies can reduce costs and ease administration, and can also result in improved pharmacodynamic properties of the drug. Half-life extension strategies have therefore attracted increasing attention by the pharmaceutical industry. This has already led to the establishment of a variety of different strategies to modulate the half-life of therapeutic proteins, including chemical but also genetic modifications. Some of these modifications such as PEGylation and hyperglycosylation are clinically approved strategies and many more are in clinical testing (see Table 1.3). It is expected that half-life extension strategies will be an integral part of many future protein drugs. The following 16 chapters of this book provide in-depth information of the most important half-life modulation strategies, also including background information on general pharmacokinetic aspects and considerations.

References

1 Walsh, G. (2010) Biopharmaceutical benchmarks 2010. *Nat. Biotechnol.*, **28**, 917–924.

2 Tang, L., *et al.* (2004) Pharmacokinetic aspects of biotechnology products. *J. Pharm. Sci.*, **93**, 2184–2196.

3 Tryggvason, K., and Wartiovaara, J. (2005) How does the kidney filter plasma? *Physiology*, **20**, 96–101.

4 Rodewald, R., and Karnovsky, M.J. (1974) Porous substructures of the glomerular slit diaphragm in the rat and mouse. *J. Cell Biol.*, **60**, 423–433.

5 Haraldsson, B., and Sörensson, J. (2004) Why do we not all have proteinuria? An update of our current understanding of the glomerular barrier. *News Physiol. Sci.*, **19**, 7–10.

6 Wartiovaara, J., *et al.* (2004) Nephrin strands contribute to a porous slit diaphragm scaffold as revealed by electron tomography. *J. Clin. Invest.*, **114**, 1475–1483.

7 Peters, T. (1996) *All about Albumin*, Academic Press.

8 Lobo, E.D., Hansen, R.J., and Balthasar, J.P. (2004) Antibody pharmacokinetics and pharmcodynamics. *J. Pharm. Sci.*, **93**, 26452668.

9 Anderson, C.L., *et al.* (2006) Perspective– FcRn transports albumin: relevance to immunology and medicine. *Trends Immunol*, **27**, 343–348.

10 Roopenian, D.C., and Akilesh, S. (2007) FcRn: the neonatal Fc receptor comes of age. *Nat. Rev. Immunol.*, **7**, 715–725.

11 Lencer, W.I., and Blumberg, R.S. (2005) A passionate kiss, the run: exocytosis and recycling of IgG by FcRn. *Trends Cell Biol.*, **15**, 5–9.

12 Kontermann, R.E. (2009) Strategies to extend plasma half-lives of recombinant antibodies. *BioDrugs*, **23**, 93–109.

13 Kontermann, R.E. (2010) Alternative antibody formats. *Curr. Opin. Mol. Ther.*, **12**, 176–183.

14 Mahmood, I., and Green, M.D. (2005) Pharmacokinetic and pharmacodynamic considerations in the development of therapeutic proteins. *Clin. Pharmacokinet.*, **44**, 331–347.

15 Pisal, D.S., *et al.* (2010) Delivery of therapeutic proteins. *J. Pharm. Sci.*, **99**, 2557–2575.

16 Jevsevar, S., *et al.* (2010) PEGylation of therapeutic proteins. *Biotechnol. J.*, **5**, 113–128.

17 Harris, J.M., and Chess, R.B. (2003) Effect of PEGylation on pharmaceuticals. *Nat. Rev. Drug Discov.*, **2**, 214–221.

18 Fishburn, C.S. (2008) The pharmacology of PEGylation: balancing PD with PK to generate novel therapeutics. *J. Pharm. Sci.*, **97**, 4167–4183.

19 Gaberc-Porekar, V., *et al.* (2008) Obstacles and pitfalls in the PEGylation of therapeutic proteins. *Curr. Opin. Drug Discov. Devel.*, **11**, 242–250.

20 Veronese, F.M., and Mero, A. (2008) The impact of PEGylation on biological therapy. *BioDrugs*, **22**, 315–329.

21 Melmed, G.Y., *et al.* (2008) Certolizumab pegol. *Nat. Rev. Drug Discov.*, **7**, 641–642.

22 Schlapschy, M., *et al.* (2007) Fusion of a recombinant antibody fragment with a homo-amino-acid polymer: effects on biophysical properties and prolonged plasma half-life. *Protein Eng. Des. Sel.*, **20**, 273–284.

23 Schellenberger, V., *et al.* (2009) A recombinant polypeptide extends the *in vivo* half-life of peptides and proteins in a tunable manner. *Nat. Biotechnol.*, **27**, 1186–1190.

24 Sinclair, A.M., and Elliott, S. (2005) Glycoengineering: the effect of glycosylation on the properties of therapeutic proteins. *J. Pharm. Sci.*, **94**, 1626–1635.

25 Li, H., and d'Anjou, M. (2009) Pharmacological significance of glycosylation in therapeutic proteins. *Curr. Opin. Biotechnol.*, **20**, 678–684.

26 Solá, R.J., and Griebenow, K. (2010) Glycosylation of therapeutic proteins: an effective strategy to optimize efficacy. *BioDrugs*, **24**, 9–21.

27 Egrie, J.C., *et al.* (2003) Darbepoetin alfa has a longer circulating half-life and greater *in vivo* potency than recombinant human erythropoietin. *Exp. Hematol.*, **31**, 290–299.

28 Fares, F., *et al.* (2007) Development of a long-acting erythropoietin by fusing the carboxy-terminal peptide of human chorionic gonadotropin beta-subunit to the coding sequence of human erythropoietin. *Endocrinology*, **148**, 5081–5087.

29 Gregoriadis, G., *et al.* (2005) Improving the therapeutic efficacy of peptides and proteins: a role for polysialic acids. *Int. J. Pharm.*, **300**, 125–130.

30 Gregoriadis, G., *et al.* (2000) Polysialic acids: potential in improving the stability and pharmacokinetics of proteins and other therapeutics. *Cell. Mol. Life Sci.*, **57**, 1964–1969.

31 Constantinou, A., *et al.* (2008) Modulation of antibody pharmacokinetics by chemical polysialylation. *Bioconjug. Chem.*, **19**, 643–650.

32 Agreda-Vásquez, G.P., *et al.* (2008) Starch and albumin mixture as replacement fluid in therapeutic plasma exchange is safe and effective. *J. Clin. Apher.*, **23**, 163–167.

33 Brecher, M.E., *et al.* (1997) Alternatives to albumin: starch replacement for plasma exchange. *J. Clin. Apher.*, **12**, 146–153.

34 Huang, C. (2009) Receptor-Fc fusion therapeutics, traps and MIMETIBODY technology. *Curr. Opin. Biotechnol.*, **20**, 692–699.

35 Schulte, S. (2009) Half-life extension through albumin fusion technologies. *Thromb. Res.*, **124**, S6–S8.

36 Parveen, S., and Sahoo, S.K. (2008) Polymeric nanoparticles for cancer therapy. *J. Drug Target.*, **16**, 108–123.

37 Allen, T.M., and Cullis, P.R. (2004) Drug delivery systems: entering the mainstream. *Science*, **303**, 1818–1822.

38 Peer, D., *et al.* (2007) Nanocarriers as an emerging platform for cancer therapy. *Nat. Nanotechnol.*, **2**, 751–760.

39 Chuang, V.T., *et al.* (2002) Pharmaceutical strategies utilizing recombinant human serum albumin. *Pharm. Res.*, **19**, 569–577.

40 Jazayeri, J.A., and Carroll, G.J. (2008) Fc-based cytokines: prospects for engineering superior therapeutics. *BioDrugs*, **22**, 11–26.

41 Petkova, S.B., *et al.* (2006) Enhanced half-life of genetically engineered human IgG1 antibodies in a humanized FcRn mouse model: potential applications in humorally mediated autoimmune disease. *Int. Immunol.*, **18**, 1759–1769.

42 Zalevsky, J., *et al.* (2010) Enhanced antibody half-life improves *in vivo* activity. *Nat. Biotechnol.*, **28**, 157–159.

43 Vaccaro, C., *et al.* (2005) Engineering the Fc region of immunoglobulin G to modulate *in vivo* antibody levels. *Nat. Biotechnol.*, **23**, 1283–1288.

44 Kratz, F. (2008) Albumin as drug carrier: design of prodrugs, drug conjugates and nanoparticles. *J Control. Release*, **132**, 171–183.

45 Rustgi, V.K. (2009) Albinterferon alfa-2b, a novel fusion protein of human albumin and human interferon alfa-2b, for chronic hepatitis C. *Curr. Med. Res. Opin.*, **25**, 991–1002.

46 Thibaudeau, K., *et al.* (2005) Synthesis and evaluation of insulin–human serum albumin conjugates. *Bioconjug. Chem.*, **16**, 1000–1008.

47 Xie, D., *et al.* (2010) An albumin-conjugated peptide exhibits potent anti-HIV activity and long *in vivo* half-life. *Antimicrob. Agents Chemother.*, **54**, 191–196.

48 Morales, J. (2007) Defining the role of insulin detemir in basal insulin therapy. *Drugs*, **67**, 2557–2584.

49 Nguyen, A., *et al.* (2006) The pharmacokinetics of an albumin-binding Fab (AB.Fab) can be modulated as a function of affinity for albumin. *Protein Eng. Des. Sel.*, **19**, 291–297.

50 Stork, R., *et al.* (2007) A novel tri-functional antibody fusion protein with improved pharmacokinetic properties generated by fusing a bispecific single-chain diabody to an albumin-binding domain from streptococcal protein G. *Protein Eng. Des. Sel.*, **20**, 569–576.

51 Trüssel, S., *et al.* (2009) New strategy for the extension of the serum half-life of antibody fragments. *Bioconjug. Chem.*, **20**, 2286–2292.

52 Smith, B.J., *et al.* (2001) Prolonged *in vivo* residence times of antibody fragments associated with albumin. *Bioconjug. Chem.*, **12**, 750–756.

53 Tijink, B.M., *et al.* (2008) Improved tumor targeting of anti-epidermal growth factor receptor nanobodies through albumin binding: taking advantage of modular nanobody technology. *Mol. Cancer Ther.*, **7**, 2288–2297.

54 Walker, A., *et al.* (2010) Anti-serum albumin domain antibodies in the development of highly potent, efficacious and long-acting interferon. *Protein Eng. Des. Sel.*, **23**, 271–278.

55 Kontermann, R.E. (2011) *Bispecific Antibodies*, Springer, Heidelberg.

56 Fischer, N., and Léger, O. (2007) Bispecific antibodies: molecules that enable novel therapeutic strategies. *Pathobiology*, **74**, 3–14.

57 Segal, D.M., *et al.* (2001) Introduction: bispecific antibodies. *J. Immunol. Methods*, **248**, 1–6.

58 Müller, D., and Kontermann, R.E. (2010) Bispecific antibodies for cancer immunotherapy: current perspectives. *BioDrugs*, **24**, 89–98.

59 Müller, D., and Kontermann, R.E. (2007) Recombinant bispecific antibodies for cellular cancer immunotherapy. *Curr. Opin. Mol. Ther.*, **9**, 319–326.

60 Müller, D., *et al.* (2007) Improved pharmacokinetics of recombinant bispecific antibody molecules by fusion to human serum albumin. *J. Biol. Chem.*, **282**, 12650–12660.

61 Stork, R., *et al.* (2008) *N*-glycosylation as novel strategy to improve the pharmacokinetic properties of bispecific single-chain diabodies. *J. Biol. Chem.*, **283**, 7804–7812.

62 Hopp, J., *et al.* (2010) The effects of affinity and valency of an albumin-binding domain (ABD) on the half-life of a single-chain diabody-ABD fusion protein. *Protein Eng. Des. Sel.*, **23**, 827–834.

63 Stork, R., *et al.* (2009) Biodistribution of a bispecific single-chain diabody and its half-life extended derivatives. *J. Biol. Chem.*, **284**, 25612–25619.

2
Pharmacokinetics and Half-Life of Protein Therapeutics

Bernd Meibohm

2.1
Introduction

The basic paradigm of clinical pharmacology is the fact that drug effects, desired as well as undesired, are a function of drug concentrations within different organs and tissues in the human body. Thus, drug concentrations are the driving force for the spectrum of drug responses observed in a drug-treated patient. The discipline of pharmacokinetics describes the time course of drug concentration in a body fluid, preferably plasma or blood, that results from the administration of a certain dosage regimen. In simple words, pharmacokinetics is *"what the body does to the drug"*.

Drug concentrations in blood or plasma are used as a surrogate for drug concentrations at the effect site, or site of drug action, under the assumption that under pharmacokinetic steady-state conditions there is a constant relationship between the free, unbound drug concentration in plasma and at the site of action. Even though this assumption is frequently not accurate, it has proven to be generally a very useful approximation to achieve the desired effect levels via modulation of their plasma concentration, especially during prolonged pharmacotherapy with multiple dose regimens. However, it is usually limited to small molecule drugs, for which passive diffusion is a major mechanism in the drug disposition process. In contrast, this concept might be less applicable for protein therapeutics, as their distribution is largely determined by convective transport rather than diffusion due to their high molecular weight and charge. Nevertheless, drug concentrations in blood and plasma are frequently also used for protein therapeutics as surrogates for effect site concentration as the latter are usually not easily accessible in human subjects.

The plasma concentration–time profile resulting from drug administration is determined by pharmacokinetic parameters and the administered dosage regimen. While the pharmacokinetic parameters are characteristic for the disposition or handling of a drug in a specific patient and can thus usually not be modulated during pharmacotherapy, the dosage regimen is the clinician's tool to affect drug concentrations for maximum therapeutic benefit. The dosing regimen is the

Therapeutic Proteins: Strategies to Modulate Their Plasma Half-Lives, First Edition. Edited by Roland Kontermann.
© 2012 Wiley-VCH Verlag GmbH & Co. KGaA. Published 2012 by Wiley-VCH Verlag GmbH & Co. KGaA.

manner in which a drug is taken. It entails the individual dose and the dosing frequency, that is, "How much?" and "How often?" a drug is taken [1].

2.2
Basic Principles of Pharmacokinetics

Pharmacokinetic parameters are characteristic for the disposition and uptake of drug into the body of one specific drug in a specific patient. As drug property, pharmacokinetic parameters are usually not accessible for therapeutic manipulation by the clinician, but may be modulated by physiological or pathophysiological processes in the patient as well as concomitant drug therapy (drug–drug interactions) and environmental factors. Structural modifications in the drug molecule, for example for protein therapeutics, may also change specific pharmacokinetic parameters and thus alter their disposition behavior [2].

2.2.1
Primary Pharmacokinetic Parameters

The most important pharmacokinetic parameters are clearance, volume of distribution and bioavailability (Figure 2.1). Clearance (CL) is reflective for the drug eliminating capacity of the body, volume of distribution (*V*) refers to the distribution of drug within the body including uptake into specific organs and tissues as

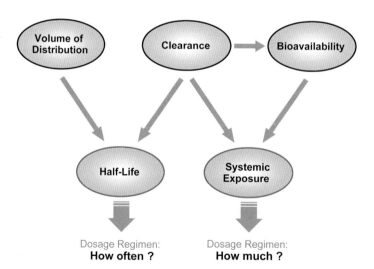

Figure 2.1 Interrelationship of primary pharmacokinetic parameters (clearance, volume of distribution, and bioavailability), secondary pharmacokinetic parameters (half-life and systemic exposure), and their relevance for determining dosage regimens. Adapted with permission from [3] Copyright © (2003) Macmillan Publishers Ltd.

well as binding to endogenous proteins and other macromolecules. Based on these underlying physiologic processes, clearance and volume of distribution are independent of each other and are called primary pharmacokinetic parameters. Bioavailability (F) refers to the extent of drug uptake into the systemic circulation.

Clearance (CL) quantifies the elimination of a drug. It is the volume of body fluid, blood or plasma, that is cleared from the drug per time unit. Thus, it measures the removal of drug from the plasma or blood. Clearance does not indicate how much drug is being removed, but it represents the volume of blood or plasma from which the drug is completely removed in a given time period. The unit of clearance is volume per time, for example, Lh^{-1} or $mLmin^{-1}$. Clearance (CL) is an independent pharmacokinetic parameter, and is the most important pharmacokinetic parameter because it determines the dose rate, that is, the amount of drug given per time to maintain a drug concentration.

Volume of distribution (V) quantifies the extent of distribution of a drug throughout the body. Drug distribution means the reversible transfer of drug from one location to another within the body. The concentration achieved in plasma after distribution depends on the dose and the extent of distribution. The volume of distribution relates the amount of drug in the body to the plasma concentration. It is an apparent volume, which is calculated upon the simplifying assumption that the plasma concentration is present in all body compartments (an assumption that is certainly not true for protein therapeutics). The unit of volume of distribution is volume, for example, L or mL. The larger the volume of distribution, the smaller the fraction of the dose that resides in the plasma.

For most drugs, distribution throughout the body is not instantaneous, but a time-consuming process. Thus, the initial drug distribution volume after intravenous (i.v.) bolus administration is frequently smaller than that after distribution equilibrium throughout the body has been reached. The initial volume of distribution is frequently referred to as the volume of the central compartment V_C, representing well-perfused organs and tissues for which drug distribution for a specific drug is nearly instantaneous. In contrast, the volume of distribution at steady state Vss is observed once a distribution equilibrium between the multiple organs and tissues in the body has been achieved.

Bioavailability (F) is the fraction of the administered dose that reaches the systemic circulation. Bioavailability can be viewed as the result of a combination of processes that reduce the amount of extravascularly administered drug that reaches the systemic circulation, for example degradation at the site of administration as frequently observed for protein therapeutics. By definition, F is 1 (or 100%) for intravascular administrations, for example, i.v. dosing.

2.2.2
Secondary Pharmacokinetic Parameters

The primary pharmacokinetic parameters clearance, volume of distribution and bioavailability are major determinants for the plasma concentration–time profile resulting from administration of a dosing regimen. The clinically most useful characteristics of the resulting concentration–time profile are the elimination half-life $t\frac{1}{2}$, as well as the average steady-state concentration $C_{ss,av}$ and the area under the plasma concentration–time curve (AUC) as measures of systemic exposure (Figure 2.1).

Half-life ($t\frac{1}{2}$) characterizes the monoexponential decline in drug concentration after drug input processes have been completed assuming a first-order elimination process. Half-life is the time required for the plasma concentration to decrease by one-half. It is a transformation of the first-order elimination rate constant K that characterizes drug removal from the body if the elimination process follows first-order kinetics. Drug concentration C at any time t during a monoexponential decrease can be described by

$$C = C_0 \cdot e^{-K \cdot t} \tag{2.1}$$

where C_0 is the initial drug concentration at time $t = 0$h. Half-life is then given as

$$t_{1/2} = \frac{\ln 2}{K} \quad \text{or} \quad t_{1/2} = \frac{0.693}{K} \tag{2.2a,b}$$

The elimination rate constant K is the negative slope of the plasma concentration–time profile in a plot of the natural log (ln) of the concentration versus time. Half-life can thus be calculated from two concentrations C_1 and C_2 during the monoexponential decline of drug concentration via the relationship

$$K = \frac{\ln\left(\dfrac{C_1}{C_2}\right)}{t_2 - t_1} \tag{2.3}$$

Half-life is a secondary pharmacokinetic parameter that is defined by the primary parameters clearance and volume. The elimination rate constant K as a transform of half-life can be seen as a proportionally factor between clearance and volume of distribution:

$$CL = K \cdot V \quad \text{or} \quad K = \frac{CL}{V} \tag{2.4a,b}$$

Thus, half-life is given by

$$t_{1/2} = \frac{0.693 \cdot V}{CL} \tag{2.5}$$

Since clearance and volume are both determined by unrelated underlying physiological processes as described earlier, they are independent of each other. If volume, for example is increased by a pathophysiologic process, then clearance still remains unaffected. According to Eq. (2.4), a change in volume of distribution V would thus result a compensatory change in the elimination rate constant K without affecting CL. Vice versa, an increase or decrease in CL will only result in a corresponding change in the elimination rate constant K, but volume of distribution V would remain unaffected.

Half-life provides important information about specific aspects of a drug's disposition, such as how long it will take to reach steady state once maintenance dosing is started and how long it will take for "all" the drug to be eliminated from the body once dosing is stopped (usually considered five half-lives). As half-life also determines the fluctuation between minimum and maximum concentrations during a multiple dose regimen, it is the major determinant of the dosing interval between two consecutive doses, that is, how often a dose should be given.

Systemic exposure to a drug in the blood circulation is a time-integrated or time-averaged measure of drug concentration that is secondary to the administered dosage regimen and the primary parameters clearance CL and bioavailability F.

The area-under-the-concentration-time curve (AUC) is the integrated concentration over time as a measure of overall exposure to a drug resulting from a specific dosage regimen. It is given by

$$AUC = \frac{F \cdot D}{CL} \tag{2.6}$$

where D is the administered dose.

The average steady state concentration $C_{ss,av}$ is the average concentration over one dosing interval in a multiple dose regimen. It is related to clearance CL and bioavailability F via

$$C_{ss,av} = \frac{F \cdot D}{\tau \cdot CL} = \frac{AUC}{\tau} \tag{2.7}$$

where τ is the dosing interval between two consecutive doses of the multiple dose regimen. The ratio D/τ is also referred to as "dose rate".

As indicated in Eqs. (2.6) and (2.7), systemic exposure assessed either as AUC or $C_{ss,av}$ is only dependent on the bioavailable dose or dose rate and clearance, but not the extent of drug distribution as quantified by volume of distribution V. Thus, clearance and bioavailability as determinants of systemic exposure determine how much of a drug needs to be given, while clearance and volume of distribution determine via the half life, how often a discrete drug dose needs to be administered (Figure 2.1).

2.3
Pharmacokinetics of Protein Therapeutics

Similar to conventional small molecule drugs, protein therapeutics are characterized by well-defined pharmacokinetic properties that form the basis for the design of therapeutic dosing regimens as well as drug delivery strategies. Potential differences, caveats and pitfalls, however, may arise from their similarity to endogenous and/or dietary molecules with which they share common drug disposition pathways, as well as their interaction with endogenous regulatory feedback pathways, especially if they are analogs of hormones or other tightly regulated endogenous substances [4].

The widespread application of pharmacokinetic concepts in drug development has repeatedly been promoted by industry, academia, and regulatory authorities [5–7]. It is believed that the application of pharmacokinetically-based concepts in all preclinical and clinical drug development phases may substantially contribute to a more scientifically-driven, evidence-based development process. In addition, it provides the pharmacologic basis for dosage selection and dosage regimen design. Thus, in-depth knowledge of a compound's pharmacokinetic characteristics will, for protein therapeutics, also continue to form a crucial element to achieve its fullest therapeutic potential for an optimal use with regard to efficacy and safety in the target patient population.

The *in vivo* disposition of endogenous peptide and protein drugs may often be predicted to a large degree from their physiological function. Peptides, for example, which frequently have hormone activity, usually have short elimination half-lives, which is desirable for a close regulation of their endogenous levels and thus function. Insulin, for example shows dose-dependent elimination with a relatively short half-life of 26 and 52 minutes at 0.1 and $0.2\,U\,kg^{-1}$, respectively. Contrary to that, proteins that have transport tasks such as albumin or long-term immunity functions such as antibodies have elimination half-lives of several days, which enables and ensures the continuous maintenance of necessary concentrations in the blood stream [2]. This is for example reflected by the elimination half-life of antibody drugs like the anti-epidermal growth factor receptor antibody cetuximab, for which a half-life of approximately 7 days was reported [8].

2.3.1
Absorption of Protein Therapeutics

Drug delivery by oral application is the preferred route of administration for the majority of small molecule drugs because of its convenience, cost-effectiveness and painlessness. Unfortunately, oral administration is not a viable delivery pathway for most protein therapeutics because of their extremely low oral bioavailability. The lack of oral bioavailability is mainly caused by two factors, high gastrointestinal enzyme activity and the function of the gastrointestinal mucosa as absorption barrier. The gastrointestinal tract is the most efficient body compartment for peptide and protein metabolism due to substantial peptidase and protease activity. Dietetic as well as therapeutic proteins undergo equally effective catabolism, and thus the fraction of the administered dose of a protein therapeutic available for absorption is rapidly reduced. In addition, the gastrointestinal mucosa presents a major absorption barrier for water-soluble, charged macromolecules such as peptides and proteins, resulting in a low membrane permeability for most protein drugs. Thus, even if protease activity could be overcome, for example by co-administration of protease inhibitors to prevent enzymatic degradation [9], oral bioavailability would remain limited for most peptides and proteins, and molecular size is generally considered the ultimate obstacle [10].

Several promising strategies have recently emerged from intensive research into methods to overcome the obstacles associated with oral drug delivery of proteins. Approaches to increase the oral bioavailability of protein drugs include absorption enhancers for increasing the amount of drug that is able to cross absorption barriers as well as encapsulation into micro- or nanoparticles thereby protecting proteins from intestinal degradation [9–11]. Other strategies include chemical modifications such as amino acid backbone modifications and chemical conjugation to improve the resistance to degradation and permeability of protein drugs.

The lack of systemic activity after oral administration for most proteins resulted besides the frequently used, but invasive i.v. application into the utilization of numerous non-oral administration pathways, for example, nasal, buccal, rectal, vaginal, percutaneous, ocular or pulmonary drug delivery. In addition, drug delivery by subcutaneous (s.c.) or by intramuscular (i.m.) administration are frequently used alternatives for administering protein drugs. Presystemic degradation processes, however, are also frequently associated with these administration routes, resulting in a reduced bioavailability of numerous proteins after s.c. or i.m. administration compared to their i.v. administration. Other potential factors that may limit bioavailability of proteins after s.c. or i.m. administration including variable local blood flow, injection trauma, and limitations of uptake into the systemic circulation related to effective capillary pore size and diffusion.

After s.c. administration, large molecule drugs like proteins may, dependent on their molecular weight, either enter the systemic circulation via blood capillaries or through lymphatic vessels. Macromolecules larger than 16 kD are predominantly absorbed into the lymphatics whereas those under 1 kD are mostly absorbed into the blood circulation. This is of particular importance for those agents whose therapeutic targets are lymphoid cells, for example interferons and interleukins, as it allows achieving high exposure in regional lymph nodes after s.c. administration since the lymphatic vessels drain into the regional lymph nodes [12].

2.3.2
Distribution of Protein Therapeutics

The distribution volume of proteins is determined largely by their molecular weight, physiochemical properties (e.g., charge, lipophilicity), protein binding, and their dependency on active transport processes. Since most therapeutic proteins have high molecular weight and are thus large in size, their apparent volume of distribution is usually small and limited to the volume of the plasma or the extracellular space, predominantly because of their limited mobility secondary to impaired passage through biomembranes [13].

In contrast to small molecule drugs, protein transport from the vascular space into the interstitial space of tissues is largely mediated by convection rather than diffusion, following the unidirectional fluid flux from the vascular space through paracellular pores into the interstitial tissue space (Figure 2.2). The subsequent removal from the tissues is accomplished by lymph drainage back into the systemic circulation [14]. This underlines the unique role the lymphatic system

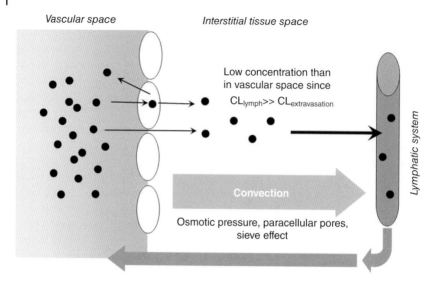

Figure 2.2 Distribution mechanisms of therapeutic proteins: convective extravasation rather than diffusion.

plays in the disposition of protein therapeutics as already mentioned in the previous section. Another, but much less prominent pathway for the movement of protein molecules from the vascular to the interstitial space is transcellular migration via endocytosis [15].

The plasma concentration–time profiles for proteins after i.v. administration are usually biphasic and can best be described by a two-compartment pharmacokinetic model [12]. The central compartment in this two-compartment model represents primarily the vascular space and the interstitial space of well-perfused organs with permeable capillary walls, including the liver and the kidneys. The peripheral compartment is more reflective of concentration-time profile in the interstitial space of poorly-perfused tissues like skin and inactive muscle.

The central compartment in which peptides and proteins initially distribute after intravenous administration has thus typically a volume of distribution equal to or slightly larger than the plasma volume, that is, 3–8 L. The total volume of distribution, 14–20 L, frequently comprises not more than twice the initial volume of distribution [16, 17].

It should be stressed that pharmacokinetic calculations of volume of distribution may be problematic for many protein therapeutics [2, 18]. Noncompartmental determination of volume of distribution at steady-state (V_{ss}) using statistical moment theory assumes first-order disposition processes with elimination occurring from the rapidly equilibrating or central compartment [19, 20]. These basic assumptions, however, are not fulfilled for numerous protein therapeutics, as proteolysis and receptor-mediated elimination in peripheral tissues may constitute a substantial fraction of the overall elimination process. Thus, V_{ss} values reported for biologics in the literature are frequently underestimating the real distribution space and should be interpreted with caution.

Another factor that can influence the distribution of peptide and protein drugs is binding to other proteins in plasma. Physiologically active, endogenous peptides and proteins frequently interact with specific binding proteins involved in their transport and regulation, for example growth hormone. It is a general pharmacokinetic principle, which is also applicable to proteins, that only the free, unbound fraction of a drug substance is accessible to distribution and elimination processes as well as interactions with its target structure at the site of action, for example a receptor. Protein binding not only affects the unbound fraction of a protein drug and thus the fraction of a drug available to exert pharmacological activity, but many times it may also either prolong protein circulation time by acting as a storage depot or it may enhance protein clearance by triggering protein complex degradation through the reticulo-endothelial system.

Aside from physicochemical properties and protein binding of protein-based drugs, site-specific target receptor-mediated uptake can also influence biodistribution. Therefore, there is often a close interrelationship between distribution, elimination and pharmacodynamics for protein therapeutics in contrast to conventional small molecule drugs. The generally low volume of distribution of protein drugs should not necessarily be interpreted as low tissue penetration. Receptor-mediated specific uptake into the target organ, as one mechanism, can result in therapeutically effective tissue concentrations despite a relatively small volume of distribution.

2.3.3
Elimination of Protein Therapeutics

2.3.3.1 Proteolysis
Protein therapeutics are nearly exclusively metabolized via the same catabolic pathways as endogenous or dietetic proteins, leading to amino acids that are reutilized in the endogenous amino acid pool for the *de novo* biosynthesis of structural or functional body proteins. Detailed investigations on the metabolism of proteins are relatively difficult because of the myriad of potential molecule fragments that may be formed, and are therefore generally not conducted. Non-metabolic elimination pathways such as renal or biliary excretion are negligible for most protein therapeutics. If biliary excretion of proteins occurs, however, it generally results in subsequent metabolism of these compounds in the gastrointestinal tract.

The metabolic rate for protein degradation generally increases with decreasing molecular weight from large to small proteins to peptides, but is also dependent on other factors like secondary and tertiary structure as well as the level of glycosylation. The elimination of proteins by proteolytic degradation can occur unspecifically nearly everywhere in the body or can be limited to a specific organ or tissue. Locations of intensive protein metabolism are liver, kidneys, gastrointestinal tissue, but also blood and other body tissues. Endothelial cells with their large surface area (>1000 m^2 in adult humans) constitute a major site for protein degradation via endocytotic uptake and subsequent lysosomal degradation. Thus, intracellular uptake is *per se* more an elimination rather than a distribution process [12].

While proteolytic enzymes like peptidases and proteases in the gastrointestinal tract and in lysosomes are relatively unspecific, soluble peptidases in the interstitial space and exopeptidases on the cell surface have a higher selectivity and determine the specific metabolism pattern of an organ.

For orally administered peptides and proteins, the gastrointestinal tract is the major site of metabolism due to its high proteolytic activity as primary digestion site for dietary proteins. The metabolic activity of the gastrointestinal tract, however, is not limited to orally administered proteins. Parenterally administered proteins may also be metabolized in the intestinal mucosa following intestinal secretion. At least 20% of the degradation of endogenous albumin, for example, takes place in the gastrointestinal tract [16].

2.3.3.2 Renal Protein Metabolism

While proteins are usually not excreted unchanged in the urine, the kidneys may serve as a major site of protein metabolism for smaller sized proteins that undergo glomerular filtration. The size-selectivity cut-off for glomerular filtration is approximately 60 kD, although the effective molecule radius based on molecular weight and conformation is probably the limiting factor. In addition to size-selectivity, charge-selectivity has also been observed for glomerular filtration where anionic macromolecules pass through the capillary wall less readily than neutral macromolecules, which in turn pass through less readily than cationic macromolecules [21].

For renal protein metabolism, glomerular filtration is the dominant, rate-limiting step as subsequent degradation processes are not saturable under physiologic conditions. Due to this limitation of renal elimination, the renal contribution to the overall elimination of proteins is dependent on the proteolytic activity for these proteins in other body regions. If metabolic activity for these proteins is high in other body regions, there is only minor renal contribution to total clearance, and it becomes negligible in the presence of unspecific degradation throughout the body. If the metabolic activity is low in other tissues or if distribution to the extravascular space is limited, however, the renal contribution to total clearance may approach 100%.

Several different pathways have been described for protein metabolism subsequent to glomerular filtration (Figure 2.3). Small linear peptides such as bradykinin or glucagon undergo intraluminal metabolism, predominantly by exopeptidases in the luminal brush border membrane of the proximal tubules. The resulting amino acids are transcellularly transported back into the systemic circulation with contribution of the proton driven peptide transporters PEPT2 and to a lesser degree PEPT1 [4].

After glomerular filtration, larger peptides and proteins such as interleukin-2, interleukin-11 and insulin are actively reabsorbed in the proximal tubules via

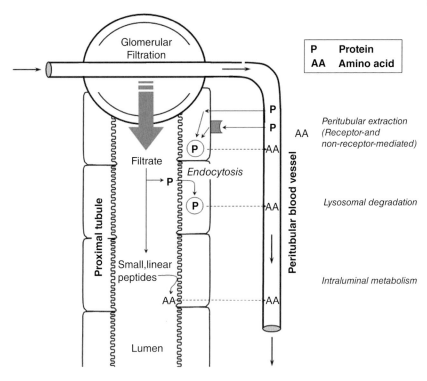

Figure 2.3 Renal elimination processes of peptides and proteins: glomerular filtration followed by either (a) intraluminal metabolism or (b) tubular reabsorption with intracellular lysosomal metabolism, and (c) peritubular extraction with intracellular lysosomal metabolism. From [4].

endocytosis. This cellular uptake is followed by lysosomal degradation to peptide fragments and amino acids that are transported into the systemic circulation to be re-utilized in the endogenous amino acid pool.

Besides intraluminal metabolism and tubular reabsorption with intracellular lysosomal metabolism, peritubular extraction from post-glomerular capillaries with subsequent intracellular metabolism is an additional renal elimination mechanism for proteins. This process can either be receptor-mediated or non-receptor-mediated and has for example been described for human growth hormone and calcitonin, but usually contributes only a small fraction to the protein's elimination.

2.3.3.3 Hepatic Protein Metabolism

Apart from proteolysis and the kidneys, the liver substantially contributes to the metabolism of peptide and protein drugs. The rate of hepatic metabolism is largely dependent on specific amino acid sequences in the protein [22].

An important first step in the hepatic metabolism of proteins and peptides is the uptake into the hepatocytes. Small peptides may cross the hepatocyte membrane via passive diffusion if they have sufficient hydrophobicity. Uptake of larger peptides and proteins is facilitated via various carrier- or receptor-mediated, energy-dependent transport processes. Receptor systems involved in the uptake process include for example the LDL-receptor, the low density lipoprotein receptor-related protein (LRP) and sugar (e.g., mannose, fucose)-recognizing receptors. Hepatic uptake and clearance of tissue plasminogen activator (tPA) as a large protein (65 kD), for example, is facilitated by mannose and asialoglycoprotein receptors. For therapeutic monoclonal antibodies and fractions thereof with intact Fc-region, Fc-γ receptors may substantially contribute to their intracellular uptake.

2.3.3.4 Receptor-Mediated Protein Metabolism and Target-Mediated Drug Disposition

Intracellular uptake of protein therapeutics can not only be facilitated by unspecific cell surface receptors such as LRP, but also by the cell surface receptors that are the pharmacologic targets of protein therapeutics. This receptor-mediated endocytosis is an additional mechanism that may mediate the uptake of a protein therapeutic into hepatocytes in the liver, tubular cells in the kidneys, and endothelial cells throughout the body, but also any other cell type that expresses the target receptor [2]. Receptor binding is usually negligible compared to the total amount of drug in the body for conventional small molecule drugs and does rarely affect their pharmacokinetic profile. In contrast to that, a substantial fraction of a protein dose can be bound to receptors and subsequently eliminated through receptor-mediated uptake and intracellular metabolism. Thus, target receptor binding may serve as the initial step in the degradation and elimination of a protein therapeutic and thus affect the compound's disposition. The phenomenon that a pharmacodynamic target structure affects the pharmacokinetics of a drug compound and results in elimination processes has been termed "target-mediated drug disposition" [23].

Since the number of receptors is limited, their binding and the related drug uptake can usually be saturated within therapeutic concentrations. Thus, receptor-mediated elimination constitutes a major source for nonlinear pharmacokinetic behavior of numerous biologics, that is, a lack of dose proportionality in pharmacokinetics. Thus, for many protein therapeutics that exhibit target-mediated drug disposition, systemic clearance at concentrations below the saturation of the target-mediated drug disposition process is much higher than at higher concentrations when the target-mediated drug disposition process is saturated. This leads to an over-proportional increase in systemic exposure with increasing dose once the capacity-limited, target-mediated drug disposition pathway has been saturated.

Macrophage colony-stimulating factor (M-CSF), for example, undergoes besides linear renal elimination a nonlinear elimination pathway that follows Michaelis–Menten kinetics and is linked to a receptor-mediated uptake into macrophages. At low concentrations, M-CSF follows linear pharmacokinetics, while at high concentrations, nonrenal elimination pathways are saturated resulting in nonlinear pharmacokinetic behavior [24].

2.3.3.5 The Role of the Neonatal Fc Receptor (FcRn) in the Disposition of Protein Therapeutics

Immunoglobulin G (IgG)-based monoclonal antibodies and their derivatives constitute one of the most important class of protein therapeutics with many members currently being under development or in therapeutic use. Interaction with the Brambell receptor, or neonatal Fc receptor (FcRn), constitutes are major component in the drug disposition of IgG molecules [25]. FcRn has been well described in the transfer of passive humoral immunity from a mother to her fetus by transferring IgG across the placenta and the proximal small intestine via transcytosis. More importantly, interaction with FcRn in a variety of cells, including endothelial cells and monocytes, macrophages and other dendritic cells, protects IgG from lysosomal catabolism and thus constitutes a salvage pathway for IgG molecules that have been internalized in these cells types. This is facilitated by intercepting IgG in the endosomes and recycling it to the systemic recirculation [26]. The interaction with the FcRn receptor thereby prolongs the elimination half-life of IgG, with a more pronounced effect the stronger the binding of the Fc fragment of the antibody is to the receptor: Based on the affinity of this binding interaction, human IgG1, IgG2, and IgG4 have a half life in humans of 18–21 days, whereas the less strongly bound IgG3 has a half-life of only seven days and murine IgG in humans has a half-life of one to two days [27]. Similar to IgG, FcRn is also involved in the disposition of albumin molecules. The kinetics of IgG and albumin recycling is illustrated in Figure 2.4. For IgG1, approximately 60% of the molecules taken up into lysosomes are recycled, for albumin 30%. As FcRn is responsible for the extended presence of IgG, albumin and other Fc- or albumin-conjugated proteins in the systemic circulation, modulation of the IgG-FcRn interaction or the albumin-FcRn interaction will deliberately control the half-life of these molecules [28].

2.4
Summary and Conclusions

The same pharmacokinetic principles underlie protein therapeutics as small molecule drugs, including volume of distribution and clearance as independent pharmacokinetic parameters and drivers for the secondary parameter half-life. The terminal half-life in the plasma concentration time profile, together with the systemic exposure that is determined by bioavailability and clearance, determine the dosing amount and dosing frequency to achieve and maintain therapeutically active drug concentrations in the human body. Thus, modulation of distribution and elimination processes of protein therapeutics affecting their volume of distribution and clearance allows changing dosing regimen requirements with regard to dose amount and dosing interval.

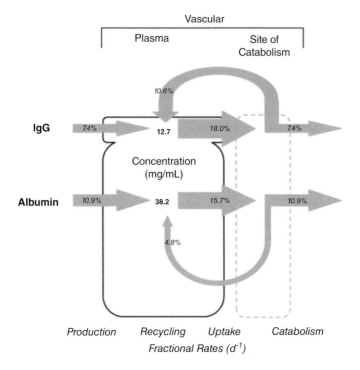

Figure 2.4 Effect of FcRn-mediated recycling on IgG and albumin turnover in humans expressed as fractional rates. Shown are homeostatic plasma concentrations (12.1 and 38.2 mg mL^{-1}), fractional catabolic rates (7.4 and 10.9%/day), the FcRn-mediated fractional recycling rates (10.6 and 4.8%/day), and the fractional production rates (7.4 and 10.9%/day). The figure is to scale: areas for plasma amount and arrow widths for rates. Reproduced with permission from [28]. Copyright © (2007) Elsevier.

Distribution of protein therapeutics is largely facilitated by convective transport, a process predominantly determined by the size, dimension and physicochemical properties of the protein. Elimination comprises multiple pathways identical to those of dietetic and/or endogenous protein. Again, size, charge and physico-chemical properties are major determinants, as well as the interaction and affinity with multiple receptor systems, including drug target receptors as well more class-specific receptor systems (e.g., LRP, Fc-γ receptor subtypes). Any intervention that modifies structural features of the protein therapeutic as well as its degree of interaction with receptor systems involved in its disposition will ultimately affect clearance and/or volume of distribution and thus the elimination half-life of the drug. The subsequent chapters in this textbook will highlight a diverse array of structural modifications of protein therapeutics and the application delivery systems that ultimately result through different pathways into a modulation of the elimination half life of the drug and thus the dosing requirements for the success-ful clinical application of protein therapeutics.

References

1 Meibohm, B., and Evans, W.E. (2006) Clinical pharmacodynamics & pharmacokinetics, in *Textbook of Therapeutics: Drug and Disease Management* (eds R.A. Helms and D.J. Quan), Lippincott Williams Wilkins, Philadelphia, PA, pp. 1–30.

2 Tang, L., Persky, A.M., Hochhaus, G., and Meibohm, B. (2004) Pharmacokinetic aspects of biotechnology products. *J. Pharm. Sci.*, **93**, 2184–2204.

3 van de Waterbeemd, H., and Gifford, E. (2003) ADMET in silico modelling: towards prediction paradise? *Nat. Rev. Drug Discov.*, **2**, 192–204.

4 Tang, L., and Meibohm, B. (2006) Pharmacokinetics of peptides and proteins, in *Pharmacokinetics and Pharmacodynamics of Biotech Drugs* (ed. B. Meibohm), Wiley-VCH Verlag GmbH, Weinheim, pp. 17–44.

5 Lesko, L.J. (2007) Paving the critical path: how can clinical pharmacology help achieve the vision? *Clin. Pharmacol. Ther.*, **81**, 170–177.

6 Meibohm, B., and Derendorf, H. (2002) Pharmacokinetic/pharmacodynamic studies in drug product development. *J. Pharm. Sci.*, **91**, 18–31.

7 Suryawanshi, S., Zhang, L., Pfister, M., and Meibohm, B. (2010) The current role of model-based drug development. *Expert Opin. Drug Discov.*, **5**, 311–321.

8 Herbst, R.S., Kim, E.S., and Harari, P.M. (2001) IMC-C225, an anti-epidermal growth factor receptor monoclonal antibody, for treatment of head and neck cancer. *Expert Opin. Biol. Ther.*, **1**, 719–732.

9 Mahato, R.I., Narang, A.S., Thoma, L., and Miller, D.D. (2003) Emerging trends in oral delivery of peptide and protein drugs. *Crit. Rev. Ther. Drug Carrier Syst.*, **20**, 153–214.

10 Shen, W.C. (2003) Oral peptide and protein delivery: unfulfilled promises? *Drug Discov. Today*, **8**, 607–608.

11 Lee, H.J. (2002) Protein drug oral delivery: the recent progress. *Arch. Pharm. Res.*, **25**, 572–584.

12 Meibohm, B., and Braeckman, R.A. (2007) Pharmacokinetics and pharmacodynamics of peptides and proteins, in *Pharmaceutical Biotechnology: Concepts and Applications* (eds D.J.A. Crommelin, R.D. Sindelar, and B. Meibohm), Informa Healthcare, New York, pp. 95–123.

13 Zito, S.W. (1997) *Pharmaceutical Biotechnology: A Programmed Text*, Technomic Pub. Co., Lancaster, PA.

14 Flessner, M.F., Lofthouse, J., and Zakaria, R. (1997) *In vivo* diffusion of immunoglobulin G in muscle: effects of binding, solute exclusion, and lymphatic removal. *Am. J. Physiol.*, **273**, H2783–H2793.

15 Reddy, S.T., Berk, D.A., Jain, R.K., and Swartz, M.A. (2006) A sensitive *in vivo* model for quantifying interstitial convective transport of injected macromolecules and nanoparticles. *J. Appl. Physiol.*, **101**, 1162–1169.

16 Colburn, W. (1991) Peptide, peptoid, and protein pharmacokinetics/ pharmacodynamics, in *Petides, Peptoids, and Proteins*, vol. **3**, (eds P. Garzone, W. Colburn, and M. Mokotoff), Harvey Whitney Books, Cincinnati, OH, pp. 94–115.

17 Kageyama, S., Yamamoto, H., Nakazawa, H., Matsushita, J., Kouyama, T., Gonsho, A., Ikeda, Y., and Yoshimoto, R. (2002) Pharmacokinetics and pharmacodynamics of AJW200, a humanized monoclonal antibody to von Willebrand factor, in monkeys. *Arterioscler. Thromb. Vasc. Biol.*, **22**, 187–192.

18 Straughn, A.B. (2006) Limitations of noncompartmental pharmacokinetic analysis of biotech drugs, in *Pharmacokinetics and Pharmacodynamics of Biotech Drugs* (ed. B. Meibohm), Wiley-VCH Verlag GmbH, Weinheim, pp. 181–188.

19 Perrier, D., and Mayersohn, M. (1982) Noncompartmental determination of the steady-state volume of distribution for any mode of administration. *J. Pharm. Sci.*, **71**, 372–373.

20 Straughn, A.B. (1982) Model-independent steady-state volume of distribution. *J. Pharm. Sci.*, **71**, 597–598.

21 Deen, W.M., Lazzara, M.J., and Myers, B.D. (2001) Structural determinants of glomerular permeability. *Am. J. Physiol. Renal Physiol.*, **281**, F579–F596.

22 Meibohm, B., and Derendorf, H. (2004) Pharmacokinetics and pharmacodynamics of biotech drugs, in *Pharmaceutical Biotechnology: Drug Discovery and Clinical Appications* (eds R. Muller and O. Kayser), Wiley-VCH Verlag GmbH, Weinheim, pp. 147–172.

23 Mager, D.E. (2006) Target-mediated drug disposition and dynamics. *Biochem. Pharmacol.*, **72**, 1–10.

24 Bauer, R.J., Gibbons, J.A., Bell, D.P., Luo, Z.P., and Young, J.D. (1994) Nonlinear pharmacokinetics of recombinant human macrophage colony- stimulating factor (M-CSF) in rats. *J. Pharmacol. Exp. Ther.*, **268**, 152–158.

25 Roopenian, D.C., and Akilesh, S. (2007) FcRn: the neonatal Fc receptor comes of age. *Nat. Rev. Immunol.*, **7**, 715–725.

26 Wang, W., Wang, E.Q., and Balthasar, J.P. (2008) Monoclonal antibody pharmacokinetics and pharmacodynamics. *Clin. Pharmacol. Ther.*, **84**, 548–558.

27 Dirks, N.L., and Meibohm, B. (2010) Population pharmacokinetics of therapeutic monoclonal antibodies. *Clin. Pharmacokinet.*, **49**, 633–659.

28 Kim, J., Hayton, W.L., Robinson, J.M., and Anderson, C.L. (2007) Kinetics of FcRn-mediated recycling of IgG and albumin in human: pathophysiology and therapeutic implications using a simplified mechanism-based model. *Clin. Immunol.*, **122**, 146–155.

Part Two
Half-Life Extension through Chemical and
Post-translational Modifications

3
Half-Life Extension through PEGylation

Simona Jevševar and Menči Kunstelj

3.1
Introduction

Protein and peptide biopharmaceuticals have been successfully used as very efficient drugs for the treatment of many pathophysiological states for almost three decades. Rapid development of recombinant DNA technology has enabled large-scale production of human proteins and has greatly extended their availability. Although proteins are very efficient in therapy, the first generation of biopharmaceuticals lack some expected characteristics for optimal drugs because their physicochemical and pharmacokinetic properties are limited. The main limitations are physicochemical instability, limited solubility, proteolytic instability, relatively short elimination half-life, immunogenicity and toxicity. Consequently protein therapeutics are mainly administrated parenterally, often very frequently and are available in liquid or lyophilized form.

Many technologies have recently been developed focusing on the characteristics of the first generation protein drugs to improve the desired pharmacokinetic properties. One of the oldest technologies for half-life extension is post-production modification by PEGylation, which also successfully addresses other deficiencies, such as limited solubility and immunogenicity.

Abuchowski *et al.* [1, 2] conducted the first conjugations of polyethylene glycol (PEG) to protein, which resulted in improved characteristics of PEG–protein conjugates during the 1970s. The first PEGylated biopharmaceutical, a PEGylated form of adenosine deaminase, Adagen® (Enzon Pharmaceuticals, USA), for the treatment of severe combined immunodeficiency disease (SCID) [3] was approved by the Food and Drug Administration (FDA) in the early 1990s. Since then, ten different PEGylated products have been FDA approved (Table 3.1). Nine of them are PEGylated proteins and one (pegaptanib [Macugen® or Macuverse®]) is a PEGylated anti-vascular endothelial growth factor (VEGF) aptamer (an RNA oligonucleotide), for the treatment of ocular vascular disease [4]. Krystexxa® (pegloticase), a PEGylated recombinant mammalian urate oxidase, an enzyme capable of reducing plasma urate concentrations, was FDA approved in September 2010 [5, 6].

Therapeutic Proteins: Strategies to Modulate Their Plasma Half-Lives, First Edition. Edited by Roland Kontermann.
© 2012 Wiley-VCH Verlag GmbH & Co. KGaA. Published 2012 by Wiley-VCH Verlag GmbH & Co. KGaA.

Table 3.1 FDA approved PEGylated proteins.

Therapeutic and number of PEG attached	Company	PEG length shape	Therapeutic indication *(Rationale for modification)*	Year of approval
Adagen® (PEGylated Bovine Adenosine Deamidase) multiPEG	Enzon	5 kDa Linear	Severe combined immuno deficiency *(Increased serum half-life)*	1990
Oncaspar® (PEGylated asparaginase) multiPEG	Enzon	5 kDa Linear	Acute lymphoblastic leukemia *(Increased serum half-life, reduced immunogenicity)*	1994
PEG-Intron® (PEGylated IFN-α2b) monoPEG	Schering-Plough/Enzon	12 kDa Linear	Hepatitis C *(Increased serum half-life)*	2001
Pegasys® (PEGylated IFN-α2a) monoPEG	Hoffmann-La Roche	40 kDa Branched	Hepatitis C *(Increased serum half-life)*	2002
Neulasta® (PEGylated hG-CSF) monoPEG	Amgen/Nektar	20 kDa Linear	Neutropenia *(Increased serum half-life)*	2002
Somavert® (PEGylated hGH mutein) multiPEG	Pfizer/Nektar	5 kDa Linear	Acromegaly *(hGH–receptor antagonist)*	2003
MIRCERA® PEGylated epoetin-β monoPEG	Hoffmann-La Roche	30 kDa Linear	anemia associated with chronic renal failure *(Increased serum half-life)*	2007
Cimzia® (Certolizumab pegol) (PEGylated anti TNF Fab) monoPEG	UCB Pharma	40 kDa Branched	Rheumatoid arthritis and Crohn's disease *(Increased serum half-life)*	2008
Krystexxa® (pegloticase) (PEGylated recombinant mammalian urate oxidase) multiPEG	Savient Pharmaceuticals	10 kDa Linear	Chronic gout *(Reduced immunogenicity and increased serum half-life)*	2010

Market authorization for several new PEGylated biopharmaceuticals can be expected in the near future, because many new products are currently in different stages of clinical trials, for example, Biogen Idec Inc. received in July 2009 a Fast Track designation from the FDA for its PEGylated IFN β-1a for treatment of multiple sclerosis, and started a Phase III study to evaluate the efficacy and safety of less frequent administration of PEGylated IFN β-1a [7, 8]. Another example is PEG-SN38, a PEGylated SN38, the active metabolite of cancer drug Campostar®, which entered Phase II clinical trials. Although SN38 exhibits 100–1000-fold higher cytotoxic activity than Campostar® its insolubility limited its conversion into a viable drug. Enzon solved the insolubility problem with PEGylation and PEG-SN38 has already showed very promising results in preclinical and clinical trials [9].

Although several successful therapeutics available on the market with improved pharmacokinetic behavior have been generated by PEGylation [10–12], the first market authorization for a product from the protein class, for which elimination half-life extension is necessary to exert a clinically meaningful effect, a PEGylated Fab, was obtained in 2008. The next two PEGylated antibody fragments, both for treatment of various tumors in combination with existing chemotherapeutic regimes, developed by UCB Pharma, di-Fab' anti- platelet derived growth factor (PDGF) (CDP860) [13] and anti- growth factor receptor (GFR) (CDP791) are in different stages of clinical trials.

As many new generation therapeutics based on protein nanobodies and protein scaffolds having short elimination half-life are in different stages of development it is expected that PEGylation will play an important role in the next generation of therapeutics.

Engineered protein scaffolds are generated from small, soluble, stable, monomeric proteins derived from several families such as lipocalins, fibronectin III, Protein A, thioredoxin, BPTI/LACI-D1/ITI-D2 (Kunitz domain) and equipped with binding sites for the desired target [14, 15]. Several drugs based on protein scaffolds are in preclinical and clinical trials, including PEGylated scaffolds: CT-322 – Adnectin-based antagonist of VEGFR-2 and DX-1000 – a Kunitz type inhibitor for blocking breast cancer growth and metastasis [14].

Nanobodies, single domain antibody fragments without the light chain, which are found in camelids and sharks, have similar characteristics so that they may generate the next generation of biopharmaceuticals to target toxins, microbes, viruses, cytokines and tumor antigens. They are fully functional and capable of binding antigen without domain pairing and can be efficiently overexpressed in a microbial expression system. An example of successful application of PEGylation to a nanobody is PEGylated nanobody neutralizing foot and mouth disease (FMD) virus [16, 17].

The result of PEG attachment to protein is a new macromolecule with significantly changed physicochemical properties showing improved pharmacokinetic (PK) and pharmacodynamic (PD) behavior [18, 19]. Generally, increase of the PEG chain brings increase in the elimination half-life of the PEG–protein conjugate, while biological activity *in vitro* is generally significantly reduced by increase of

PEG length. In addition to PEG length, its shape greatly influences the absorption rate, elimination half-life, and biological activity *in vitro*. Branched PEGs extend elimination half-life more than linear PEGs of the same nominal molecular weight (MW), but at the same time biological activity *in vitro* is reduced more when branched PEG is employed [20, 21].

The flexible nature of PEG chains shields the protein and protects it from the environment influences on the interactions of the protein that are responsible for its biological function. This characteristic of PEG chains means that *in vitro* activity determined by cell-based bioassays for PEGylated proteins is not predictive for the *in vivo* therapeutic effect, because the major effect of PEGylation is steric hindrance caused by flexible PEG chain and not conformational changes. Although steric hindrance reduces the binding affinity to the receptor, there are plenty of opportunities for the receptor–ligand interactions because of the prolonged circulating half-life [10].

Pegasys®, a PEGylated IFN α-2a, is a typical example of a very efficient PEGylated protein drug that displays an *in vitro* activity of only a few percent of the level of the unmodified IFN α, while its efficacy justifies replacement of the first generation IFN α in therapy [22].

3.2
Preparation and Physico-Chemical Characterization of PEGylated Proteins

PEGylation of a protein of interest starts with the selection of an appropriate PEG reagent, followed by suitable chemistry for PEG attachment and subsequent purification and characterization of the final PEG–protein conjugates, which is illustrated in Figure 3.1.

3.2.1
Selection of PEG Reagents and PEGylation Chemistry

PEG reagents suitable for pharmaceutical use are commercially available in different lengths, shapes and chemistries. Commercial suppliers are for example, NOF Corporation (Japan); SunBio (South Korea); Dr. Reddys (UK); JenKem (China); and Creative PEGWorks (USA).

PEG is regarded as a nontoxic, nonimmunogenic, hydrophilic, uncharged and nondegradable polymer and has been approved by the FDA [23]. Although there are many new PEG formats such as forked, multi-arm and comb-shaped PEGs available with certain potential for the future, linear and branched PEGs with molecular masses up to 40 kDa, which bring the desired improvement of pharmacokinetic properties, are still the first choice. One of the main limitations for the employment of traditional long linear PEG reagent with MW above 30 kDa is the high viscosity of the resulting PEG–protein conjugates as well as potential accumulation of PEG reagents in the body. In this sense multi-arm PEGs and especially comb-shaped PEGs bearing numerous short PEG chains attached to the polymer

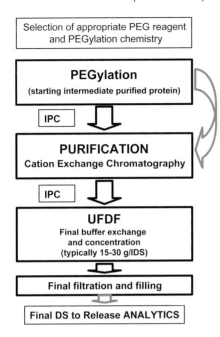

Figure 3.1 Typical flow chart diagram for manufacturing of PEG–protein conjugates starting from purified protein intermediate.

backbone offer the additional advantage of significantly reduced viscosity of final PEG–protein conjugates compared with classical PEG–protein conjugates [12]. A promising approach is suggested by releasable PEGylation – an attachment of PEG reagent with releasable linker to the protein. Despite lacking the major benefits of traditional PEGylation for example, long elimination half-life, reduced immunogenicity, reduced proteolysis, easier formulation, and analytics of stable PEGylated proteins, this approach shows that PEGylation can be used to improve the solubility of poorly soluble drugs and deposit such drugs at the target; it allows random PEGylation and by appropriate selection of the linker also control of PK parameters [24].

3.2.1.1 Random PEGylation

Chemical reaction between the protein and suitably activated PEGylation reagents results in a PEG–protein conjugate. Various chemical groups in the amino acid side-chains can be exploited for the reaction with PEG moiety, such as –NH2, –NH–, –COOH, –OH, –SH groups as well as disulfide (–S–S–) bonds. For successful drug candidate generation based on PEGylation, several aspects have to be considered besides the PEG attachment site on the protein, such as activation chemistry of PEG reagent, stable or releasable PEG attachment to protein, length and shape of the linker, length, shape and structure of the PEG reagent, and end capping of the PEG chains.

The majority of the PEGylation reagents for random PEGylation targets amino groups on the protein, most frequently the ε-amino groups on the side-chains of lysine residues, which are often exposed and available for chemical attachment of PEG reagents. Reaction proceeds quickly and results in complex mixtures of isoforms, differing in the PEG attachment site to the protein. Although the reaction can be directed to some extent by the pH of the reaction mixture, this type of conjugation reaction usually leads to complex PEGylation mixtures. Although a variety of other chemistries have been tried in the past [2, 3, 25–27], the most frequently used random PEGylation reagents nowadays possess the activated carbonyl group in the form of *N*-hydroxy-succinimide (NHS) esters that form stable protein–PEG conjugates via amide linkages. Stable linkage between PEG and protein enables easier production, characterization and formulation of the final drug. The first PEGylated pharmaceuticals Adagen® and Oncaspar® were complex mixtures of various PEGylated species and were launched by Enzon in the early 1990s. The most prominent examples of randomly PEGylated drugs available on the market are PegIntron®, Pegasys® and Mircera®, approved 2001, 2002 and 2007, respectively. All three are mixtures of mono-PEGylated positional isomers possessing substantially longer elimination half-lives than their unmodified versions enabling less frequent administration [28–30].

Another interesting PEGylated biopharmaceutical drug, produced by random PEGylation, is Somavert®, genetically modified and multiPEGylated human growth hormone (hGH) acting as an antagonist of the hGH receptor (HGHR) for the treatment of acromegaly [31]. Somavert® is a good example showing certain potential of PEGylation to convert agonist to antagonist.

An example of modified protein, where lysines were introduced into the molecule to achieve more selective attachment of the PEG moiety, is an improved version of PEGylated G-CSF (Maxy G34), multiPEGylated G-CSF analog, that has successfully completed Phase IIa clinical trials [32, 33]. A similar approach by eliminating two of three lysines present in the original molecule, enabling more selective attachment of PEG to a single lysine in combination with removal of N-terminal fusion protein after PEGylation, is described by Roche [34–36] for the preparation of homogeneous monoPEGylated IGF1 for the treatment of neuromuscular disorders, in particular amyotrophic lateral sclerosis (ALS), which is in the early stages of clinical trials.

3.2.1.2 Site-Specific PEGylation

Although purification would allow homogenous product preparation, one of the obvious trends in the development of the PEGylation technology is a shift from random to site-specific PEGylation reactions, leading to better defined products and higher yields of the manufacturing process. Examples of classical approaches of site-specific PEGylation reactions are reductive alkylation for N-terminal PEGylation and cysteine-specific PEGylation.

N-terminal PEGylation, performed as a reductive alkylation step with a PEG–aldehyde reagent [37], was applied to generate Neulasta® [38], a N-terminally mono-PEGylated G-CSF bearing a 20 kDa PEG.

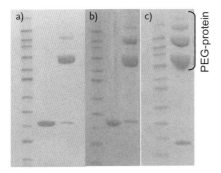

Figure 3.2 Complexity of PEGylation mixture after reductive alkylation directed to N-terminus of different proteins containing different number of exposed lysines (a) protein with low number of exposed lysines; (b) protein with high number of exposed lysines; (c) peptide with high number of exposed lysines.

Reductive alkylation is a general tool which can be applied to any protein. The method is said to be rather selective for the amino group of N-terminus, the selectivity being claimed to depend on the pH value [37–39]. Although the reaction may primarily occur at the amino group of N-terminus, amino groups of exposed lysines are accessible for reaction as well, leading to higher complexity of PEGylation mixture when many exposed lysines are present in protein which is shown in Figure 3.2.

PEGylation of thiol groups in natural or genetically introduced free cysteines is another well-known approach to site-specific PEGylation. Although a variety of thiol-specific reagents are available, maleimide PEG reagents are the most frequently used, leading to stable PEG–protein conjugates. In native proteins, exposed free cysteines are rarely available, therefore free Cys has to be genetically introduced into protein in order to be available for PEGylation. Genetically introduced cysteines can be carefully selected far from the receptor binding site to retain high *in vitro* bioactivity. Additionally, higher selectivity of reaction can be achieved than for reductive alkylation directed to the N-terminus when many exposed lysines are present in the protein. Preparation of PEGylated cysteine analogs of IFN-α2a and Factor VIII with high preserved *in vitro* activity can be found in the literature [40, 41]. Although engineered cysteines appear as a very elegant solution to selective PEGylation, their use is often limited with complications caused in the refolding and purification process.

Cysteine-specific PEGylation is usually employed for Fab' fragments. Cysteine residues in the hinge region of Fab' fragments that are far from the antigen-binding region are exploited for specific conjugation, resulting in a well-defined product. Although short elimination half-life of Fab' fragments is substantially prolonged by PEGylation, PEG–Fab's do not fully compete with full antibody regarding elimination half-life. However PEGylation appears to be an ideal method of reducing the antigenicity of Fab's, and PEGylated Fab's can be superior to full antibodies when elimination of undesired side effects originating from the Fc region is required. Cimzia®, a PEGylated Fab' fragment bearing a 40 kDa branched PEG site specifically attached to a free hinge cysteine is the result of the approach described. More efficient PEGylation of Fab' fragments can be achieved by exploiting an interchain disulfide bond for cysteine-specific PEGylation and removal of free Cys in the hinge region. The interchain disulfide bond has to be reduced prior to PEGylation, but contrary to careful application of mild reduction conditions required for free cysteines, quite harsh reduction conditions can be applied, resulting in selective reduction of only interchain S–S bonds. Final PEGylated Fab' molecules devoid of the covalent linkage between the two antibody chains usually retain high levels of chemical and thermal stability and good behavior in pharmacokinetic and efficacy models [42].

A promising approach to the PEGylation of protein disulfide bonds is a PEGylation technology of PolyTherics: TheraPEG™ using PEG monosulfone reagents [43, 44]. The strategy is appropriate for specific PEGylation of Fab' fragments where the interchain disulfide bond can be exploited. Due to the nature of the TheraPEG™ reagent the two chains of Fab' fragments remain linked through the PEG reagent instead of the interchain disulfide bond.

Another well-known site-specific PEGylation is offered by so-called Amber technology via non-natural amino acids [45]. Several products now in different stages of preclinical and clinical development have been developed by the Ambrx company applying Amber technology (e.g., mono-PEGylated hGH molecules with improved pharmacological properties in Phase II clinical trial; IFN-β, FGF21, Leptine, all still in preclinical trials) [46].

Another interesting approach to site-specific conjugation of synthetic molecules to the C-terminus of recombinant proteins has been developed by the Almac company and can be applied to a wide range of proteins including cytokines, antibody fragments and enzymes [47].

3.2.2
Purification of PEGylated Proteins

After selection of suitable PEGylation chemistry leading to a rather complex PEGylation mixture, purification of PEGylated proteins is required to obtain the final product. The desired PEG–protein conjugate has to be separated from unreacted protein, multi-PEGylated proteins, unreacted PEG reagent and from other reagents that might be added to the PEGylation mixture. Differences in charge, hydrodynamic radius, hydrophobicity and in some cases also affinity are exploited to separate the desired PEG–protein conjugate [48, 49]. Size-exclusion chromatography (SEC) has been widely used to separate PEG conjugates in the past as increase of MW is one of the most evident changes caused by PEGylation. SEC has several limitations, in addition to low throughput and high costs making it highly undesired on the industrial scale, its resolution power for PEG and PEG–

protein conjugate is not very high. Additionally, SEC is not able to separate positional isomers of the same MW.

Generally the method of choice for isolation of PEGylated proteins is cation-exchange chromatography (CEX) enabling an efficient single-step purification of the target PEG–protein conjugate from the PEGylation mixture, although other types of chromatography can be successfully used [50]. CEX also possesses the ability to separate positional isomers of the same MW based on their charge difference. The elution order of PEG–protein conjugates is determined by the PEG-to-protein mass ratio [51, 52]. Additionally retention times on CEX also depend on MW of PEG attached to the protein meaning that conjugates with PEGs of higher MW exhibit lower retention times as a consequence of stronger shielding effect of PEG. The PEG shielding effect can be exploited for efficient CEX separation, even in cases where no charge difference between PEGylated and un-PEGylated protein exists [48, 52–54].

3.2.3
Physico-Chemical Characterization of PEG Reagents and PEG–Protein Conjugates

Analytical methods capable of good characterization of PEG–protein conjugates and PEG reagents at various stages are crucial for the successful development of an efficient PEGylation and subsequent purification process as well as for the safety of the final product.

3.2.3.1   Characterization of PEG Reagents
The quality of PEG reagents is reflected by the consistency of the molecular mass, polydispersity index, presence of activated and nonactivated impurities and degree of activation. The most critical quality attributes of PEG reagents affecting the final quality of drug substance are polydispersity and absence of activated impurities in PEG reagent.

Polydispersity has been significantly improved with the development of PEG reagent synthesis and purification and several PEG suppliers are able to provide PEG reagents of relatively narrow distribution. Polydispersity, the average MW, and amount of the main peak fraction in PEGs are routinely controlled by SEC. However, a more precise determination of MW of PEGs as well as PEG–protein conjugates is possible by the matrix assisted laser desorption/ionization time-of-flight (MALDI-TOF) technique [55].

PEG reagents are ultraviolet (UV) transparent and nonfluorescent, therefore they can be detected with evaporative light scattering or corona discharge charged aerosol detectors that detect particulate matter in the gas phase [56]. For UV or fluorescence detection a derivatization method is needed to produce UV absorbance, for example, PEG aldehyde can be derivatized with 4-aminobenzoic acid and analyzed using reverse phase (RP) high performance liquid chromatography (HPLC) or size-exclusion SE-HPLC [52]. Such an alternative method is very powerful in detecting activated impurities in PEG reagent.

3.2.3.2 **Characterization of PEG–Protein Conjugates**

The characterization of PEGylated proteins is influenced by the fact that the PEG molecule attached to the protein changes and masks the characteristics of the protein substantially. This leads to limited characterization power compared with unmodified protein and is still the main reason why highly purified protein usually enters into the PEGylation process. Standard analytical techniques used to characterize proteins are successfully employed to characterize PEG–protein conjugates. However the behavior of unmodified protein and its PEGylated counterpart is significantly different. This usually requires different conditions, often also a different column for the characterization of PEG–protein conjugates with the same power as unmodified protein by HPLC analytical methods.

Comparison of the behavior of PEG reagents, PEG–protein conjugates and proteins of approximately the same nominal MW on SEC and sodium dodecyl sulfate polyacrylamide gel electrophoresis (SDS-PAGE) show distinctly different retention times and mobility in gel; PEGs and PEG–protein conjugates are in complete disagreement with the protein MW standard [21, 57, 58].

The RP-HPLC method is a powerful and robust method for the determination of purity and content of PEGylated proteins, including the amount of higher-PEGylated and un-PEGylated species, protein oxidation, deamidation, and cleavage of the protein backbone [59] as well as for RP-HPLC peptide mapping [26, 54, 60]. The resolution power for the detection of oxidized and deamidated species is often significantly affected by the masking effect of PEGs, which is more pronounced when longer PEGs are attached to protein.

Cation exchange (CE)-HPLC is the only analytical method with separation power for positional isoforms and it has been employed for characterization of commercially available products, Pegasys® [60] and PegIntron® [61]. Additionally CE-HPLC is able to distinguish between different length of PEG chain, as the PEG size affects the retention time as a consequence of the shielding effect of PEG, which influences the interaction between the protein and the chromatographic matrix [54].

Various modes of mass spectroscopy are already in use for the characterization of biopharmaceuticals [62] and with further development it is likely to be introduced as a standard technique in the routine production of modern biopharmaceuticals [63]. It is a useful tool for the identification and quantification of PEGylation sites [63–65], as well as for characterization of impurities, which may sometimes not be resolved and detected by simpler techniques due to the masking effect of PEG.

3.3
Pharmacokinetic (PK) Behavior of PEGylated Proteins

The modulation of protein pharmacokinetic properties focused on prolongation of circulating half-life, change of elimination, and biodistribution properties is the main result of protein modification, including PEGylation. Size enlargement of

the protein by PEGylation is the main reason for altered pharmacokinetics properties, although better proteolytic stability and generally reduced immunogenicity
cannot be neglected.

3.3.1
Administration Route of PEGylated Proteins

The administration route greatly influences the pharmacokinetic properties of the
PEGylated drugs. These drugs are usually administrated by the intravenous (i.v.)
or subcutaneous (s.c.) route. Different routes of administration have been studied
by injecting PEGs of different sizes [66]. The slowest absorption has been achieved
by intramuscular (i.m.) application, followed by s.c. and intraperitoneal (i.p.) application. PEGs of lower MW (e.g., 6 kDa) disappear very quickly from the injection
site, while the absorption of larger PEGs is substantially slower. For larger PEGs
(e.g., 50 kDa) administered s.c. and i.m., the subcutis and muscles work as a reservoir that slowly releases PEG into the blood circulation. Consequently the
absorption is significantly slower than elimination and the absorption becomes
the parameter controlling the elimination rate of the PEG from the body. Similar
pharmacokinetic behavior has also been determined for PEG–protein
conjugates.

Bioavailability of s.c. and i.m administered PEGylated proteins is reduced compared with i.v. application [19]. The bioavailability after i.m., s.c. and i.p. application has been studied with PEGylated superoxide dismutase. Compared with i.v.
application with 100% bioavailability the i.p., i.m. and s.c. application resulted in
71%, 54%, and 29% bioavailability, respectively [67]. The bioavailability after s.c.
application of a PEGylated protein increases with PEG length. The bioavailability
of human granulocyte colony stimulating factor (hG-CSF) PEGylated with 30 kDa
linear PEG compared with 20 kDa linear PEG has been reported to be higher by
60% [68].

3.3.2
Elimination Properties of PEGylated Proteins

Longer PEG chains, which are used for PEGylation of proteins, are not subjected
to metabolism, and the elimination mechanism depends on their MW. Study of
elimination properties of PEG after i.v. administration revealed that PEGs smaller
than 20 kDa are predominantly removed by the kidneys. As the renal elimination
decreases with increasing PEG length the liver uptake becomes a predominant
way of elimination of larger PEGs and PEGs above 50 kDa are eliminated only by
liver uptake [69]. The relationship between PEG MW and elimination half-life after
i.v. application is sigmoidal. PEGs smaller than 8 kDa are filtrated freely. The renal
filtration of PEGs with MW between 8 and 30 kDa is size-dependent with significant increase of elimination half-life between 20 and 30 kDa. Renal filtration of
PEGs larger than 30 kDa is reduced to a minimum and the half-life of larger PEGs
increases very slowly [66, 67]. It is now well established that at least 20 kDa PEG

is needed to achieve clinically significant reduction of renal clearance [10]. The renal elimination properties of PEGs can be explained with glomerular filtration cut-off value, which for globular proteins is around 70 kDa. This MW cannot be directly extrapolated to PEGs, since PEGs are very well hydrated and depending on the MW exhibits up to 10 times larger hydrodynamic volumes (Table 3.2). The pore size of the glomerular membrane is reported to be between 3 and 5 nm [67].

Table 3.2 Hydrodynamic radius of protein standards, linear PEG reagents and PEGylated interferon alpha2b conjugates (PEG-IFN) determined with DLS and corresponding elimination half-life after i.v. application – the data were obtained from Kusterle *et al.* [21].

Hydrodynamic radius [nm]	MW of Proteins [kDa]	MW of PEGs [kDa]	MW of PEG-IFN with 10, 20, 30 and 45 kDa PEG [kDa]	$t_{1/2}$ after i.v administration [h]
2.1	13.7 (Ribonuclease A)			
2.2	19.5 (Interferon alpha2b)			
2.3		5		0.3[a]
2.6	29 (Carbonic anhydrase)			
3.6	43 (Ovalbumin)			
3.7		10		
4.1	75 (Conalbumin)			
4.7	158 (Aldolaza)			
5.0		20		3[a]
5.7			29.5 (PEG-IFN-10L)	7.3
6.6		30		
7.2		45 B		~14[a]
7.4			39.5 (PEG-IFN-20L)	10.5
9.1			49.5 (PEG-IFN-30L)	19.9
9.6	440 (Feritin)			
9.6			64.5 (PEG-IFN-45B)	23.9

a) The data for elimination half-life of PEG reagents were obtained from the paper of Caliceti and Veronese [67].

The hydrodynamic radius of 20 and 30 kDa PEG has been determined to be 5.0 nm and 6.6 nm, respectively (Table 3.2). The hydrodynamic radius of 20 and 30 kDa PEG are in the range of glomerular membrane pore size which nicely explains the significant increase of PEG elimination half-life between 20 and 30 kDa. PEG with hydrodynamic radius larger than glomerular membrane pore size can still be filtrated at low rates, due to high flexibility of the linear polymer chain [67].

PEGylated conjugates are predominantly eliminated by a combination of renal and hepatic pathways. Renal clearance is the predominant route of elimination for low MW conjugates, as with increasing MW the elimination pathway shifts from renal to liver uptake. For IFN alpha2a PEGylated with 40 kDa branched PEG it was determined that the primary route of elimination is hepatic [70].

The renal filtration properties of PEGylated conjugates depend on their hydrodynamic radius, which is determined by the molecular mass of the protein and the length and the number of PEG attached. In the case of N-terminally PEGylated IFN alpha2b analogs with 10, 20, 30 and 45 kDa PEG the hydrodynamic radius of the PEG–protein conjugate is a sum of the PEG and the protein hydrodynamic radius (Table 3.2). However this is not always the case. It has been shown that attachment of a few small size PEGs does not significantly alter the hydrodynamic radius [71] or that single long PEG chain increases elimination half-life more than attachment of a few short PEGs with the same net MW [72]. The basic rule that elimination half-life increases with the increased PEG length or with the number of PEG attached was confirmed in a i.v. study of N-terminally PEGylated IFN alpha2b analogs with 10, 20, 30 and 45 kDa PEG, where elimination half-life increased from 7.3 h for 10 kDa analog to 23.9 h for 45 kDa analog (Table 3.2). The same study also revealed that the elimination half-life also increased linearly for 30 and 45 kDa PEG conjugate, both with hydrodynamic radius of 9.1 nm and 9.6 nm, respectively, clearly above the glomerular filtration cut-off limit. This indicates that beside renal filtration other mechanisms, like reduced proteolysis and clearance by the immune system, are involved in the elimination process of IFN alpha2b analogs. There are some results from the literature, that support the kidney filtration model with a hydrodynamic cut-off limit after which the elimination half-life does not increase significantly [71, 73].

Proteolysis, one of the protein elimination pathways, is usually significantly reduced after PEGylation due to the shielding effect of the highly hydrated polyethylene glycol chain. PEGylation protection from trypsin digestion has been studied on two differently PEGylated conjugates of hGH [own unpublished results]. Unmodified hGH was digested in a few minutes, while 14% of N-terminally PEGylated hGH with 30 kDa linear PEG and 61% of randomly multiPEGylated hGH with four 5 kDa PEG were found undigested after 30 min (Table 3.3). The value of 61% of undigested hGH with four 5 kDa PEG compared with 14% of undigested hGH with one 30 kDa PEG attached is in agreement with literature data reporting that attachment of a higher number of small PEGs and employment of branched PEGs offers better proteolytic protection than monoPEGylation with linear PEG of the same net MW [20].

Table 3.3 Time profile of trypsin digestion of native hGH, monoPEGylated hGH with 30 kDa PEG and multiPEGylated hGH with 5 kDa PEG.

Time (min)	% of undigested hGH	% of undigested PEG-hGH-30L	% of undigested multiPEG-hGH-5L
0	100	100	100
1	2	36	89
2	0	29	88
10	0	18	70
30	0	14	61

3.3.3
Biodistribution Properties of PEGylated Proteins

PEGs and PEG–protein conjugates administered i.v. exhibit two-compartment pharmacokinetic behavior, meaning that they are distributed in to extravascular tissues [66, 67]. The vascular wall functions as a permeation barrier for PEG molecules. PEGs of low MW permeate to extravascular tissues more rapidly and in a larger amount than PEGs of high MW. Small PEGs (e.g., 6 kDa) display a clear peak in extravascular accumulation-time curve, while for larger PEGs accumulation to a constant level is observed. This can be explained by the fact that low MW PEGs exhibit high vascular permeability and are distributed rapidly into extravascular tissues, but can also return to the blood circulation when the concentration in blood drops. For PEGs of larger MW (above 20 kDa) the clearance from the tissues is reduced resulting in gradual accumulation in the extravascular space [66].

The PEG–protein conjugates display the same behavior as PEGs. The clearance after hepatic uptake of PEGylated catalase was much slower than for the unmodified or glycosylated form, which resulted in more extensive liver accumulation. The same behavior can also be expected for other organs rich in reticuloendothelial cells, such as spleen, lymph nodes, lung, and kidneys. It has been demonstrated that PEGylation can promote lymph localization and that lymphatic bioavailability can be increased with the PEG size [67].

PEGylation can also be used for passive targeting of solid tumors, as it is known that solid tumors are hypervascular with defective vascular architecture and exhibit enhanced permeability and retention for macromolecules [10].

3.3.4
Increased Circulation Lifespan of PEGylated Proteins

Protein therapeutics have very limited lifespan in the circulation because the eliminating mechanisms are efficient. Increased residence time in the bloodstream of PEG–protein conjugates is a result of the changes in pharamacokinetic properties achieved by PEGylation. The PEG shielding effect reduces proteolysis, opsonization, and removal of PEG–protein conjugates by the immune system. Increased MW of PEG–protein conjugate reduces renal elimination and at the same time lowers diffusion into extravascular tissues. Typically, elimination half-life of plasma proteins, such as cytokines and growth factors, is in the range of few minutes to a few hours [74], which clearly indicates the necessity for half-life extension to ensure less frequent administration, more efficient and patient-friendly therapies. Contrary to most plasma proteins, the elimination half-life of antibodies is much longer, ranging from one to four weeks, and the desirability of half-life strategies is to extend the elimination half-life of proteins close to the half-life of antibodies. Interestingly, the elimination half-life of Fab' fragments is very short, only up to a few hours, despite their MW of 50 kDa being larger than the MW of cytokines and growth hormones.

3.4
Safety of PEGylated Proteins

PEGylation normally reduces immunogenicity of proteins by the steric hindrance mechanism. There are examples of transforming immunogenic proteins into a tolerogen by PEGylation [75]. PEGaspargase is a good example, where PEGylation solved the problem of neutralizing the formation of antibodies associated with the use of native asparaginase. Another example is a PEGylated recombinant mammalian urate oxidase, an enzyme capable of reducing plasma urate concentrations in patients with hyperuricemia [6, 76]. However, a few reports on induction of anti-PEG immune responses in the case of repeated administration of PEGylated liposomes [77, 78] or PEG-glucuronidase can be found in the literature [79].

PEG is generally recognized as a nonbiodegradable polymer, but some reports show that it can be oxidatively degraded by various enzymes as alcohol and aldehyde dehydrogenases [80] and cytochrome P450-dependent oxidases [81]. *In vivo* cleavage of branched PEGs at the junction site of two linear PEG chains was observed for 40 kDa branched PEG attached to IFN alpha2a [70]. All the aforementioned degradation pathways occur only to a limited extent and therefore cannot be considered as a significant degradation route for PEGylated proteins.

Despite slow kidney elimination, very limited biodegradability of PEG, limited proteolysis and excretion, all the elimination mechanisms can prevent high mass accumulation of PEGylated proteins if they are administered in low doses [67].

Toxicological experiments with very high doses of PEG–protein conjugates induced reversible renal tubular vacuolization which has not caused functional abnormalities. However, equivalent doses of PEG alone or the non-PEG-linked protein did not cause light microscopic evidence of vacuolation suggesting that the combination of PEG and protein in PEG–protein conjugates was necessary to induce changes of sufficient magnitude to be detected [82]. A few warnings referring to significant PEG–protein accumulation in the liver which may increase the risk of toxicity can be found in the literature [67, 83, 84]. At significantly

1000–10000-fold lower quantities of PEG–protein conjugates administered in therapy, side effects such as renal tubular vacuolization or a significant immunogenic response are not expected [67, 75, 85–87].

Although no severe side effects have been reported after use of PEGylated therapeutics for a more than decade, the potential consequences of lifelong treatment with high dosages of PEG–protein conjugates bearing PEGs of higher molecular mass (above 30 kDa) should be considered as certain PEG accumulation can be expected in such cases.

3.5
Conclusions

Several aspects of PEGylation technology which should be considered in order to generate successful PEG–protein therapeutics are discussed in this review, starting with the selection of appropriate PEG length and shape to gain the desired pharmacokinetic characteristics. In addition to pharmacokinetic PEG length and shape, PEGylation chemistry influences production efficiency and characterization of final PEG–protein conjugates. Although protein modification technologies primarily focus on half-life extension, other characteristics such as solubility, biodistribution, and safety aspects should not be neglected. Reduction of immunogenicity of the proteins by the masking effect of the PEG chain has been demonstrated in several examples that show the certain potential of this technology in preparation of less immunogenic biotherapeutics.

PEGylation is a proven technology, which has already resulted in ten FDA approved therapeutics, confirming its safety and applicability. Since its introduction PEGylation has been focused more on existing therapeutic proteins and their life-cycle management than generating of completely new drugs. With the emergence of therapeutics based on protein nanobodies and scaffolds wider employment of PEGylation technology can be expected. In addition to half-life extension, broader use of this technology could solve solubility problems for many proteins by employing either traditional PEGylation, which also brings half-life extension, or releasable PEGylation, which enables release of fully active drug.

References

1 Abuchowski, A., Vanes, T., Palczuk, N.C., and Davis, F.F. (1977) Alteration of immunological properties of bovine serum-albumin by covalent attachment of polyethylene-glycol. *J. Biol. Chem.*, **252**, 3578–3581.

2 Abuchowski, A., Mccoy, J.R., Palczuk, N.C., Vanes, T., and Davis, F.F. (1977) Effect of covalent attachment of polyethylene-glycol on immunogenicity and circulating life of bovine liver catalase. *J. Biol. Chem.*, **252**, 3582–3586.

3 Levy, Y., Hershfield, M.S., Fernandez-Mejia, C., Polmar, S.H., Scudiery, D., Berger, M., and Sorensen, R.U. (1988) Adenosine deaminase deficiency with late onset of recurrent infections: response to treatment with polyethylene glycol-modified adenosine deaminase. *J. Pediatr.*, **113**, 312–317.

4 Ng, E.W., Shima, D.T., Calias, P., Cunningham, E.T., Jr, Guyer, D.R., and Adamis, A.P. (2006) Pegaptanib, a targeted anti-VEGF aptamer for ocular vascular disease. *Nat. Rev. Drug Discov.*, 5, 123–132. doi: nrd1955 [pii];10.1038/nrd1955

5 Sherman, M.R., Saifer, M.G., and Perez-Ruiz, F. (2008) PEG-uricase in the management of treatment-resistant gout and hyperuricemia. *Adv. Drug Deliv. Rev.*, 60, 59–68. doi: S0169-409X(07)00138-X [pii];10.1016/j.addr.2007.06.011

6 U.S. Food and Drug Administration, (2010) FDA approves new drug for gout, http://www.fda.gov/NewsEvents/Newsroom/PressAnnouncements/ucm225810.htm (accessed 26 October 2010).

7 Medical News Today, (2009) Biogen Idec receives fast track designation from FDA for PEGylated interferon beta-1a for relapsing multiple sclerosis, http://www.medicalnewstoday.com/articles/156976.php (accessed 26 October 2010).

8 Baker, D.P., Lin, E.Y., Lin, K., Pellegrini, M., Petter, R.C., Chen, L.L., Arduini, R.M., Brickelmaier, M., Wen, D., Hess, D.M., Chen, L., Grant, D., Whitty, A., Gill, A., Lindner, D.J., and Pepinsky, R.B. (2006) N-terminally PEGylated human interferon-beta-1a with improved pharmacokinetic properties and *in vivo* efficacy in a melanoma angiogenesis model. *Bioconjug. Chem.*, 17, 179–188. doi: 10.1021/bc050237q

9 Zhao, H., Rubio, B., Sapra, P., Wu, D., Reddy, P., Sai, P., Martinez, A., Gao, Y., Lozanguiez, Y., Longley, C., Greenberger, L.M., and Horak, I.D. (2008) Novel prodrugs of SN38 using multiarm poly(ethylene glycol) linkers. *Bioconjug. Chem.*, 19 (4), 849–859. Epub 2008 Mar 28.

10 Bailon, P., and Won, C.Y. (2009) PEG-modified biopharmaceuticals. *Expert. Opin. Drug Deliv.*, 6, 1–16. doi: 10.1517/17425240802650568

11 Kang, J.S., DeLuca, P.P., and Lee, K.C. (2009) Emerging PEGylated drugs. *Expert Opin. Emerg. Drugs*, 14, 363–380. doi: 10.1517/14728210902907847

12 Ryan, S.M., Mantovani, G., Wang, X., Haddleton, D.M., and Brayden, D.J. (2008) Advances in PEGylation of important biotech molecules: delivery aspects. *Expert Opin. Drug Deliv.*, 5, 371–383. doi: 10.1517/17425247.5.4.371

13 Serruys, P.W., Heyndrickx, G.R., Patel, J., Cummins, P.A., Kleijne, J.A., and Clowes, A.W. (2003) Effect of an anti-PDGF-beta-receptor-blocking antibody on restenosis in patients undergoing elective stent placement. *Int. J. Cardiovasc. Intervent.*, 5, 214–222. doi: 10.1080/14628840310017177; VXUDJ9RWG3MN9XHU

14 Gebauer, M., and Skerra, A. (2009) Engineered protein scaffolds as next-generation antibody therapeutics. *Curr. Opin. Chem. Biol.*, 13, 245–255. doi: S1367-5931(09)00068-4 [pii];10.1016/j.cbpa.2009.04.627

15 Skerra, A. (2007) Alternative non-antibody scaffolds for molecular recognition. *Curr. Opin. Biotechnol.*, 18, 295–304. doi: S0958-1669(07)00080-8 [pii];10.1016/j.copbio.2007.04.010

16 Harmsen, M.M., and De Haard, H.J. (2007) Properties, production, and applications of camelid single-domain antibody fragments. *Appl. Microbiol. Biotechnol.*, 77, 13–22. doi: 10.1007/s00253-007-1142-2

17 Wesolowski, J., Alzogaray, V., Reyelt, J., Unger, M., Juarez, K., Urrutia, M., Cauerhff, A., Danquah, W., Rissiek, B., Scheuplein, F., Schwarz, N., Adriouch, S., Boyer, O., Seman, M., Licea, A., Serreze, D.V., Goldbaum, F.A., Haag, F., and Koch-Nolte, F. (2009) Single domain antibodies: promising experimental and therapeutic tools in infection and immunity. *Med. Microbiol. Immunol.*, 198, 157–174. doi: 10.1007/s00430-009-0116-7

18 Fishburn, C.S. (2007) The pharmacology of PEGylation: balancing PD with PK to generate novel therapeutics. *J. Pharm. Sci.*, doi: 10.1002/jps.21278

19 Hamidi, M., Rafiei, P., and Azadi, A. (2008) Designing PEGylated therapeutic molecules: advantages in ADMET properties. *Expert Opin. Drug Discov.*, doi: 10.1517/17460441.3.11.1293

20 Veronese, F.M., Caliceti, P., and Schiavon, O. (1997) Branched and linear poly(ethylene glycol): influence of the polymer structure on enzymological,

pharmacokinetic, and immunological properties of protein conjugates. *J. Bioact. Compat. Polym.*, **12**, 196–207. doi: 10.1177/088391159701200303

21 Kusterle, M., Jevsevar, S., and Porekar, V.G. (2008) Size of pegylated protein conjugates studied by various methods. *Acta Chim. Slov.*, **55**, 594–601.

22 Bailon, P., Palleroni, A., Schaffer, C.A., Spence, C.L., Fung, W.J., Porter, J.E., Ehrlich, G.K., Pan, W., Xu, Z.X., Modi, M.W., Farid, A., Berthold, W., and Graves, M. (2001) Rational design of a potent, long-lasting form of interferon: a 40 kDa branched polyethylene glycol-conjugated interferon alpha-2a for the treatment of hepatitis C. *Bioconjug. Chem.*, **12**, 195–202. doi: bc000082g [pii]

23 Pasut, G., and Veronese, F.M. (2007) Polymer-drug conjugation, recent achievements and general strategies. *Prog. Polym. Sci.*, **32**, 933–961. doi: 10.1016/j.progpolymsci.2007.05.008

24 Filpula, D., and Zhao, H. (2008) Releasable PEGylation of proteins with customized linkers. *Adv. Drug Deliv. Rev.*, **60**, 29–49. doi: S0169-409X(07)00136-6 [pii];10.1016/j.addr.2007.02.001

25 Abuchowski, A., Kazo, G.M., Verhoest, C.R., Jr, Van, E.T., Kafkewitz, D., Nucci, M.L., Viau, A.T., and Davis, F.F. (1984) Cancer therapy with chemically modified enzymes. I. Antitumor properties of polyethylene glycol-asparaginase conjugates. *Cancer Biochem. Biophys.*, **7**, 175–186.

26 Zalipsky, S., Seltzer, R., and Menon-Rudolph, S. (1992) Evaluation of a new reagent for covalent attachment of polyethylene glycol to proteins. *Biotechnol. Appl. Biochem.*, **15**, 100–114.

27 Zalipsky, S., Seltzer, R., and Nho, K. (1991) Succinimidyl carbonates of polyethylene glycol, in *Polymeric Drugs and Drug Delivery Systems* (eds R.L. Dunn and R.M. Ottenbrite), American Chemical Society, Washington, DC, pp. 91–100.

28 European Medicines Agency, (2009) PegIntron: EPAR product information: Summary of product characteristics, http://www.ema.europa.eu/docs/en_GB/document_library/EPAR_-_Product_

Information/human/000280/WC500039388.pdf (accessed 04 August 2011).

29 European Medicines Agency, (2009) Pegasys: EPAR product information: Summary of product characteristics, http://www.ema.europa.eu/docs/en_GB/document_library/EPAR_-_Product_Information/human/000395/WC500039195.pdf (accessed 19 August 2011).

30 European Medicines Agency, (2009) MIRCERA: EPAR product information: Summary of product characteristics, http://www.ema.europa.eu/docs/en_GB/document_library/EPAR_-_Product_Information/human/000739/WC500033672.pdf (accessed 7 March 2011).

31 European Medicines Agency, (2009) SOMAVERT: EPAR product information: Summary of product characteristics, http://www.ema.europa.eu/docs/en_GB/document_library/EPAR_-_Product_Information/human/000409/WC500054629.pdf (accessed 7 June 2011).

32 Nissen, T.L., Andersen, K.V., Hansen, C.K., Mikkelsen, J.M., and Schambye, H.T. (2002) hG-CSF conjugates. WO 03/006501, filed Jul. 10, 2002 and published Jan. 23, 2003.

33 Maxygen (2011) Next-generation G-CSF for the treatment of neutropenia and acute radiation syndrome http://www.maxygen.com/products-mye.php (accessed 26 October 2010).

34 Amrein, B., Foeser, S., Lang, K., Metzger, F., Regula, J., Schaubmar, A., Hesse, F., Kuenkele, K.-P., and Lanzendoerfer, M. (2005) Conjugates of insulin-like growth factor-I and poly(ethylene glycol). WO2006/066891 filed Dec., 21, 2005, published Jun. 29, 2006..

35 Fisher, S., Hesse, F., Knoetgen, H., Lang, K., Metzger, F., Regula, J.T., Schantz, C., Schaubmar, A., Schoenfeld, H.J. (2007) Method for the production of conjugates of insulin-like growth factor-I and poly(ethylene glycol). WO2008/025528 filed Aug., 29, 2007, published Mar. 6, 2008..

36 Holtmann, B., Metzger, F., and Sendtner, M. (2009) Use of pegylated IGF-I variants

for the treatment of neuromuscular disorders. WO2009/121759 filed Mar., 24, 2009, published Oct. 8, 2009.

37 Kinstler, O., Molineux, G., Treuheit, M., Ladd, D., and Gegg, C. (2002) Mono-N-terminal poly(ethylene glycol)-protein conjugates. *Adv. Drug Deliv. Rev.*, **54**, 477–485. doi: S0169409X02000236 [pii]

38 Molineux, G. (2004) The design and development of pegfilgrastim (PEG-rmetHuG-CSF, Neulasta). *Curr. Pharm. Des*, **10**, 1235–1244.

39 Means, G.E., and Feeney, R.E. (1995) Reductive alkylation of proteins. *Anal. Biochem.*, **224**, 1–16.

40 Rosendahl, M.S., Doherty, D.H., Smith, D.J., Carlson, S.J., Chlipala, E.A., and Cox, G.N. (2005) A long-acting, highly potent interferon alpha-2 conjugate created using site-specific PEGylation. *Bioconjug. Chem.*, **16**, 200–207. doi: 10.1021/bc049713n

41 Mei, B., Pan, C., Jiang, H., Tjandra, H., Strauss, J., Chen, Y., Liu, T., Zhang, X., Severs, J., Newgren, J., Chen, J., Gu, J.M., Subramanyam, B., Fournel, M.A., Pierce, G.F., and Murphy, J.E. (2010) Rational design of a fully active, long-acting PEGylated factor VIII for hemophilia A treatment. *Blood*, **116**, 270–279. doi: blood-2009-11-254755 [pii];10.1182/blood-2009-11-254755

42 Humphreys, D.P., Heywood, S.P., Henry, A., Ait-Lhadj, L., Antoniw, P., Palframan, R., Greenslade, K.J., Carrington, B., Reeks, D.G., Bowering, L.C., West, S., and Brand, H.A. (2007) Alternative antibody Fab' fragment PEGylation strategies: combination of strong reducing agents, disruption of the interchain disulphide bond and disulphide engineering. *Protein Eng. Des. Sel.*, **20**, 227–234. doi: gzm015 [pii];10.1093/protein/gzm015

43 Balan, S., Choi, J.W., Godwin, A., Teo, I., Laborde, C.M., Heidelberger, S., Zloh, M., Shaunak, S., and Brocchini, S. (2007) Site-specific PEGylation of protein disulfide bonds using a three-carbon bridge. *Bioconjug. Chem.*, **18**, 61–76. doi: 10.1021/bc0601471

44 Shaunak, S., Godwin, A., Choi, J.W., Balan, S., Pedone, E., Vijayarangam, D., Heidelberger, S., Teo, I., Zloh, M., and Brocchini, S. (2006) Site-specific PEGylation of native disulfide bonds in therapeutic proteins. *Nat. Chem. Biol.*, **2**, 312–313. doi: nchembio786 [pii];10.1038/nchembio786

45 Deiters, A., and Schultz, P.G. (2005) *In vivo* incorporation of an alkyne into proteins in Escherichia coli. *Bioorg. Med. Chem. Lett.*, **15**, 1521–1524. doi: S0960-894X(04)01527-6 [pii];10.1016/j.bmcl.2004.12.065

46 Ambrx, Inc (2010) Ambrx Pipeline, http://www.ambrx.com/wt/page/technology (accessed 26 October 2010).

47 Cotton, G. (2003) Ligation method WO 2005014620 filed Aug., 5, 2003, published Feb. 2, 2005.

48 Fee, C.J., and Van Alstine, J.M. (2006) PEG–proteins: Reaction engineering and separation issues. *Chem. Eng. Sci.*, **61**, 924–939.

49 Gaberc-Porekar, V., Zore, I., Podobnik, B., and Menart, V. (2008) Obstacles and pitfalls in the PEGylation of therapeutic proteins. *Curr. Opin. Drug Discov. Devel.*, **11**, 242–250.

50 Jevsevar, S., Kunstelj, M., and Porekar, V.G. (2010) PEGylation of therapeutic proteins. *Biotechnol. J.*, **5**, 113–128. doi: 10.1002/biot.200900218

51 Seely, J.E., and Richey, C.W. (2001) Use of ion-exchange chromatography and hydrophobic interaction chromatography in the preparation and recovery of polyethylene glycol-linked proteins. *J. Chromatogr. A*, **908**, 235–241.

52 Seely, J.E., Buckel, S.D., Green, P.D., and Richey, C.W. (2005) Making site-specific PEGylation work. *Biopharm. Int.*, **18**, 30–35.

53 Chapman, A.P., Antoniw, P., Spitali, M., West, S., Stephens, S., and King, D.J. (1999) Therapeutic antibody fragments with prolonged *in vivo* half-lives. *Nat. Biotechnol.*, **17**, 780–783. doi: 10.1038/11717

54 Caserman, S., Kusterle, M., Kunstelj, M., Milunovic, T., Schiefermeier, M., Jevsevar, S., and Porekar, V.G. (2009) Correlations between *in vitro* potency of polyethylene glycol-protein conjugates and their chromatographic behavior. *Anal. Biochem.*, **389**, 27–31.

doi: S0003-2697(09)00183-3 [pii];10.1016/
j.ab.2009.03.023

55 Montaudo, G., Samperi, F., and
Montaudo, M.S. (2006) Characterization
of synthetic polymers by MALDI-MS.
Prog. Polym. Sci., **31**, 277–357.
doi: 10.1016/j.progpolymsci.
2005.12.001

56 Kou, D., Manius, G., Zhan, S., and
Chokshi, H.P. (2009) Size exclusion
chromatography with Corona charged
aerosol detector for the analysis of
polyethylene glycol polymer.
J. Chromatogr. A, **1216**, 5424–5428. doi:
S0021-9673(09)00770-5 [pii];10.1016/j.
chroma.2009.05.043

57 Fee, C.J., and Van Alstine, J.M. (2004)
Prediction of the viscosity radius and the
size exclusion chromatography behavior
of PEGylated proteins. *Bioconjug. Chem.*,
15, 1304–1313.

58 Kurfurst, M.M. (1992) Detection and
molecular-weight determination of
polyethylene glycol-modified hirudin by
staining after sodium dodecyl-sulfate
polyacrylamide-gel electrophoresis. *Anal.
Biochem.*, **200**, 244–248.

59 Piedmonte, D.M., and Treuheit, M.J.
(2008) Formulation of Neulasta(R)
(pegfilgrastim). *Adv. Drug Deliv. Rev.*, **60**,
50–58.

60 Foser, S., Schacher, A., Weyer, K.A.,
Brugger, D., Dietel, E., Marti, S., and
Schreitmuller, T. (2003) Isolation,
structural characterization, and antiviral
activity of positional isomers of
monopegylated interferon alpha-2a
(PEGASYS). *Protein Expr. Purif.*, **30**,
78–87.

61 Wang, Y.S., Youngster, S., Grace, M.,
Bausch, J., Bordens, R., and Wyss, D.F.
(2002) Structural and biological
characterization of PEGylated
recombinant interferon alpha-2b and its
therapeutic implications. *Adv. Drug Deliv.
Rev.*, **54**, 547–570.

62 Srebalus Barnes, C.A., and Lim, A. (2007)
Applications of mass spectrometry for the
structural characterization of
recombinant protein pharmaceuticals.
Mass Spectrom. Rev., **26**, 370–388. doi:
10.1002/mas.20129

63 Kaltashov, I.A., Bobst, C.E., Abzalimov,
R.R., Berkowitz, S.A., and Houde, D.

(2010) Conformation and dynamics of
biopharmaceuticals: transition of
mass spectrometry-based tools from
academe to industry. *J. Am. Soc. Mass
Spectrom.*, **21** (3), 323–337. Epub 2009
Oct 29.

64 Cindric, M., Cepo, T., Galic, N.,
Bukvic-Krajacic, M., Tomczyk, N.,
Vissers, J.P., Bindila, L., and Peter-
Katalinic, J. (2007) Structural
characterization of PEGylated rHuG-CSF
and location of PEG attachment sites.
J. Pharm. Biomed. Anal., **44**, 388–395. doi:
S0731-7085(07)00151-3 [pii];10.1016/j.
jpba.2007.02.036

65 Mero, A., Spolaore, B., Veronese, F.M.,
and Fontana, A. (2009)
Transglutaminase-mediated PEGylation
of proteins: direct identification of the
sites of protein modification by mass
spectrometry using a novel monodisperse
PEG. *Bioconjug. Chem.*, **20**, 384–389. doi:
10.1021/bc800427n;10.1021/
bc800427n [pii]

66 Yamaoka, T., Tabata, Y., and Ikada, Y.
(1995) Fate of water-soluble polymers
administered via different routes.
J. Pharm. Sci., **84**, 349–354.

67 Caliceti, P., and Veronese, F.M. (2003)
Pharmacokinetic and biodistribution
properties of poly(ethylene glycol)-protein
conjugates. *Adv. Drug Deliv. Rev.*, **55**,
1261–1277.

68 Zhai, Y., Zhao, Y., Lei, J., Su, Z., and Ma,
G. (2009) Enhanced circulation half-life
of site-specific PEGylated rhG-CSF:
optimization of PEG molecular weight.
J. Biotechnol., **142**, 259–266. doi:
S0168-1656(09)00230-2 [pii];10.1016/j.
jbiotec.2009.05.012

69 Yamaoka, T., Tabata, Y., and Ikada, Y.
(1994) Distribution and tissue uptake of
poly(ethylene glycol) with different
molecular weights after intravenous
administration to mice. *J. Pharm. Sci*, **83**,
601–606.

70 Modi, M.W., Fulton, J.S., and Buchmann,
D.K. (2000) Clearance of PEGylated
(40 kDa) interferon alpha-2a (Pegasys®)
is primary hepatic. *Hepatology*, **32**, 371.

71 Knauff, M.J., Bell, D.P., Hirtzer, P., Luo,
Z.P., Young, J.D., and Kartre, N.V. (1988)
Relationship of effective molecular size to
systemic clereance in rats of recombinant

interleukin-2 chemically modified with water soluble polymers. *J. Biol. Hem.*, **263**, 1506–15070.

72 Lee, L.S., Connover, C., Shi, C., Whiltlow, M., and Fipula, D. (1999) Prolonged circulating lives of single-chain Fv proteins conjugated with polyethylene glycol: a comparison of conjugation chemistries and compounds. *Bioconjug. Chem.*, **10**, 973–981.

73 Clark, R., Olson, K., Fuh, G., Marian, M., Mortensen, D., Teshima, G., Chang, S., Chu, H., Mukku, V., Canova-Davis, E., Somers, T., Cronin, M., Winkler, M., and Wells, J.A. (1996) Long-acting growth hormones produced by conjugation with polyethylene glycol. *J. Biol. Chem.*, **271**, 21969–21977.

74 Kontermann, R.E. (2009) Strategies to extend plasma half-lives of recombinant antibodies. *BioDrugs.*, **23** (2), 93–109. doi: 10.2165/00063030-200923020-00003

75 Sehon, A.H. (1991) Suppression of antibody responses by conjugates of antigens and monomethoxypoly(ethylene glycol). *Adv. Drug Deliv. Rev.*, **6**, 203–217. doi: 10.1016/0169-409X(91)90041-A

76 Caliceti, P., Schiavon, O., and Veronese, F.M. (2001) Immunological properties of uricase conjugated to neutral soluble polymers. *Bioconjug. Chem.*, **12**, 515–522.

77 Judge, A., McClintock, K., Phelps, J.R., and MacLachlan, I. (2006) Hypersensitivity and loss of disease site targeting caused by antibody responses to PEGylated liposomes. *Mol. Ther.*, **13**, 328–337. doi: S1525-0016(05)01620-5 [pii];10.1016/j.ymthe.2005.09.014

78 Wang, X., Ishida, T., and Kiwada, H. (2007) Anti-PEG IgM elicited by injection of liposomes is involved in the enhanced blood clearance of a subsequent dose of PEGylated liposomes. *J. Control Release*, **119**, 236–244. doi: S0168-3659(07)00116-2 [pii];10.1016/j.jconrel.2007.02.010

79 Cheng, T.L., Chen, B.M., Chern, J.W., Wu, M.F., and Roffler, S.R. (2000) Efficient clearance of poly(ethylene glycol)-modified immunoenzyme with anti-PEG monoclonal antibody for prodrug cancer therapy. *Bioconjug. Chem.*, **11**, 258–266. doi: bc990147j [pii]

80 Mehvar, R. (2000) Modulation of the pharmacokinetics and pharmacodynamics of proteins by polyethylene glycol conjugation. *J. Pharm. Pharm. Sci.*, **3**, 125–136.

81 Beranova, M., Wasserbauer, R., Vancurova, D., Stifter, M., Ocenaskova, J., and Mara, M. (1990) Effect of cytochrome P-450 inhibition and stimulation on intensity of polyethylene degradation in microsomal fraction of mouse and rat livers. *Biomaterials*, **11**, 521–524.

82 Bendele, A., Seely, J., Richey, C., Sennello, G., and Shopp, G. (1998) Short communication: renal tubular vacuolation in animals treated with polyethylene-glycol-conjugated proteins. *Toxicol. Sci.*, **42**, 152–157. doi: S1096-6080(97)92396-9 [pii];10.1006/toxs.1997.2396

83 Bukowski, R., Ernstoff, M.S., Gore, M.E., Nemunaitis, J.J., Amato, R., Gupta, S.K., and Tendler, C.L. (2002) PEGylated interferon alfa-2b treatment for patients with solid tumors: a phase I/II study. *J. Clin. Oncol.*, **20**, 3841–3849.

84 Gregoriadis, G., Jain, S., Papaioannou, I., and Laing, P. (2005) Improving the therapeutic efficacy of peptides and proteins: a role for polysialic acids. *Int. J. Pharm.*, **300**, 125–130. doi: S0378-5173(05)00366-2 [pii];10.1016/j.ijpharm.2005.06.007

85 Richter, A.W., and Akerblom, E. (1983) Antibodies against polyethylene glycol produced in animals by immunization with monomethoxy polyethylene glycol modified proteins. *Int. Arch. Allergy Appl. Immunol.*, **70**, 124–131.

86 Roberts, M.J., Bentley, M.D., and Harris, J.M. (2002) Chemistry for peptide and protein PEGylation. *Adv. Drug Deliv. Rev.*, **54**, 459–476. doi: S0169409X02000224 [pii]

87 Harris, J.M., and Chess, R.B. (2003) Effect of pegylation on pharmaceuticals. *Nat. Rev. Drug Discov.*, **2**, 214–221. doi: 10.1038/nrd1033;nrd1033 [pii]

4

Half-Life Extension of Therapeutic Proteins via Genetic Fusion to Recombinant PEG Mimetics

Uli Binder and Arne Skerra

4.1
Introduction

Rapid clearance from blood circulation by renal filtration is a typical property of small proteins and peptides. However, by expanding the apparent molecular dimensions beyond the pore size of the kidney glomeruli, the plasma half-life of therapeutic proteins can be extended to a medically useful range of several days in humans. One established strategy to achieve such an effect is chemical conjugation of the biologic with the synthetic polymer polyethylene glycol (PEG; see Chapter 3). This has led to several approved drugs, for example PEG-interferon α2a (Pegasys®), PEG-G-CSF (Neulasta®) and, recently, a PEGylated anti-TNFα Fab fragment (Cimzia®). Nevertheless, "PEGylation" technology has several drawbacks: clinical grade PEG derivatives are expensive and their covalent coupling to a recombinant protein requires additional downstream processing and purification steps, thus lowering yield and raising the costs. Furthermore, PEG is not biodegradable, which can cause side effects such as vacuolation of kidney epithelium upon continuous treatment [1–3].

To overcome these caveats, recombinant polypeptide mimetics of PEG have been invented, meaning specific amino acid sequences that can be fused to the pharmaceutically active protein using recombinant DNA technology and which retard kidney filtration in a similar manner to a chemically attached synthetic polymer chain, based on size and/or charge effects (Figure 4.1, Table 4.1). A number of such amino acid polymers have been investigated *in vitro* or developed to the preclinical stage and, in a few cases, already entered clinical trials. These approaches will be described and discussed in this chapter.

Therapeutic Proteins: Strategies to Modulate Their Plasma Half-Lives, First Edition. Edited by
Roland Kontermann.
© 2012 Wiley-VCH Verlag GmbH & Co. KGaA. Published 2012 by Wiley-VCH Verlag GmbH & Co. KGaA.

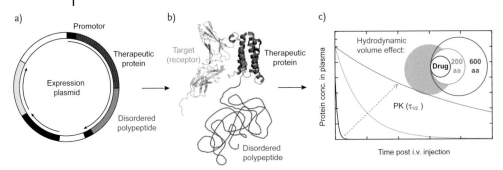

Figure 4.1 Principle of producing recombinant biopharmaceuticals as fusion with a disordered polypeptide chain to extend their plasma half-life. (a) A therapeutic protein is genetically coupled with the polypeptide and (b) produced in one step. (c) The hydrodynamic volume of the biologic grows with increasing length of the PEG mimetic, leading to a strong rise in plasma half-life.

Table 4.1 Overview of recombinant PEG mimetics.

PEG mimetic	Amino acid composition	Company	Reference
SAPA repeats	[DSSAHSTPSTPA]$_n$	–	[7]
Antigen 13 repeats	[EPKSA]$_n$	–	[7]
Elastin-like polymers	[VPGXG]$_n^{a)}$	PhaseBio (www. phasebio.com)	[20, 21]
Gelatin-like polymers	[GXZ]$_n^{b)}$	–	[15]
Polyanionic polymers	poly E or poly D	Cell Therapeutics (www. celltherapeutics.com)	[26]
Genetic Polymers	G, N, Q plus A, S, T, D, E	Aequus BioPharma (www. aequusbiopharma. com)	[27]
HAP	[GGGGS]$_n$	–	[36]
HRM	P, D, S, T, A	–	[10]
XTEN	P, E, S, T, A, G	Amunix (www. amunix.com)	[31]
PAS	P, A, S	XL-protein (www. xl-protein.com)	[40]

a) X represents every amino acid except for P.
b) X designates E, K, N, P, Q or S; Z is an amino acid selected from the group E, K, N, P or Q.

4.2
Mechanisms to Retard Kidney Filtration Using Conjugates of Drugs with Polymers

To mimic PEG–or other natural or synthetic polymers–for the purpose of biological drug half-life extension, suitable amino acid sequences are designed to be (i) hydrophilic, in some cases even charged, and (ii) structurally disordered (Figure 4.1). Absence of hydrophobic interaction sites and a high propensity of aqueous solvation are crucial to avoid nonspecific association with biological components of the organism, for example cell surfaces or extracellular matrix including their embedded proteins. High solubility of the biopharmaceutical substance also prevents aggregation upon storage and permits formulation at elevated concentration, as it is needed for bolus injection. On the other hand, the lack of secondary or tertiary structure, in conjunction with random Brownian motion, leads the conformationally flexible polypeptide chain to adopt an expanded hydrodynamic volume. This provides a physical mechanism to prevent rapid diffusion through the pores of the glomerular basement membrane, which is the limiting step during kidney filtration (Figure 4.2).

It is well known from kidney physiology [4] that the filtration efficiency of a macromolecular plasma solute depends on its size and charge. This can be explained by the defined pore size distribution of the glomeruli with an upper threshold of around 60 Å (corresponding to a globular protein of approximately 70 kDa) and its overall negative surface charge caused by anionic proteoglycans

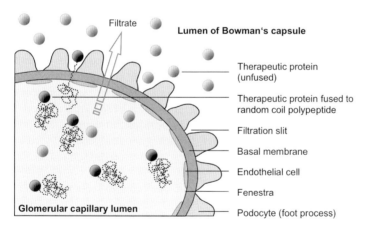

Figure 4.2 Principle of renal elimination of circulating biological drugs. The glomerular capillary provides a three-layer filter: dissolved plasma molecules permeate the capillary endothelium via narrow pores, traverse the negatively charged mesh-like basement membrane and finally enter the lumen of the Bowman capsule via filtration slits between the foot appendages of podocytes. Coupling of a therapeutic protein with a conformationally flexible polypeptide or polymer chain increases its hydrodynamic diameter beyond the glomerular pore size and thus retards filtration process if compared with the naked protein.

present in the extracellular matrix of the basement membrane. Thus, large acidic proteins such as human serum albumin (HSA), with a diameter of 90 Å and pI = 5, are mostly (though not totally) resistant to kidney filtration. The same is the case for the even larger immunoglobulins (Igs) while for both HSA and Igs the "biological" mechanism of endosomal recycling also comes into play, which specifically prolongs their circulation further (see Chapter 8 dealing with FcRn biology). On the other hand, proteins and peptides that are smaller than a lower threshold diameter of approximately 36 Å show a sieving coefficient of 1, which means that their concentration becomes fully equilibrated with the primary urinary filtrate upon each kidney passage, leading to an extremely short plasma-half life, in the range of minutes, for humans.

The purely physical phenomenon of retarding kidney filtration of a solute by increasing its molecular size can be compared with the well known laboratory techniques of dialysis or ultrafiltration. In this case dialysis tubings or ultrafiltration membranes with different pore sizes are available and usually a choice is made such that the cut-off is rather smaller than the size of the protein to be studied in order to avoid its leakage into the dialysis buffer or filtrate. In the physiological situation discussed here one faces the inverse task, that is, the apparent size of the therapeutic protein has to be enlarged to an extent that its diffusion through the glomerular filter becomes slowed down by a useful factor. In principle, this problem can be solved in an elegant way by attaching a highly solvated and structurally disordered polypeptide chain to the drug. In fact, it is well known from polymer biophysics that an unfolded amino acid chain adopts a much larger hydrodynamic volume than a polypeptide with the same number of residues in a tertiary structure and/or with a globular shape [5, 6].

Owing to thermal motion the conformation of such a "random chain" constantly fluctuates, which leads to a more or less uniform, on average wadding-like structure, whose molecular dimensions only depend on the number of monomer building blocks (here amino acid residues), their size/diameter, and conformational stiffness. In general, the radius of gyration for such a random chain scales with the square root of residue number and, in principle, in aqueous solution there is no difference between a conventional linear chemical polymer, such as PEG, and a disordered polypeptide chain, as long as there are no intramolecular attractive or repulsive forces.

However, one has to consider that most natural amino acid sequences do not behave like an ideal random chain in physiological solution because they either tend to adopt a folded conformation (secondary structure) or, if unfolded, they usually are insoluble and form aggregates. In fact, most of the classical experiments to investigate the random chain behavior of polypeptides were conducted under denaturing conditions, that is, in the presence of chemical denaturants like urea or guanidinium chloride [6]. Hence, the technologies described here generally rest upon peculiar amino acid sequences that resist folding, aggregation, and unspecific adsorption, and thus provide stable random chains under physiological buffer conditions and temperature even if genetically fused to a folded therapeutic protein domain. Under these circumstances, such recombinant PEG mimetics can confer a size increase much larger than one would normally expect on the basis of their molecular mass alone – as will be exemplified below – eventually retarding kidney filtration and effectively extending plasma half-life of the attached biologic by considerable factors.

Generally, polypeptides made of natural amino acids offer several advantages over chemical polymers such as PEG. First, they are usually biodegradable by (intracellular) proteases and, consequently, toxicity of the polymer or accumulation in organs (e.g., kidney, spleen or liver) or cells (in particular macrophages) is not to be expected. Second, the amino acid polymer can be genetically encoded and, therefore, the pharmaceutically active protein can be directly produced as translational fusion with the recombinant polymer using genetic engineering. The resulting chimeric protein, which has the somewhat unusual feature of comprising a folded and a distinct unfolded part, can simply be produced in one step in microbial hosts or also in eukaryotic cells and recovered as a soluble protein, thus avoiding laborious chemical modification steps.

4.3
Naturally Occurring Repetitive Amino Acid Sequences

Considering the concept of utilizing naturally disordered polypeptide sequences to extend the plasma half-life of biologically active proteins as explained above, there are hints that protozoan parasites employ such a strategy for certain proteinogenic virulence factors in their mammalian hosts. The trans-sialidase of *Trypanosoma cruzi* contains a 680-amino acid residue catalytic domain followed by a C-terminal repetitive domain, dubbed "shed acute phase antigen" (SAPA), which comprises a variable number of 12mer amino acid repeats. Pharmacokinetic (PK) studies in mice of the trans-sialidase containing 13 of such hydrophilic and (at physiological pH) negatively charged amino acid repeats, having the natural sequence DSSAHSTPSTPA, revealed a fivefold longer plasma half-life compared with the recombinant enzyme from which the C-terminal repetitive sequence had been deleted [7]. A similar half-life extending effect was observed after fusion of the same trans-sialidase, that is, its 76 kDa catalytic domain, with 13 charged amino acid repeats of the sequence EPKSA that were found in the *T. cruzi* protein antigen 13.

Furthermore, the repeats from both SAPA and the antigen 13 were able to prolong the plasma half-life of the heterologous protein gluthatione S-transferase (GST) from *Schistosoma japonicum* by a factor of 7–8 after genetic fusion to both C-termini of this homo-dimeric enzyme [7]. The precise mechanism for the improved plasma stability was not investigated; however, it is conceivable that these repeat sequences are intrinsically unstructured and hence retard kidney filtration primarily via the resulting hydrodynamic volume effect. In addition, the overall negative charge may contribute via electrostatic repulsion from the glomerular basement membrane.

While in principle it may appear attractive to use these naturally occurring repetitive amino acid sequences from human pathogens to optimize the pharmacokinetics of therapeutic proteins they were found to be highly immunogenic [8, 9]. This problem was addressed in a protein engineering approach by designing an artificial sequence derived from the SAPA and antigen 13 repeats, which was shown to have lower immunogenicity in animal experiments [10]. Indeed, the engineered amino acid repeat with the simplified sequence PSTAD, termed hybrid repetitive motif (HRM), was still able to extend the plasma half-life of the catalytic domain of *trypanosomal* trans-sialidase and also of rat tyrosine aminotransferase by a factor of 4.5–6. Interestingly, repeat sequences with similar half-life extending properties were also identified in bacteria [10].

4.4
Gelatin-Like Protein Polymers

Gelatin, hydrolyzed and denatured animal collagen, contains long stretches of Gly-Xaa-Yaa repeats, wherein Xaa and Yaa mostly constitute proline and 4-hydroxyproline, respectively. Succinylation of gelatin, primarily via the ε-amino groups of naturally interspersed lysine side chains, increases the hydrophilicity of this biopolymer and lowers its isoelectric point (pI). The intramolecular electrostatic repulsion between the negatively charged carboxylate groups of the modified side chains supposedly spreads out the molecule into a more or less extended, yet disordered conformation [11, 12]. The resulting expanded volume makes succinylated gelatin an ideal macromolecule for use as plasma expander which is well tolerated in humans [13], for example marketed as Volplex® (Beacon Pharmaceuticals Ltd.; www.beaconpharma.co.uk) or Gelofusine® (B. Braun Melsungen AG; www.bbraun.com).

Furthermore, it was shown that chemical conjugation of succinylated gelatins to proteins having a short plasma half-life, such as the soybean trypsin inhibitor or Cu/Zn-superoxide dismutase, can prolong circulation by a factor of 6–7 [11, 14]. Notably, a similar half-life extending effect has been achieved by genetic fusion of granulocyte-colony-stimulating factor (G-CSF) to an artificial gelatin-like polypeptide [15]. To this end, all hydrophobic side chains in a natural gelatin were exchanged by hydrophilic residues, resulting in a 116 amino acid gelatin-like protein (GLK) comprising the amino acids G, P, E, Q, N, S, and K in varying order. G-CSF was fused at its N-terminus with four copies of this GLK sequence and secreted in *Pichia pastoris*. The fusion protein retained biological activity, exhibited decreased aggregate formation after incubation at 40 °C for 48 h and showed a 5.7-fold longer terminal plasma half-life after subcutaneous administration to rats. An antibody response after four-times weekly immunization of mice was only detected against G-CSF but not against the GLK part.

According to the authors of this study, the underlying mechanism for half-life extension may comprise reduced proteolysis due to the outspread structure of the GLK, its slower renal clearance due to the increased hydrodynamic volume, or its reduced extravasation as a consequence of the net negative charge [15]. *Pichia pastoris* appeared as a favorable production organism for GLK fusion proteins; it remains to be shown whether GLK polymers can also be produced in other organisms, as it is known that recombinant gelatin fragments can only be expressed with low yield in *E. coli*, for example [16].

4.5
Elastin-Like Polypeptides

Elastin is a component of the extracellular matrix in many tissues like skin, lung, arteries, ligament, and cartilage. It is formed from the soluble precursor tropoelastin, which consists of a hydrophilic Lys/Ala-rich domain and a hydrophobic, elastomeric domain with repetitive sequence. Enzymatic crosslinking of lysine side chains within the hydrophilic domain leads to insoluble elastin formation. Elastin-like polypeptides (ELPs) are artifically designed, repetitive amino acid sequences derived from the hydophobic domain of tropoelastin [17]. The most common repeat sequence motif of ELPs is VPGXG, wherein "X" can be any amino acid except Pro [17, 18].

Interestingly, soluble ELPs undergo a reversible thermal phase transition, forming insoluble microaggregates above a characteristic temperature. The chemical and physical polymer properties, including the transition temperature, depend on ELP sequence and length and can be adjusted to desired applications. Employing a genetic library, an ELP was selected that remains soluble at room temperature but aggregates above 32 °C, that is, at mammalian body temperature. As demonstrated for an isolated 47 kDa ELP by intra-articular injection into the rat knee, the resulting microaggregates can form a depot with a half-life prolonged by a factor of 25 if compared with a 61 kDa nonaggregating ELP having a transition temperature above 50 °C [19].

Suitable ELPs can be fused with therapeutic proteins and produced in *E. coli*. Consequently, the ability of ELPs to form gel-like depots after injection can significantly prolong the *in vivo* half-life of an attached biologic, albeit by a mechanism different from the other unstructured polypeptides discussed in this chapter. For example, a vasoactive intestinal peptide fused to ELP showed a terminal plasma half-life in monkeys of 9.9 to 45.8 h after subcutaneous injection. These PK data likely reflect a slow (and somewhat variable) absorption rate rather than the actual elimination [20]. Unfortunately, ELP attachment can hamper the bioactivity of the fusion partner as demonstrated for the interleukin-1 receptor antagonist in an IL-1-induced lymphocyte proliferation bioassay. Bioactivity of $ELP(V_5G_3A_2)_{90}$-IL-1RA was significantly diminished as indicated by the increase of the IC_{50} from 1.4 ± 0.3 nM, measured for the commercially available unfused IL1-RA, to 730 ± 240 nM.

Yet, ELPs are also subject to degradation by endogenous proteases such as collagenase, and resulting truncated proteins can nearly exhibit a comparable bioactivity to the unmodified biologic [21]. Several therapeutic ELP fusion proteins with prolonged action, for example of phenylalanine hydroxylase [22], oxyntomodulin

(Oxymera™) as well as a GLP-1 analog (Glymera™) [23], are currently under development by PhaseBio Inc. (www.phasebio.com).

4.6
Polyanionic Polymers

Synthetic polyglutamate has been chemically coupled to poorly soluble cytotoxic small molecule drugs for cancer treatment. The most advanced polyglutamate (PG) product is Opaxio™ (formerly known as XYOTAX™), a paclitaxel drug conjugate currently in clinical phase III studies sponsored by Cell Therapeutics Inc. (www.celltherapeutics.com). The biodegradable PG increases hydrophilicity, reduces toxicity and, notably, enhances tumor accumulation of the cytotoxic drug. The latter phenomenon, also known as enhanced permeability and retention (EPR) effect [24], can be explained by impaired lymphatic drainage in cancer tissues and by the aberrant structure and higher permeability of tumor vasculature. Apart from these beneficial drug targeting properties, coupling of paclitaxel to PG also improves its pharmacokinetics. In human studies the plasma half-life of a paclitaxel PG conjugate was prolonged by a factor of 3 to 14 in comparison with the unmodified compound [25].

Based on these findings the idea was proposed that the polyglutamate technology could be applied to therapeutic proteins, promising less frequent and/or lower dosing of the biologic. Several fusion proteins, for example G-CSF fused at its N-terminus with a stretch of 175 consecutive Glu residues or IFNα2 carrying at its C-terminus a PG tail of 84 residues, were successfully produced in a soluble state in the cytoplasm of *E. coli* [26]. For efficient translation, the N-terminal fusion required a leader peptide, which was later removed by tobacco etch virus (TEV) protease cleavage. Polyglutamate fusions of G-CSF and INFα2 showed bioactivity in cell culture assays [26]. However, to date no PK data have been reported.

4.7
Genetic Polymers™

A less massively charged, unstructured polypeptide sequence suitable for the half-life extension of biologics was also invented at Cell Therapeutics Inc. [27]. Their Genetic Polymer™ comprises an amino acid sequence with 2 to 500 repeat units of three to six amino acids, wherein G, N or Q represent the major constituents while minor constituents can be A, S, T, D or E. This amino acid composition allows integration of the glycosylation sequon Asn-Xaa-Ser/Thr (where Xaa is any amino acid except Pro) for *N*-linked glycosylation of the Asn side chain in eukaryotic expression systems. The increased macromolecular size of the resulting fusion protein, including post-translational modification with bulky solvated carbohydrate structures, can extend the pharmacokinetics of a genetically conjugated protein. In fact, it is well known that oligosaccharide attachment in general can both reduce susceptibility to proteolysis and increase the hydrodynamic volume, thus prolonging the *in vivo* half-life of therapeutic proteins [28].

As an example, G-CSF was fused at its C-terminus with a Genetic Polymer™ comprising 155 NNT repeats, expressed in CHO-K1 cells, and secreted into the culture medium. No proteolytic degradation of the amino acid polymer was detectable while a substantial degree of glycosylation was observed. When intravenously or subcutaneously injected into mice G-CSF-(NNT)$_{115}$ was able to increase white blood cells and neutrophils, hence demonstrating bioactivity. Furthermore, its plasma half-life in mice, after single intravenous application, was fourfold increased [27]. Aequus Biopharma Inc. (www.aequusbiopharma.com) was launched as a spin-off to commercialize the Genetic Polymer™ technology.

4.8
XTEN Polypeptides

So-called "Xtended" recombinant polypeptides, developed by Amunix Inc. (www.amunix.com), comprise an unstructured nonrepetitive amino acid polymer encompassing the residues P, E, S, T, A, G [29, 30]. This set of amino acids, which shows a composition not unlike the PSTAD repeat described above, was systematically screened for sequences to yield a solvated polypeptide with large molecular size, suitable for biopharmaceutical development, by avoiding hydrophobic side chains – in particular F, I, L, M, V and W – that can give rise to aggregation and may cause an HLA/MHC-II mediated immune response. Also, potentially crosslinking Cys residues, the cationic amino acids K, R and H, which could interact with negatively charged cell membranes, and the amide side chains of N and Q, which are potentially prone to hydrolysis, were excluded [31]. Synthetic gene libraries encoding nonrepetitive sequences comprising the PESTAG set of residues, which were fused to the green fluorescent protein (GFP), were screened with respect to soluble expression levels in *E. coli*, and a resulting subset was further investigated for genetic stability, protein solubility, thermostability, aggregation tendency, and contaminant profile. Eventually, an 864 amino acid sequence containing 216 Ser residues (25.0 mole %), 72 Ala residues (8.3 mole %) and 144 residues (16.7 mole %) of each Pro, Thr, Glu, and Gly was further tested for fusion to the GLP-1 receptor agonist Exendin-4 (E-XTEN) and a few other biologics.

The fusion proteins – typically carrying a cellulose binding domain, which was later cleaved off – were produced in a soluble state in the cytoplasm of *E. coli* and isolated. Investigation of E-XTEN by circular dichroism (CD) spectroscopy revealed lack of secondary structure while during size exclusion chromatography (SEC) the fusion protein showed substantially less retention than expected for a 84 kDa protein, thus demonstrating an increased hydrodynamic volume [31]. The disordered structure of the PESTAG polypeptide and the associated increase in hydrodynamic radius may be favoured by the electrostatic repulsion between amino acids carrying a high net negative charge which are distributed across the XTEN sequence [29]. When mice were weekly injected for six weeks with E-XTEN plus adjuvant only 5 of 20 animals developed a low detectable antibody titer. Of those, two animals were crossreactive with Exendin-4; thus, XTEN seems to show almost negligible immunogenicity [31]. Following subcutaneous injection, the terminal half-life of Exendin-4 fused to the 864 residue XTEN sequence was increased 65-fold up to 32 h in rats if compared with the original peptide and 125-fold in monkeys [31]. This pronounced PK effect is probably not only caused by the expanded hydrodynamic volume of the XTEN polypeptide, but also by its high overall negative charge, which leads to a very low pI value of 2.3 for the XTEN moiety alone (due to its altogether 144 Glu residues).

Furthermore, it was demonstrated that the *in vivo* half-life of the fusion partner can be tuned by variation of the polypeptide length. To this end the plasma clearance of glucagon was optimized for an overnight dosing regime – to prevent nocturnal hypoglycemia during treatment of diabetes – by C-terminal fusion with a shortened XTEN version of 144 residues [32]. However, this study also demonstrated that XTEN decreases potency of its therapeutic fusion partner. In a cell culture assay, the same glucagon XTEN fusion showed merely 15% bioactivity of the unmodified peptide. An even stronger loss in receptor affinity (17-fold decreased EC$_{50}$) was described for an XTEN fusion of human growth hormone (hGH) [33], which has been subject to Phase I clinical trials sponsored by Versartis Inc. (www.versartis.com). This reduction in receptor affinity is probably also a result of the high intrinsic negative charge of the XTEN polypeptide but could provide an advantage if dose level toxicity or receptor-mediated clearance is an issue.

4.9
Glycine-Rich Homo-Amino-Acid Polymers

Glycine is the smallest and structurally simplest amino acid, which lacks a side chain, is nonchiral and, hence, has long been considered – based on theoretical grounds – as the conformationally most flexible amino acid [5, 34]. Furthermore, computer simulations have indicated that Gly polymers lack secondary structure and are likely to form a random coil in solution [35]. From a chemical perspective, polyglycine is a linear unbranched polyamide that shows certain resemblance to the polyether PEG in so far as both are essentially one-dimensional macromolecules with many rotational degrees of freedom along the chain, comprising repeated short hydrocarbon units that are regularly interrupted by hydrogen-bonding and highly solvated polar groups. Consequently, polyglycine should constitute the simplest genetically encodable PEG mimetic with prospects for extending the plasma half-life of therapeutic proteins. This concept was independently proposed by two groups and called homo-amino-acid polymer (HAP) [36] or glycine-rich sequence (GRS) [30], respectively.

However, it has long been known that chemically synthesized pure polymers of Gly show poor solubility in water [37]. Hence, different attempts were made to increase hydrophilicity, either by introducing hydrogen-bonding serine alcohol side chains [30, 36] or, in addition, negatively charged glutamate residues [30]. Notably, peptide spacers with the composition $(Gly_4Ser)_n$ have already become popular to link domains in fusion proteins in a flexible manner, for example, the pair of variable antibody domains to yield so-called single-chain Fv fragments and as such have entered clinical trials [38, 39]. Related sequences also occur in human proteins (e.g., zinc finger DNA-binding protein 99 and mitogen-activated protein kinase 4) and, thus, can be expected to show low immunogenicity.

After developing a suitable gene synthesis strategy, taking into consideration the high G/C nucleotide content, the rare occurrence of some involved tRNAs and the highly repetitive nature of the polyglycine-encoding DNA sequences, long $(Gly_4Ser)_n$ repeat polypeptides (HAPs) were successfully produced as soluble fusion proteins in E. coli [36]. In this manner the recombinant Fab fragment of the therapeutic antibody Trastuzumab (Herceptin®) was fused at the C-terminus of its light chain to HAPs of 100 and 200 residues and functionally secreted in E. coli, followed by affinity purification from the periplasmic cell extract. The isolated $Fab\text{-}(Gly_4Ser)_{20}$ and $Fab\text{-}(Gly_4Ser)_{40}$ fusion proteins appeared monodisperse in the ESI mass spectrum, without signs of premature translational termination. In addition, they showed $1:1$ light:heavy chain stoichiometry, indicating that the presence of the long HAP tag did not interfere with Ig chain hetero-dimer association and/ or disulfide-crosslinking. Also, an essentially unchanged Her2 antigen-binding activity compared with the unfused Fab fragment was observed in ELISA.

Importantly, a significantly increased hydrodynamic volume was detected for these fusion proteins in analytical SEC. In the case of the 200 residue HAP version the apparent size increase was 120% compared with the unfused Fab fragment, whereas the true mass was only bigger by 29%, hence revealing the effect of an enhanced hydrodynamic volume due to the solvated random coil structure of the polyglycine tag. Furthermore, CD difference spectra were characteristic for disordered secondary structure for the HAP moiety. Finally, terminal plasma half-life of the Fab fragment carrying the 200 residue HAP in mice was prolonged by approximately a factor of 3. Though moderate, this effect could be appropriate for specialized medical applications, such as in vivo imaging [36]. Unfortunately, the production of fusion proteins with longer $(Gly_4Ser)_n$ repeat sequences appeared less feasible due to an increasing tendency to form aggregates, thus posing a natural limitation to the use of–more or less pure–glycine polymers as PEG mimetics.

4.10
PASylation® Technology

Based on the previous experience with polyglycine sequences and in an attempt to abolish the general aggregation propensity of an amino acid polymer without introducing charged side chains, the so-called PAS sequences were invented [40]. The design of these intrinsically unstructured polypeptides arose from the notion

that small amino acids other than Gly all have a certain secondary structure preference: for example, α-helix for Ala, β-sheet for Ser, and the poly(Pro) II helix for
Pro [5]. The hypothesis was that amino acid sequences with an appropriate mixture
of Pro, Ala, and Ser (i.e., PAS) might lead to mutual cancellation of their distinct
conformational preferences and, thus, result in a stably disordered polypeptide.
Additional conformational variability should arise from the slow *cis/trans* isomeri-
zation of the N-terminal prolyl peptide bonds in such PAS polypeptides. Long
PAS-encoding gene cassettes were synthesized by ligation of hybridized oligode-
oxynucleotide building blocks for 20mer or 24mer amino acid sequence stretches.
Hence, PAS polymers have defined sequences and, due to the repetition of these
long building blocks, show uniform biophysical properties over their entire length.
So far, several fusion proteins with PAS polymers of up to 600 residues were successfully produced in *E. coli* and data have been published demonstrating that PAS
tags with different sequences exhibit stable random coil conformation in physiological solution [40].

For example, the Fab fragment of a therapeutic antibody, interferon α-2b, inter-
leukin-1 receptor antagonist and human growth hormone were fused with PAS
sequences of differing lengths between 200 and 600 residues. These therapeutic
proteins were produced in a soluble and active form via secretion into the periplasm of *E. coli*, where their structural disulfide bonds were efficiently formed.
The unhindered translocation of long PAS polypeptides across biological membranes is a general advantage of these uncharged, yet hydrophilic sequences. Mass
spectrometry demonstrated that the purified fusion proteins were always monodisperse. Under physiological buffer conditions all of the produced fusion proteins
showed stable random conformation for the PAS moiety as well as high solubility,
while retaining correct folding and biochemical activity of the associated therapeutic protein (Skerra and coworkers, to be published).

Biophysical characterization using analytical SEC demonstrated drastically
increased hydrodynamic volumes of more than 0.5 MDa, whereas the actual mass
of the protein was increased by merely up to 50 kDa (for 600 PAS residues). Investigation of the plasma half-life in mice and rats performed for PAS fusion proteins
of a humanized Fab fragment, human interferon α-2b, and human growth
hormone revealed considerably prolonged circulation by factors in the range of
10–100, depending on the length of the tag. Furthermore, PAS sequences proved
resistant to serum proteases and showed no detectable immunogenicity in mice.
In contrast, they were rapidly degraded during incubation with a kidney enzyme
extract; hence, organ accumulation should not be expected.

These results suggest that PASylation® should be useful to develop biologics
with comparable pharmacological properties to those achievable by PEGylation,
but with potentially less side effects. In principle, this technology for extending
the plasma half-life is applicable to all recombinant proteins in which the N-
or C-terminus of the polypeptide chain is freely accessible and not directly
involved in the interaction with the disease-relevant target receptor or soluble
signaling factor, while being compatible with both microbial and cell culture
expression. PAS fusion is even possible with bioactive peptides, and a correspond-

ing collaborative project was recently announced by XL-protein GmbH (www.xl-protein.com), the company that develops PASylation® for human therapy [41].

4.11
Conclusions and Outlook

Genetic fusion of therapeutic proteins and peptides with structurally disordered amino acid sequences currently emerges as a powerful strategy to tailor their plasma half-life without the need for either chemical or post-translational modification of the biotechnological product. Generally, this approach offers several advantages over chemical coupling with PEG, HES, polysialic acid or other polymers, in particular with regard to the lower cost of goods (COG) and by avoiding corresponding *in vitro* processing steps. Similar to HESylation® or polysialylation, most of the recombinant PEG mimetics are stable in blood plasma while being degradable by kidney (or other) enzymes. In this way organ accumulation, which is a common problem of fully synthetic polymers such as PEG, can be prevented. Another advantage for biopharmaceutical development, especially in light of quality assurance/quality control (QA/QC) and good manufacturing practice (GMP) requirements, is the monodisperse nature of most of the recombinant PEG mimetics. Due to their genetic encoding and the high processivity as well as fidelity of the cellular protein biosynthesis machinery, the hybrid proteins discussed here typically show a single peak in mass spectrometric analysis, which significantly simplifies bioanalytics and batch-to-batch control. Exceptions are amino acid polymers that require post-translational modification such as glycosylation in the case of the Genetic Polymers™, for example.

Genetic coupling to conformationally flexible polypeptides should also provide benefits over fusion with large globular proteins such as serum albumin (Chapter 12) or Ig Fc fragments (Chapter 9). For example, fusion proteins with amino acid polymers can be produced in a cost-effective way in *E. coli* or other bacterial microorganisms, whereas albumin with its 17 disulfide bridges or Fc fusion proteins with their pair of glycosylated polypeptide chains require eukaryotic production systems. Structurally disordered polypeptide sequences prolong plasma half-life of the fusion partner primarily via a physical effect, as detailed above. In contrast, albumin and Fc fragments bind to the neonatal Fc receptor (FcRn; cf. Chapter 8), which leads to endosomal recycling [42]. Another benefit of recombinant PEG mimetics is their easily variable length – just by trimming the coding DNA region – thus allowing customized PK properties of the biological drug according to clinical requirements.

Apart from that, the Ig Fc fragment can elicit undesired immunological effects and it imposes bivalence on the drug, which may lead to side effects via receptor clustering and agonistic signaling [43]. Albumin binding domains or peptides may be a preferred technology over direct albumin fusion because the decrease in activity of the fusion partner is expected to be much lower (Chapters 13–15). However, especially if these binding domains are derived from nonhuman organisms, immunogenicity may be a problem. On the other hand, at least for some of the structurally disordered polypeptides described in this chapter immunogenicity seems to be negligible. Anti-drug antibodies (ADAs) are also a risk factor for albumin fusion proteins, especially because of potential crossreactivity with endogenous albumin. Such an effect cannot arise for artificially designed amino acid polymers.

In assessing the differential properties of the amino acid polymers one should be aware that the unstructured nature of these tags does not totally prevent kidney filtration because their conformational flexibility still allows diffusional penetration through the glomerular pores (also depending on the size of the attached therapeutic protein or peptide), albeit at reduced speed (cf. Figure 4.2). At the same time, this sterical flexibility is an advantage when considering complex formation with the biological target (e.g., cell surface receptor or soluble factor) as well as tissue penetration, if compared with fusion to large globular and bulky proteins. Nevertheless, a compromise between long plasma half-life on one hand and quick tissue penetration on the other has to be found for each entity as both are inversely related to the molecular size. Finally, when interpreting PK data from animal studies (cf. Chapter 2) one has to be aware of the rules of allometric scaling, which can lead to differences in plasma half-life by more than one order of magnitude between mouse and man [44].

All disordered polypeptides proposed so far, except for the elastin-like polymers, recruit physical effects of size and charge in order to retard kidney filtration. However, their distinct amino acid compositions also result in a differing behavior of corresponding fusion proteins with respect to immunogenicity, tissue distribution, biological receptor-binding activity, and options for biotechnological manufacturing. For example, naturally occurring sequences such as the *trypanosomal* SAPA or antigen 13 repeats are probably not useful for clinical applications due to their known high immunogenicity.

However, less immunogenic amino acid polymers like the HRM, GLK or glycine-rich HAP, in principle, may offer viable alternatives to PEGylation. This would be true if GLK polymers, for example, were compared with first generation PEGs with a size of around 5 kDa [45]; yet, a less than eightfold prolonged plasma half-life as reported for these polypeptides is not enough to compete with either the newer PEG generations (e.g. 40 kDa branched PEG) or the more advanced recombinant polymer technologies like XTEN or PASylation®. Likewise, for the Genetic Polymers™ further improvement of plasma half-life is clearly required, which in this case may also involve optimization of the cellular glycosylation efficiency.

Except for the neutral HAP and PAS sequences, all other (nonglycosylated) amino acid polymers described up to now carry negative charges. This is beneficial in so far as repulsion from the negatively charged glomerular basement membrane can additionally contribute to an increased plasma half-life. On the other hand, the repulsion from the glycocalyx of endothelial cells, which also is negatively charged, will reduce tissue penetration of the fused drug. Indeed, such an effect has been anticipated for cytotoxic anticancer drugs coupled, for example, to polyglutamate sequences in order to prevent damage of healthy tissue. However,

Table 4.2 Comparison of the characteristics of recombinant PEG mimetics and related polymer technologies.

	PEG	HES	Polysialic acid	PAS	XTEN	Genetic polymer	HAP	HRM	PolyE	GLP	ELP
Genetic fusion	−	−	−	+	+	+	+	+	+	+	+
Biodegradability	−	+	+	+	+	+	+	+	+	+	+
PK effect	++	++	++	++	++	+	+	+	+	+	?
No/Low immunogenicity	−	+	+	+	+	?	?	+	+	+	+
Monodisperse product	−	−	−	+	+	−	+	+	+	+	+
Uncharged	+	+	−	+	−	−	+	−	−	−	−
Manufacturability	−	−	−	+	+	−	−	+	+	?	+
Biochemical activity	(−)	?	?	+	−	?	+	?	?	?	−
Product in clinic	+	−	−	−	+	−	−	−	(+)[a]	−	+
Product on market	+	(+)[b]	−	−	−	−	−	−	−	−	−

a) So far, only a small molecule conjugated to polyglutamate is in clinical development.
b) While HES is approved as plasma expander, a HESylated therapeutic protein is not yet on the market.

most biologics must enter the tissue and interstitial space in order to exert their full therapeutic effect. Furthermore, as most targets are receptors presented at the negatively charged cell surface, electrostatic repulsion will also decrease binding activity and, thus, efficacy of the drug. Consequently, uncharged disordered polypeptides should generally provide an advantage.

At present, the field of recombinant PEG mimetics is still in its infancy, with only few examples that have reached the clinical trial stage (Table 4.2). Nevertheless, preclinical data are encouraging. While a lot of data about the biotechnological manufacturing, biochemistry, biophysics, and PK properties of corresponding chimeric drug candidates are already available, future research and development will have to clarify mostly two topics: pharmacodynamics and immunogenicity. In theory, by fusion to a structurally disordered polypeptide – similar to conjugation with PEG – pharmacodynamics should be influenced not only in a positive manner – that is, by the enlarged area under the curve (AUC) as a result of the retarded kidney filtration – but also in a negative way – via potential reduction of target-binding activity. It seems that uncharged random coil polypeptides with a maximally expanded hydrodynamic volume, such as the PAS sequences, should be optimal in both respects, in particular if fused to the chain terminus that is most distant from the active site of the therapeutic protein. Regarding immunogenicity, the majority of the amino acid polymers discussed here lack the characteristic T-cell epitopes that are a prerequisite for efficient antigen presentation via HLA/MHC-II. On top of that, animal immunization studies with HRM, GLK, XTEN and PAS polypeptides have indicated lack of pronounced antibody responses. Hence, the risk of ADAs arising in patients can be considered low, although potential immunogenicity effects will probably also depend on the biopharmaceutically active fusion partner [46].

Taken together, although yet at an early stage, some of these novel concepts, in particular PASylation® or XTEN, possess the potential to deliver superior biopharmaceuticals with an improved safety profile and increased patient's compliance. Chances are high that in the long term such biotechnologies will replace traditional chemical polymers, such as PEG, in biological drug development.

References

1 Gaberc-Porekar, V., et al. (2008) Obstacles and pitfalls in the PEGylation of therapeutic proteins. *Curr. Opin. Drug Discov. Devel.*, **11**, 242–250.

2 Knop, K., et al. (2010) Poly(ethylene glycol) in drug delivery: pros and cons as well as potential alternatives. *Angew. Chem. Int. Ed. Engl.*, **49**, 6288–6308.

3 Armstrong, J.K. (2009) The occurrence, induction, specificity and potential effect of antibodies against poly(ethylene glycol, in *PEGylated Protein Drugs:*

Basic Science and Clinical Applications (ed. F.M. Veronese), Birkhäuser Verlag, Basel, pp. 147–168.

4 Maddox, D.A., Deen, W.M., and Brenner, B.A. (1992) Glomerular filtration, in *Handbook of Physiology* (ed. E.E. Windhager), Oxford University Press, New York, pp. 545–638.

5 Creighton, T.E. (1992) *Proteins: Structures and Molecular Properties*, W.H. Freeman and Company, New York.

6 Cantor, C.R., and Schimmel, P.R. (1980) *Biophysical Chemistry*, W.H. Freeman and Company, New York.

7 Buscaglia, C.A., *et al.* (1999) Tandem amino acid repeats from *Trypanosoma cruzi* shed antigens increase the half-life of proteins in blood. *Blood*, **93**, 2025–2032.

8 Affranchino, J.L., *et al.* (1989) Identification of a *Trypanosoma cruzi* antigen that is shed during the acute phase of Chagas' disease. *Mol. Biochem. Parasitol.*, **34**, 221–228.

9 Buscaglia, C.A., *et al.* (1998) The repetitive domain of *Trypanosoma cruzi* trans-sialidase enhances the immune response against the catalytic domain. *J. Infect. Dis.*, **177**, 431–436.

10 Alvarez, P., *et al.* (2004) Improving protein pharmacokinetics by genetic fusion to simple amino acid sequences. *J. Biol. Chem.*, **279**, 3375–3381.

11 Kojima, Y., *et al.* (1993) Conjugation of Cu,Zn-superoxide dismutase with succinylated gelatin: pharmacological activity and cell-lubricating function. *Bioconjug. Chem.*, **4**, 490–498.

12 Habeeb, A.F., *et al.* (1958) Molecular structural effects produced in proteins by reaction with succinic anhydride. *Biochim. Biophys. Acta*, **29**, 587–593.

13 Schortgen, F., *et al.* (2004) Preferred plasma volume expanders for critically ill patients: results of an international survey. *Intensive Care Med.*, **30**, 2222–2229.

14 Shin, Y.-H., *et al.* (1996) Conjugation of succinylated gelatin to soybean trypsin inhibitor. *J. Bioact. Compat. Polym.*, **11**, 3–16.

15 Huang, Y.S., *et al.* (2010) Engineering a pharmacologically superior form of granulocyte-colony-stimulating factor by fusion with gelatin-like-protein polymer. *Eur. J. Pharm. Biopharm.*, **74**, 435–441.

16 Olsen, D., *et al.* (2003) Recombinant collagen and gelatin for drug delivery. *Adv. Drug Deliv. Rev.*, **55**, 1547–1567.

17 MacEwan, S.R., and Chilkoti, A. (2010) Elastin-like polypeptides: biomedical applications of tunable biopolymers. *Biopolymers*, **94**, 60–77.

18 Kim, W., and Chaikof, E.L. (2010) Recombinant elastin-mimetic biomaterials: emerging applications in medicine. *Adv. Drug Deliv. Rev.*, **62**, 1468–1478.

19 Betre, H., *et al.* (2006) A thermally responsive biopolymer for intra-articular drug delivery. *J. Control. Release*, **115**, 175–182.

20 Sadeghi, H., *et al.* (2011) Modified vasoactive intestinal peptides. WO2011/020091 A1.

21 Shamji, M.F., *et al.* (2007) Development and characterization of a fusion protein between thermally responsive elastin-like polypeptide and interleukin-1 receptor antagonist: sustained release of a local antiinflammatory therapeutic. *Arthritis Rheum.*, **56**, 3650–3661.

22 Turner, A., and Sadeghi, H. (2010) Phenylalanine hydroxylase fusion protein and methods for treating phenylketonuria. WO2010/124180 A1.

23 Dagher, S., *et al.* (2010) Pharmaceutical formulations comprising elastin-like proteins. WO2010/014689 A1.

24 Matsumura, Y., and Maeda, H. (1986) A new concept for macromolecular therapeutics in cancer chemotherapy: mechanism of tumoritropic accumulation of proteins and the antitumor agent Smancs. *Cancer Res.*, **46**, 6387–6392.

25 Singer, J.W. (2005) Paclitaxel poliglumex (XYOTAX, CT-2103): a macromolecular taxane. *J. Control. Release*, **109**, 120–126.

26 Leung, D.W., *et al.* (2002) Recombinant production of polyanionic polymers and uses thereof. US 2002/0169125 A1.

27 Besman, M., *et al.* (2009) Conjugates of biological active proteins having a modified *in vivo* half-life. US 2009/0298762 A1.

28 Sinclair, A.M., and Elliott, S. (2005) Glycoengineering: the effect of glycosylation on the properties of therapeutic proteins. *J. Pharm. Sci.*, **94**, 1626–1635.

29 Schellenberger, V., *et al.* (2010) Extended recombinant polypeptides and compositions comprising same. WO 2010/091122 A1.

30 Schellenberger, V., *et al.* (2007) Unstructured recombinant polymers and uses thereof. WO 2007/103515 A2.

31 Schellenberger, V., *et al.* (2009) A recombinant polypeptide extends the *in vivo* half-life of peptides and proteins in a tunable manner. *Nat. Biotechnol.*, **27**, 1186–1190.

32 Geething, N.C., *et al.* (2010) Gcg-XTEN: an improved glucagon capable of preventing hypoglycemia without increasing baseline blood glucose. *PLoS ONE*, **5**, e10175.

33 Schellenberger, V., *et al.* (2010) Growth hormone polypeptides and methods of making and using same. WO2010/144502 A1.

34 Schulz, G.E., and Schirmer, R.H. (1979) *Principles of Protein Structure*, Springer, New York.

35 Shental-Bechor, D., *et al.* (2005) Monte Carlo studies of folding, dynamics, and stability in α-helices. *Biophys. J.*, **88**, 2391–2402.

36 Schlapschy, M., *et al.* (2007) Fusion of a recombinant antibody fragment with a homo-amino-acid polymer: effects on biophysical properties and prolonged plasma half-life. *Protein Eng. Des. Sel.*, **20**, 273–284.

37 Bamford, C.H., *et al.* (1956) *Synthetic Polypeptides – Preparation, Structure, and Properties*, Academic Press, New York.

38 Holliger, P., and Hudson, P.J. (2005) Engineered antibody fragments and the rise of single domains. *Nat. Biotechnol.*, **23**, 1126–1136.

39 Nagorsen, D., *et al.* (2009) Immunotherapy of lymphoma and leukemia with T-cell engaging BiTE antibody blinatumomab. *Leuk. Lymphoma*, **50**, 886–891.

40 Skerra, A., *et al.* (2008) Biological active proteins having increased *in vivo* and/or *vitro* stability. WO 2008/155134 A1.

41 Skerra, A. (2011) Extending plasma half-life. *EuroBiotechNews*, **10**, 30–31.

42 Roopenian, D.C., and Akilesh, S. (2007) FcRn: the neonatal Fc receptor comes of age. *Nat. Rev. Immunol.*, **7**, 715–725.

43 Labrijn, A.F., *et al.* (2008) When binding is enough: nonactivating antibody formats. *Curr. Opin. Immunol.*, **20**, 479–485.

44 Mahmood, I. (2005) *Interspecies Pharmacokinetic Scaling – Principles and Application of Allometric Scaling*, Pine House Publishers, Rockville, MD.

45 Kozlowski, A., and Harris, J.M. (2001) Improvements in protein PEGylation: pegylated interferons for treatment of hepatitis C. *J. Control. Release*, **72**, 217–224.

46 Hermeling, S., *et al.* (2004) Structure-immunogenicity relationships of therapeutic proteins. *Pharm. Res.*, **21**, 897–903.

5
Half-Life Extension through *O*-Glycosylation

Fuad Fares

5.1
Introduction

Recombinant DNA technology has been used to develop long-acting therapeutic proteins. One strategy is to add *O*-linked or *N*-linked oligosaccharide chains to the backbone of the protein.

The interest in carbohydrate (CHO) chains of glycoproteins has grown because of accumulating data concerning the importance of these chains in a wide array of biological processes. The CHO chains of glycoprotein hormones can be classified into two groups. The first group contains an *N*-acetylgalactosamine (GalNAc) residue which is linked to the hydroxyl group of either a serine or threonine residue of a polypeptide; these are called *O*-linked oligosaccharides. The recognition signal of *O*-linked oligosaccharides is unknown. *O*-linked glycosylation occurs at a later stage during protein processing, probably in the Golgi apparatus. This is the addition of GalNAc to serine or threonine residues by the enzyme *UDP-N-acetyl-D-galactosamine:polypeptide N-acetylgalactosaminyltransferase* followed by other carbohydrates (such as galactose and sialic acid). This process is important for certain types of proteins such as proteoglycans, which involves the addition of glycosaminoglycan chains to an initially unglycosylated "proteoglycan core protein".

The second group, which are called *N*-linked oligosaccharides, contains an *N*-acetylglucosamine (GlcNAc) residue at its reducing terminal and is linked to an amide group of an asparagine (Asn) residue of a polypeptide. Three major classes of *N*-linked saccharides result from this core: high-mannose oligosaccharides, complex oligosaccharides, and hybrid oligosaccharides. High-mannose contains two *N*-acetylglucosamines with many mannose residues. Complex oligosaccharides contain almost any number of the other types of saccharides including more than the original two *N*-acetylglucosamines. Hybrid oligosaccharides possess structural features characteristic of both types. The addition of *N*-linked glycans starts in the lumen of the endoplasmic reticulum (ER), with the transfer of oligosaccharides to the asparagine residues in the sequence, Asn-X-Ser/Thr (NXS/T), in the nascent polypeptide chain where X could be any amino acid except proline.

Therapeutic Proteins: Strategies to Modulate Their Plasma Half-Lives, First Edition. Edited by Roland Kontermann.
© 2012 Wiley-VCH Verlag GmbH & Co. KGaA. Published 2012 by Wiley-VCH Verlag GmbH & Co. KGaA.

The N-linked glycosylation process occurs in eukaryotes and widely in archaea, but very rarely in bacteria. In eukaryotes, most N-linked oligosaccharides begin with addition of a 14-sugar precursor to the asparagine in the polypeptide chain of the target protein. A complex set of reactions attaches this branched chain to a carrier molecule called dolichol, and then it is transferred to the appropriate point on the polypeptide chain, as it is translocated into the ER lumen.

The structures of N- and O-linked oligosaccharides are different where different sugar residues are found in each type. O-linked oligosaccharides are generally short and contain one to four sugar residues. In contrast, N-linked oligosaccharides usually have several branches each terminating with negatively charged sialic acid residue (Figure 5.1).

Protein glycosylation is believed to be the most complicated post-translational event. Carbohydrates are added to proteins in two organelles, the ER and the Golgi apparatus. Carbohydrates addition to proteins occurs both co- and post-translationally. The RNA that codes the protein sequence enters the cytoplasm where it binds to ribosomes which are the site for protein synthesis. Ribosomes bind to the ER, the nascent protein chain enters the lumen, and a core oligosaccharide is added to the protein. Further additions of monosaccharides are performed in the lumen until a final core mannose structure has been added. The ER lumen contains high concentrations of molecular chaperones to assist protein folding. Additional carbohydrate modifications (post-translational) are made as the protein moves from the lumen of the ER to the Golgi apparatus. Here terminal carbohydrate modification is completed. The Golgi does not contain molecular chaperones since protein folding is complete when the proteins arrive. Rather they have high concentrations of membrane bound enzymes, including glycosidases, and glycosyltransferases.

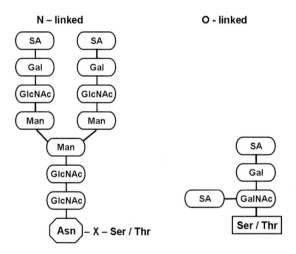

Figure 5.1 Structure of typical N-linked and O-linked oligosaccharide chains.

5.2
The Role of *O*-Linked Oligosaccharide Chains in Glycoprotein Function

Many reports suggest that *O*-GlcNAc might be implicated in pathological conditions such as cancer, cardiovascular or neurodegenerative disorders. Several reports closely link *O*-GlcNAc to glucose toxicity and insulin resistance. This close connection between *O*-GlcNAc and the development of insulin resistance involves the hexosamine biosynthesis pathway (HBP), which is the metabolic pathway leading to the synthesis of UDP-GlcNAc, the direct donor substrate for *O*-GlcNAc. The role of O-GlcNAc in striated muscle must be considered in a wider extent than its implication in insulin resistance. Indeed, the activity of O-GlcNAc transferase is two- to fourfold higher in skeletal muscle and heart than in liver, therefore suggesting an important role of this post-translational modification in "normal" physiology of muscle [1–3]. More generally, different experiments have showed that blocking or reducing *O*-GlcNAc increased the sensitivity of cells to stress, causing a decrease in cell survival, whereas an increase in *O*-GlcNAc level protected cells against stress [4].

Two other main functions of *O*-GlcNAc have been described . One function involves secretion to form components of the extracellular matrix, adhering one cell to another by interactions between the large sugar complexes of proteoglycans. The other main function is to act as a component of mucosal secretions, and it is the high concentration of carbohydrates that tends to give mucus its "slimy" feel.

One apparently universal consequence of *O*-linked oligosaccharide chains is the relative resistance to proteases of *O*-glycosylated regions in glycoproteins. The most likely explanation for protease resistance is simply that the attached carbohydrate blocks access to the peptide core, since these same sequences are quite susceptible to proteases in the absence of attached carbohydrate. The second consequence of *O*-glycosylation is the induction of a specific conformation [5].

The human chorionic gonadotropin beta (hCGβ) subunit is distinguished from the other human β subunits of the glycoprotein hormone family [follitropin (FSH), thyrotropin (TSH), lutropin (LH) and human chorionic gonadotropin (hCG)] in that it contains a unique 28-amino acid carboxyl-terminal peptide bearing four *O*-linked oligosaccharide chains (Figure 5.2). Glycoprotein hormones are composed of two subunits: the α-subunit which is common to FSH, TSH, LH and

N-Terminal - Ser Ser Ser **Ser** Lys Ala Pro Pro Pro **Ser**

Leu ProI Ser Pro **Ser** Arg Leu Pro Gly Pro

Ser Asn Thr Pro Ile Leu Pro Gln – *C - Terminal*

Figure 5.2 Amino-acid sequence of hCG carboxyl-terminal peptide. The positions of *O*-linked oligosaccharide chains are indicated. *O*-linked oligosaccharides are indicated as ⟨.

hCG and a specific β-subunit, which confers the hormone bioactivity. The assembly of subunits is the rate-limiting step for secretion, receptor binding and bioactivity of the hormone.

It has been suggested that the O-linked oligosaccharide chains play an important role in the secretion of intact hCG from the cell, enhanced bioactivity *in vivo*, and extension of its circulating half-life [6]. Deletion of the carboxyl-terminal peptide (CTP) including the O-linked oligosaccharide chains from hCG using site-directed mutagenesis, did not affect assembly of the subunits or secretion of the dimer from the cell. On the other hand, it was shown that truncated hCG without the CTP is three times less potent than intact hCG *in vivo* [7]. On the other hand, the O-linked oligosaccharide chains play a minor role in receptor binding and signal transduction. These findings indicate that the CTP of hCGβ and the associated O-linked oligosaccharides are not important for receptor binding or *in vitro* signal transduction, but are critical for *in vivo* bioactivity and half-life [6]. It has been reported that the kidney is the main site of clearance for glycoprotein hormones. On the other hand much less hCG, which contains the CTP associated with the four O-linked oligosaccharide chains, is cleared by the kidney [8, 9]. Other studies indicated that sialic acids attached to the end of oligosaccharide chains play an important role in the survival of glycoproteins in the circulation [10]. It has been suggested that more negatively charged forms of glycoprotein hormones have longer half-lives, which may be related to a decrease in glomerular filtration [11]. Thus, it was hypothesized that the presence of CTP with its sialylated O-linked oligosaccharides may prolonged the circulating half-life of the hormone secondary to a decrease in renal clearance.

Previous studies have shown the important role of sialylation in the pharmacokinetics and distribution of glycoproteins. It was demonstrated that asialoceruloplasmin was cleared rapidly in rabbits, on the other hand, sialic acid capping of the glycan chains of erythropoietin (EPO) reduces the clearance *in vivo*. Similar observations have been made with other glycoproteins including orosomucoid and hCG [5, 12, 13]. The primary reason for the rapid clearance of the asialoglycoproteins in previous studies seems to be asialo (exposed) Gal or GalNAc leading to uptake by lectin receptors in the liver [14]. Circulating glycoproteins usually do not contain O-linked oligosaccharides, and previous data focused on glycoproteins bearing terminal N-linked oligosaccharides. Asialoglycoproteins bearing terminal N-linked Gal or GalNAc are internalized by the asialoglycoprotein (ASGP) receptor via the clathrin-coated pit pathway and are delivered to lysosomes for degradation [15]. The ASGP receptor has been reported to be on the plasma membrane of parenchymal cells/hepatocytes.

5.3
Designing Long-Acting Agonists of Glycoprotein Hormones

One major issue regarding the clinical use of glycoprotein hormones is their relatively short half-life *in vivo*. To address this issue, the CTP sequence of hCGβ containing four recognition sites of O-linked oligosaccharides, was fused to the carboxyl-terminal of hFSHβ, hTSHβ, and EPO (Figure 5.3). Crystallographic studies indicated that the two sides of glycoprotein hormones: N-terminal and C-terminal, are not involved in receptor binding and they are accessible (Figure 5.4). Therefore, CTP was ligated to the N-terminal

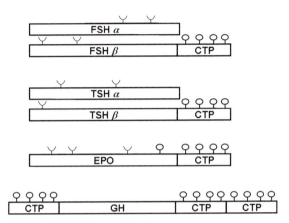

Figure 5.3 Construction of glycoprotein hormone chimeras containing the translated sequence of FSHβ subunit, TSHβ subunit, EPO, GH and hCGβ carboxyl-terminal peptide. O-linked oligosaccharides are indicated as ♀ and N-linked oligosaccharides as ⋎.

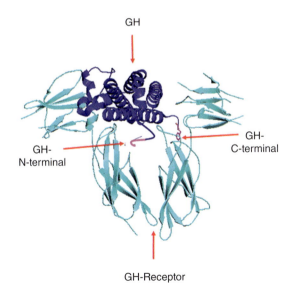

Figure 5.4 Crystallographic diagram of GH and its receptor complex. The N-terminal and C-terminal end of GH are not involved in the binding site of the hormone.

and to the C-terminal of the glycoprotein hormones coding sequence (Figure 5.3).

5.3.1
Construction of Chimeric Genes and Expression Vectors

The overlapping mutagenesis polymerase chain reaction (PCR) technique was used in order to ligate the CTP to the coding sequence of the hormone. Using this technique the cassette gene containing the recognition sites of the O-linked oligosaccharides on CTP was ligated in tandem to the coding sequence of the hormone as a single peptide chain. It was hypothesized that hormone containing the CTP would have a prolonged half-life and higher bioactivity *in vivo*.

The FSHβ wild-type (WT), FSHβ-CTP and FSHα genes were inserted into the eukaryotic expression vector, pM^2-HA, that contains the ampicillin (Amp^R) and the neomycin (Neo^R) resistance genes and a strong promoter of the HaMuSV virus, LTR. EPO-WT and EPO-CTP cassette genes were inserted into the cloning site of the eukaryotic expression vector, pGT, contains the ampicillin (Amp^R), zeocin resistance genes and a strong promoter of the CMV virus. hGH variant genes, WT and chimera, were inserted into the pCI–DHFR plasmid, a eukaryotic expression vector. TSH-WT and chimeric subunits were ligated into the expression vector pLBCMV and transfected into COS-7 and 293 cells.

The PCR-generated constructs were completely sequenced to ensure that no errors were introduced during the PCR.

5.3.2
Expression of Chimeric Genes

Expression vectors containing wild-type or CTP-chimeric genes were transfected into Chinese hamster ovary (CHO) cells. Stable clones expressing hormones were selected using antibiotics (neomycin or zeocin) and Western blot analysis. Assembly of hFSHβ-CTP with α subunit was examined by transfecting stable clones expressing the α subunit with hFSHβ-CTP. Immunoprecipitation studies prove that dimerization has occurred. It seems that α and β subunits of FSH-WT migrate to the same level, meaning that the molecular weight of the two subunits is similar (~22 kDa). However, FSH-CTP migrate more slowly in the gel and exhibit higher molecular weight (~30 kDa) compared with α subunit (Figure 5.5). Similarly, the EPO-WT migrated faster than EPO-CTP. EPO-CTP exhibited high molecular weight (~48 kDa) compared with EPO-WT (~36 kDa) (Figure 5.5). The hGH-WT migrated faster than GH-CTPs and exhibited molecular weight of ~22 kDa. GH variants containing three CTPs exhibited higher molecular weight of ~ 47.5 kDa. The increase of molecular weight of the variants containing CTP, due to the addition of 28 amino acids and the O-linked oligosaccharides. These data may indicate that the O-linked glycosylation recognition site of the C-terminal region is preserved even though the sequence is fused to different proteins.

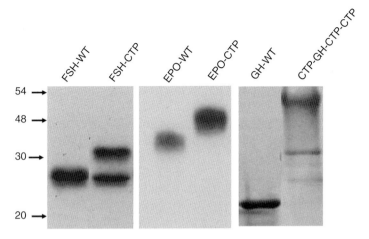

Figure 5.5 Expression of FSH, EPO and GH chimeras from transfected CHO cells. Condition: media from transfected cells were prepared for SDS-PAGE and proteins were detected by Western blot analysis using specific antisera.

5.3.3
Bioactivity of Designed Long-Acting Glycoproteins

5.3.3.1 Follicle-Stimulating Hormone (FSH)

Follitropin (follicle-stimulating hormone; FSH) is a pituitary glycoprotein hormone essential for the maintenance of ovarian follicle and testicular tubule development. Because FSH specifically induces aromatase enzyme in granulosa cells, *in vitro* signal transduction of the modified FSH dimers was assessed in the granulosa cell aromatase bioassay by measuring hormone-stimulated estrogen production. The detected steroidogenic activity *in vitro* of FSH-CTP chimeras was comparable to that of wild-type FSH. Thus, ligation of CTP to the coding sequence of FSH subunit, did not affect receptor binding affinity or *in vitro* bioactivity. The *in vivo* biopotencies of wild-type FSH and the chimeras were examined by determining ovarian weight augmentation and granulosa cell aromatase induction. It was found that ovarian weight increased significantly between animals treated with wild-type FSH and the FSH chimeras. In addition, estrogen production by granulosa cells from chimera-treated rats increased three- to fivefold over that seen in rats treated with wild-type FSH.

Because the increased biopotency of the chimeras may reflect a change in their *in vivo* longevity, the circulatory half-life of the hormones was detected. The clearance patterns of the wild-type and chimera reveal multiple phases. It seems that the clearance of the chimera is much slower than that of wild-type FSH; presumably. Radioimmunoassay (RIA) determinations show that a high level of the chimera is still detectable in serum after 24 h and yet injected wild-type hFSH reaches basal level between 8 and 24 h [16, 17].

The safety, pharmacokinetics, and pharmacodynamics of FSH-CTP were studied in hypogonadotrophic hypogonadal male subjects as a phase I in multicenter study. The results indicated that FSH-CTP use is safe and does not lead to detectable formation of antibodies. Furthermore, the pharmacokinetic and dynamic profile of FSH-CTP seemed to be promising. Compared with recombinant FSH-WT (Puregon), the half-life of FSH-CTP was increases two to three times [18].

Further studies in *in vitro* fertilization (IVF) patients, indicated that a single dose of FSH-CTP is able to induce multifollicular growth compared with daily injection of FSH-WT for seven days [19, 20].

According to the promising results described above in clinical trials, on January 28, 2010 the European Commission (EC) gave Merck & Co. marketing approval with unified labeling valid in all European Union member states for FSH-CTP, now branded as ELONVA.

5.3.3.2 Thyrotropin (TSH)

Thyrotropin (TSH) is a member of the glycoprotein hormone family which includes lutropin (LH), follitropin (FSH) and chorionic gonadotropin (hCG). These hormones are heterodimeric proteins which share a common α subunit, but differ in their hormone-β specific subunit which is unique and confers biological specificity. These hormones activate the target cells via G–protein coupled receptors which, in the case of TSH, lead to increased production of the thyroid hormones, triiodothyronine (T_3) and thyroxine (T_4). Thyroid hormones act on organs throughout the body to regulate growth and various metabolic processes.

Recombinant TSH is important in clinical studies of thyroid cancer. Because of its therapeutic potential, a longer acting analog of TSH was constructed by fusing the carboxyl-terminal extension peptide (CTP) of hCGβ onto the coding sequence of TSHβ sub unit. When co-expressed either with alpha-subunit complementary DNA or alpha minigene in African green monkey (COS-7) or in human embryonic kidney (293) cells, the chimera was fully bioactive *in vitro* and exhibited enhanced *in vivo* potency associated with a prolonged plasma half-life. Ligation of the CTP to the C-terminal of the TSHβ subunit did not affect the assembly and secretion of chimeric TSH. Wild-type and chimeric TSH secreted by COS-7 and 293 cells displayed wide differences in their plasma half-lives, presumably due to the presence of terminal sialic acid and SO_4 on their oligosaccharide chains, respectively. Chimeric and WT TSH secreted by both cell lines demonstrated similar bioactivity in cAMP production. Chimeric TSH appears to be more effective in COS-7 cells than in 293 cells, as judged by growth assay. COS-7-produced chimeric TSH showed the maximum increase in half-life, indicating the importance of sialic acid in prolonging half-life and *in vivo* potency. Sulfation of both subunits, predominantly beta and to a lesser extent alpha, appears to be responsible, at least in part, for the increased metabolic clearance of WT and chimeric TSH secreted by 293 cells. Apart from its therapeutic potential, chimeric TSH produced in various cell lines can be used as a tool to delineate the roles of sulfate and sialic acid in the *in vivo* clearance and, thereby, the *in vivo* bioactivity [21].

Assembly of the glycoprotein hormone subunits is the rate-limiting step in the production of functional heterodimers. To bypass the problem of dimerization of deglycosylated subunits, hTSHβ and α subunits were genetically fused in a single chain hormone with or without the CTP as a linker between the subunits. Single chains of hTSH retained a biologically active conformation similar to that of the wild-type heterodimer [22, 23].

5.3.3.3 Erythropoietin

Erythropoietin (EPO) is a glycoprotein hormone produced primarily by cells of the peritubular capillary endothelium of the kidney. EPO production is stimulated by reduced oxygen content in the renal arterial circulation. Circulating EPO binds to EPO receptors on the surface of erythroid progenitors resulting in replication and maturation to functional erythrocytes by an incompletely understood mechanism. Recombinant human (rhu) EPO has been used successfully in a variety of clinical situations to increase production of red blood cells. Currently, this agent is licensed for use in the treatment of the anemia of renal failure, the anemia associated with HIV infection in zidovudine (AZT) treated patients, and anemia associated with cancer chemotherapy. Administration of rhuEPO has become routine in the treatment of anemia secondary to renal insufficiency where doses of 50–75 U Kg^{-1} given three times per week are used to gradually restore hematocrit and eliminate transfusion dependency.

One major issue regarding the clinical use of EPO is its relatively short half-life *in vivo* due to its rapid clearance (~5 h) from the circulation when it is injected intravenously (i.v.). Thus, the therapeutic protocol used in the treatment of patients, required frequent injections of EPO. The recommended therapy with rhuEPO is two to three times per week by subcutaneous or intravenous injections. Therefore, it was anticipated that enhancing the *in vivo* half-life of EPO will reduce the number of injections per week.

The CTP was ligated to the coding sequence at the C-terminal end of EPO. The *in vitro* biological activity of EPO analogs was demonstrated by measuring their ability to stimulate proliferation of erythroid burst forming colonies (BFU-E) from human peripheral blood. BFU-E colonies were grown from blood of healthy donors using a microwell modification of the methylcellulose technique. The optimal formation of BFU-E colonies achieved by EPO-CTP was similar to that achieved by EPO-WT and rhuEPO using 1 U mL^{-1} of the protein. Receptor binding assay indicated that ligation of the CTP to the carboxyl-terminal of EPO has no significant effect on the affinity of the hormone to the receptor.

For further pharmacological evaluation of EPO-CTP, comparative pharmacodynamic studies of EPO-CTP and commercial rhuEPO were performed in male C57BL mice ($n = 7$/group) using different frequencies and a wide dose range. The *in vivo* efficacy was obtained by measuring the mean values of hematocrit percentage in the blood. The results indicated that EPO-CTP is significantly ($p = 0.05$) more efficient than EPO-WT when administered i.v. once a week with a dose of 5 μg kg^{-1}. EPO-CTP can successfully increase the hematocrit when administered once a week with a dose of 15 μg kg^{-1}. Once weekly dosing with the same concentration of commercial rhuEPO or EPO-WT was significantly ($p = 0.001$) less efficient than once weekly dosing of EPO-CTP.

An interesting observation from the present study was the ability of a single injection once a week of EPO-CTP ($15\,\mu g\,kg^{-1}$) to increase the levels of hematocrit, whereas the same effect was achieved by administration of the same total dose of rhuEPO administered three times a week as $5\,\mu g\,kg^{-1}$ per injection. These results indicated the importance of sustained blood levels, rather than total dose of EPO. These findings are consistent with the hypothesis that the ability of a single injection of EPO-CTP to increase hematocrit, results from its increased stability in the circulation. The increased biopotency of the chimera may reflect a change in their metabolic clearance *in vivo*. Detecting the half-lives of EPO analogs in mice indicated that a higher level of the chimera is still detectable in serum after 24 h. The half-life of EPO CTP is increased two- to threefold compared with EPO-WT. These data suggest that the mechanism of EPO metabolic clearance is affected by the presence of CTP [24].

Another variant of long-acting EPO was design by introduction of additional *N*-linked glycosylation into the protein backbone. Using site-directed mutagenesis, the DNA sequence was changed in order to design a hyperglycosylation recombinant EPO analog (novel erythropoiesis stimulating protein, NESP) which contain five *N*-linked oligosaccharide chains (two more than EPO-WT) [25, 26]. It was found that NESP is distinct from rHuEPO, having an increased molecular weight and greater negative charge. Compared with rHuEPO, it has an approximately threefold longer serum half-life, greater *in vivo* potency, and can be administered less frequently to obtain the same biological response. On the other hand, removal of the sialic acid from either native EPO or rHuEPO resulted in molecules having an increased activity *in vitro*, but very low activity *in vivo*, presumably due to removal from circulation by the asialoglycoprotein receptor in the liver. Similarly, it was shown that EPO molecules, which have been deglycosylated to remove carbohydrate (or produced in *E. coli* to allow expression of only the EPO polypeptide), are active *in vitro*, but have very low *in vivo* activity [27]. RHuEPO and NESP were tested for EPO receptor-binding activity by a radioreceptor assay. Consistent with the results obtained in this assay for the isolated isoforms and the hypothesis, the relative EPO receptor-binding affinity was inversely correlated with the carbohydrate content. The relative affinity of NESP for the EPO receptor was 4.3-fold lower than that of rHuEPO [25].

5.3.3.4 Growth Hormone (GH)

Growth hormone (GH) is secreted by the somatotrophs of the anterior pituitary gland and regulates a wide variety of physiological processes including growth and differentiation of muscle, bone and cartilage cells.

The use of GH for the treatment of children with impaired linear growth has been accepted as an important therapeutic modality for many years. In addition, GH substitution in adults increases muscle mass by 5–10%, but part of the effect is attributed to rehydration rather than protein accretion [28–31]. One major issue regarding the clinical use of GH is its short half-life due to its rapid clearance

(~12 min) from the circulation. Previous studies were carried out in order to stabilize human growth hormone (hGH) and extend its half-life. It was shown that complexation of hGH with heparin did not cause a major distribution in the tertiary structure of hGH but decreased the hydrophilic environment and stabilized the hormone. Another study demonstrated that crystals of hGH coated with positively charged poly(arginine) allowed delivery of hGH over a period of several days. Other studies designed a long-acting GH by fusing the CTP of hCGβ subunit that contains four O-linked oligosaccharide recognition sites to the coding sequence of the hormone. Crystallographic studies indicated that the N-terminal and C-terminal of GH are not important for binding of the hormone to its receptor. Therefore, CTP was ligated to the N-terminal and C-terminal of hGH. The results indicate that ligation of CTP to the coding sequence of GH did not affect secretion of the chimeric protein into the medium. *In vivo* studies in hypophysectomized rats indicated that both bioactivity and pharmacokinetic parameters, MRT, AUC, T_{max}, C_{max} and half-life, of GH bearing the CTP were dramatically enhanced.

The efficacy of GH-CTPs was assessed by weight gain in a hypophysectomized rat model. Weight gain was measured in all animals before treatment, 24 h after first injection and then twice weekly until the end of the study on day 13. Subcutaneous injections of GH-WT, or CTP–GH-CTP-CTP (50 μg/rat) once a week for two weeks or with biotropin (10 μg/rat) daily for two weeks, resulted a dramatic increase (p=0.001) in weight gain. These results indicated that accumulation of CTP-GH-CTP-CTP variant induces an incremental weight gain of 16.5 and 16.8 g, respectively. Similar results were obtained when biotropin was injected daily.

Injection of GH-CTPs (0.4, 0.8 or 4 mg kg^{-1}) once every four days significantly ($p = 0.001$) increased the weight gain compared to controls. Similar effect was achieved by daily administration of 0.1 mg kg^{-1} Biotropin. Injection of a higher amount (1.05 mg kg^{-1}) of CTP-GH-CTP-CTP, once every four days dramatically increased the weight gain. These results indicated the importance of sustained blood levels, rather than total dose of GH for higher bioactivity *in vivo*. These findings are consistent with the hypothesis that the ability of a single injection of GH–CTPs to increase weight gain, results from its increased stability in the circulation. This may support the strategy of hGH administration of at least once a week.

Pharmacokinetic studies indicated the importance of sustained blood levels, rather than total dose of GH. These findings are consistent with the hypothesis that the ability of a single injection of GH–CTP to increase weight gain, results from its increased stability in the circulation. The increased biopotency of the chimera may reflect a change in their metabolic clearance *in vivo*. Therefore, the circulatory half-lives of the hormones were determined. GH-WT or CTP-GH-CTP-CTP chimera were injected i.v. into hypophysectomized male mice. At selected intervals after injection of 50 μg kg^{-1}, blood samples were collected and the concentrations of GH were determined. The results indicated that a higher level of the chimera is still detectable in serum after 50 h where the level of Biotropin after 24 h was undetectable. The estimated half-life of CTP-GH-CTP- CTP is increased by four- to fivefold compared with Biotropin. These data suggest that the mechanism of GH metabolic clearance is affected by the presence of CTP [32]. Preliminary studies of GH-CTPs clinical trials indicated that this peptide is safe for use and clinical trials phase II is on the way.

5.4
Conclusions and Summary

In this chapter we have described the role of O-linked oligosaccharide chains on the glycoprotein function. Moreover, the design of long-acting glycoprotein hormones (FSH, TSH, EPO and GH) agonists was described. This was achieved by ligation of the sequence of hCG beta carboxyl-terminal peptide, containing the recognition signal of four O-linked oligosaccharide chains, to the coding sequence of the hormone. Ligation of the CTP to different proteins indicated that the O-linked glycosylation recognition sites of the CTP are preserved. Moreover, this ligation is not involved in the assembly of the subunits, secretion of the dimer, receptor binding affinity, nor *in vitro* bioactivity. However, both the *in vivo* bioactivity and half-life in circulation of hormones bearing the CTP were significantly enhanced. Clinical trials of proteins containing the CTP indicated that these chimeras are not immunogenic and are safe for use.

Carboxyl-terminal peptide (CTP) is a small peptide naturally found in the body as a portion of hCG and contains four recognition sites of O-linked oligosaccharide chains. CTP can readily be attached to a wide array of existing therapeutic proteins, stabilizing the therapeutic protein in the bloodstream and greatly extending its life span without additional toxicity or loss of desired biological activity. Moreover, CTP-modified proteins can be manufactured using established recombinant DNA techniques in widely used mammalian protein expression systems. This strategy may have wide applications for enhancing the *in vivo* bioactivity and half-life of diverse proteins. FSH-CTP was approved by the European Commission (EC) for marketing in all European Union member states. The safety and efficacy of other glycoprotein hormones bearing the CTP as therapeutic peptides are being evaluated in ongoing clinical trials.

References

1 Copeland, R.J., *et al.* (2008) Cross-talk between GlcNAcylation and phosphorylation: roles in insulin resistance and glucose toxicity. *Am. J. Physiol. Endocrinol. Metab.*, **295**, E17–E28.

2 Wopereis, S., *et al.* (2006) Mechanisms in protein O-glycan biosynthesis and clinical and molecular aspects of protein O-glycan biosynthesis defects: a review. *Clin. Chem.*, **52**, 574–600.

3 Yki-Järvinen, H., *et al.* (1997) UDP-N-acetylglucosamine transferase and glutamine: fructose 6-phosphate amidotransferase activities in insulin-sensitive tissues. *Diabetologia*, **40**, 76–81.

4 Zachara, N.E., and Hart, G.W. (2004) O-GlcNAc a sensor of cellular state: the role of nucleocytoplasmic glycosylation in modulating cellular function in response

to nutrition and stress. *Biochim. Biophys. Acta*, **1673**, 13–28. Review.

5 Fares, F. (2006) The role of O-linked and N-linked oligosaccharides on the structure-function of glycoprotein hormones: development of agonists and antagonists. *Biochim. Biophys. Acta*, **1760**, 560–567. Review.

6 Matzuk, M.M., *et al.* (1990) The biological role of the carboxyl-terminal extension of human chorionic gonadotropin beta-subunit. *Endocrinology*, **126**, 376–383.

7 Matzuk, M.M., *et al.* (1987) Effect of preventing O-glycosylation on the secretion of human chorionic gonadotropin in Chinese hamster ovary cells. *Proc. Natl. Acad. Sci. U.S.A.*, **84**, 6354–6358.

8 Kalyan, N.K., and Bahl, O.P. (1983) Role of carbohydrate in human chorionic gonadotropin. Effect of deglycosylated on the subunit interaction and on its *in vitro* and *in vivo* biological properties. *J. Biol. Chem.*, **258**, 67.

9 Sowers, J.R., *et al.* (1979) Metabolism of exogenous human chorionic gonadotropin in men. *J. Endocrinol.*, **80**, 83–89.

10 Van Den Hamer, C.J., *et al.* (1970) Physical and chemical studies on ceruloplasmin. IX. The role of galactosyl residues in the clearance of ceruloplasmin from the circulation. *J. Biol. Chem.*, **245**, 4397–4402.

11 Wide, L. (1986) The regulation of metabolic clearance rate of human FSH in mice by variation of the molecular structure of the hormone. *Acta Endocrinol.*, **112**, 336–344.

12 Morell, A.G., *et al.* (1971) The role of sialic acid in determining the survival of glycoproteins in the circulation. *Biol. Chem. Circ.*, **246**, 1461–1467.

13 Briggs, D.W., *et al.* (1974) Hepatic clearance of intact and desialylated erythropoietin. *Am. J. Physiol.*, **227**, 1385–1388.

14 Ashwell, G., and Kawasaki, T. (1978) A protein from mammalian liver that specifically binds galactose-terminated glycoproteins. *Meth. Enzymol.*, **50**, 287–288.

15 Yik, J.H., *et al.* (2002) Nonpalmitoylated human asialoglycoprotein receptors

recycle constitutively but are defective in coated pit-mediated endocytosis, dissociation, and delivery of ligand to lysosomes. *J. Biol. Chem.*, **277**, 40844–40852.

16 Fares, F.A., *et al.* (1992) Design of a long-acting follitropin agonist by fusing the C-terminal sequence of the β chorionic gonadotropin subunit. *Proc. Natl. Acad. Sci. U.S.A.*, **89**, 4304–4308.

17 Lapolt, P.S., *et al.* (1992) Enhanced stimulation of follicle maturation and ovulatory potential by long acting follicle-stimulating hormone agonist with extended carboxyl-terminal peptide. *Endocrinology*, **131**, 2514–2520.

18 Bouloux, P.M., *et al.* (2001) First human exposure to FSH-CTP in hypogonadotrophic hypogonadal males. *Hum. Reprod.*, **16**, 1592–1597.

19 Devroey, P., *et al.* (2004) Induction of multiple follicular development by a single dose of long-acting recombinant follicle-stimulating hormone (FSH-CTP, Corifollitropin Alfa) for controlled ovarian stimulation before *in vitro* fertilization. *J. Clin. Endocrinol. Metab.*, **89**, 2062–2070.

20 Abyholm, T., *et al.* (2008) A randomized dose-response trial of a single injection of corifollitropin alfa to sustain multifollicular growth during controlled ovarian stimulation. *Hum. Reprod.*, **23**, 2484–2492.

21 Joshi, L., *et al.* (1995) Recombinant thyrotropin containing a beta-subunit chimera with the human chorionic gonadotropin-beta carboxy-terminus is biologically active, with a prolonged plasma half-life: role of carbohydrate in bioactivity and metabolic clearance. *Endocrinology*, **136**, 3839–3848.

22 Fares, F.A., *et al.* (1998) Conversion of thyrotropin heterodimer to a biologically active single-chain. *Endocrinology*, **139**, 2459–2464.

23 Grossman, M., *et al.* (1997) Human thyroid-stimulating hormone (hTSH) subunit gene fusion produces hTSH with increased stability and serum half-life and compensates for mutagenesis-induced defects in subunit association. *J. Biol. Chem.*, **272**, 21312–21316.

24 Fares, F., *et al.* (2007) Development of a long-acting erythropoietin by fusing the

carboxyl-terminal peptide of human chorionic gonadotropin beta-subunit to the coding sequence of human erythropoietin. *Endocrinology*, **148**, 5081–5087.

25 Egrie, J.C., and Browne, J.K. (2001) Development and characterization of novel erythropoiesis stimulating protein (NESP). *Br. J. Cancer*, **84** (Suppl. 1), 3–10.

26 Nissenson, A.R. (2001) Novel erythropoiesis stimulating protein for managing the anemia of chronic kidney disease. *Am. J. Kidney Dis.*, **38**, 1390–1397.

27 Higuchi, M., *et al.* (1992) Role of sugar chains in the expression of the biological activity of human erythropoietin. *J. Biol. Chem.*, **267**, 7703–7709.

28 Aben, M.S. (1958) Treatment of a pituitary dwarf with human growth hormone. *J. Clin. Endocrinol. Metab.*, **18**, 901–903.

29 de Boer, H., *et al.* (1995) Clinical aspects of growth hormone deficiency in adults. *Endocr. Rev.*, **16**, 63–86.

30 Carroll, P., *et al.* (1998) Growth hormone deficiency in adulthood and the effects of growth hormone replacement: a review. *J. Clin. Endocrinol. Metab.*, **83**, 382–395.

31 Drake, W., *et al.* (2001) Optimizing GH therapy in adults and children. *Endocr. Rev.*, **22**, 425–450.

32 Fares, F., *et al.* (2010) Designing a long-acting human growth hormone (hGH) by fusing the carboxyl-terminal peptide of human chorionic gonadotropin beta-subunit to the coding sequence of hGH. *Endocrinology*, **151**, 4410–4417.

6

Polysialic Acid and Polysialylation to Modulate Antibody Pharmacokinetics

Antony Constantinou, Chen Chen, and Mahendra P. Deonarain

6.1
Introduction

The use of chemically inert polymers is now a common practical consideration in the formulation of drugs for therapy. Their conjugation to drugs can improve their efficacy by potentially improving solubility, decreasing immunogenicity and altering their blood clearance and tissue uptake rates. In doing so their pharmacokinetics can be modulated to minimize potential side-effects and optimize therapeutic effect.

PEGylation (conjugation to polymers of long-chain polyethylene glycol) is the "market leader" in such enhancements and its use and development is discussed at length in Chapter 3. However, the use of synthetic nonbiodegradable polymers has raised concerns. The primary driver for this is the concern that PEG used for chronic conditions generates PEGylated peptides or by-products which could accumulate in tissues and cause unforeseen toxic effects and/or generate an immune response. Degradable polymers are being developed to overcome possible toxicity issues presented by nondegradable polymers such as cell vacuolization, toxicity, renal/hepatic uptake and anti-PEG antibodies [1–4]. As such a new generation of biopolymers are now being investigated as candidates to supersede PEG technology by addressing these concerns. Such biopolymers include HEPylation (heparosan polymer), HAPylation (homoaminoacid), PASylation (proline-alanine-serine repeats) and HESylation (hydroxyethyl starch). In this chapter the use of polysialic acid as a biopolymer for conjugation purposes will be the subject in focus.

As a biopolymer the use of polysialic acid (PSA) aims to overcome the synthetic related limitations of PEG mentioned above. Polysialylation, the process whereby PSA is conjugated to drugs and proteins, is rapidly becoming a competitive candidate to rival PEGylation by addressing its synthetic concerns [5]. Unlike PEG, PSA has the advantage of being a naturally occurring polymer that can be broken down over time, hence limiting potential toxic accumulation and immunogenic concerns.

Therapeutic Proteins: Strategies to Modulate Their Plasma Half-Lives, First Edition. Edited by Roland Kontermann.
© 2012 Wiley-VCH Verlag GmbH & Co. KGaA. Published 2012 by Wiley-VCH Verlag GmbH & Co. KGaA.

As with PEGylation, polysialylation works on the principle of increasing the molecular mass of a target molecule, such that the addition of different lengths of polymer or a multiple chains to a single drug molecule can change the clearance properties of the drug moiety on the basis of size. Moreover, the attraction of water molecules to the helical polymer [6–8] that surrounds the conjugate insulates the drug with a "watery cloud" which provides it with a much larger hydrodynamic volume than could be attributed to the primary structure of the conjugate alone [9–11]. The effect of PSA hydration compares similarly to the physicochemical properties of PEG which have been the most extensively studied. Each of its ethylene oxide subunits can support up to three molecules of water [12, 13] resulting in an increased hydrodynamic volume that provides an effective molecular weight 5–10-fold greater than that of proteins or other macromolecules of a similar size [14, 15]. The biophysical properties of PSA lend itself to two important benefits. Firstly, gel filtration studies have shown that polysialic acid of various lengths can emulate a much larger apparent molecular mass and therefore display a much slower than expected clearance rate in *in vivo* systems (Deonarain *et al.*, unpublished). Therefore in theory, a relatively small PSA chain could confer a greater clearance decrease than a protein component of the same molecular weight. For example, an increase in the apparent molecular weight (as determined by size exclusion chromatography) from 30 kDa to 600 kDa effectively increases the hydrodyamic radius from around 2 nm to 7 nm, taking it above the size excluded by renal tubules (5–6 nm), thus significantly reducing its systemic clearance [15, 16], (Deonarain *et al.*, unpublished). Secondly, the hydrophilic support of PSA provides a biophysical property that can be conferred to otherwise poorly soluble drugs [9, 17, 18].

A significant physiochemical difference between PSA and other polymers is its highly charged anionic nature. The effect of charge has been found to influence the pharmacokinetics of proteins and has been particularly investigated using antibodies. Changes to the primary structure of antibodies, so as to alter their charge, have been studied as a means of modulating tissue penetration and systemic clearance. Lowering the isoelectric points of several antibodies through recombinant amino acid substitutions has resulted in lowered nonspecific uptake toxicity and decreased renal uptake [19, 20]. Similarly, acylation of anti-Tac monoclonal antibody so as to lower its isoelectric point resulted in a more favorable biodistribution than unmodified antibody by inhibiting tubular reabsorption through the kidney [21, 22]. Conversely, cationization (increased pI) of an anti-EGFR monoclonal antibody showed improved localization and increased clearance leading to improved pharmacokinetics [23]. A study of similarly sized proteins with varying charge established that positively charged molecules extravasated faster into solid tumors than similar sized ones with neutral or negative charges. Conversely, these cationic molecules had a ~twofold faster serum clearance than there anionic counterparts [24]. The increased renal clearance can be physiologically explained by the negatively charged basement membrane of the glomerulus, which may facilitate the passage of cationic proteins of high molecular weight [25, 26]. In lieu of this observation, tissue uptake, renal uptake and blood clearance can be modulated through direct modification of a protein's charge. Indirectly, nonspecific binding and renal clearance of antibodies (associated with charge) can be altered by "flooding" the system with nonreactive charged molecular decoys. Injections of cationic amino acids, sugars and peptides in mice prior to injections with tumor specific therapeutic antibodies resulted in a five- to sixfold reduction in kidney uptake compared to untreated controls [25, 27].

On the basis of these observations it is possible that proteins conjugated to a negatively charged polymer such as PSA could cause a decrease in passive extravasation through tissues and a decrease in renal uptake and blood clearance. However, while charge modifications can be used to modulate tumor uptake and the clearance pharmacokinetics of therapeutic antibodies, care must be taken to ensure that nonspecific uptake is not also increased.

6.2
Polysialic Acid in Nature

Sialic acids are nine-carbon monosaccharides ubiquitously found in nature, particularly as defined components of glycoproteins and gangliosides. In this format several receptors have been identified that recognize different sialic acid containing moieties. These lectin-recognizing receptors have been categorized within two families: the selectin family and siglec (sialic acid binding Ig-like lectins) family. A description of the role of these receptors is beyond the scope of this chapter; however, in both instances these receptors appear to recognize only the terminal sialic acid of a glycan and then only in conjunction with its linkage to its neighboring residue and/or the underlying glycan structure. For the purpose of using polysialic acid for pharmaceutical purposes it is important that there is no cross-reactivity with endogenous receptors that may lead to undesired side-effects *in vivo*.

The existence of polysialic acid in natural systems has been observed in several glycoproteins and glycolipids across a variety of organisms. In bacteria, PSA has been observed in some pathogenic strains of *Escherichia coli* such as K1, K12, K92 and *Neisseria meningitidis* C. Their capsular glycolipids include PSA chains that are immunologically similar to PSA expressed in host organisms [28–30], but differences have been observed arising from the inter-residue linkages between sialic acid units (these differences are discussed in more detail below). It is the presence of near immunologically inert PSA on the cell surface of these micro-organisms that is thought to affect bacterial virulence by mediating resistance to phagocytosis and complement-mediated killing, and influence their neurotropic invasive potential in vertebrates [28, 31–34]. Another possibly important factor in PSA acting as a particularly potent virulence factor is the ability of some of these micro-organisms to alter the acetylation status of PSA. It is believed that acetylation may lead to increased antigenicity leading to increased recognition by a host's immune system [35]; however, such modifications can also increase resistance to the effects of exo-sialidase cleavage, which may serve to promote bacterial colonization through resistance to host killing and evasion of a host's innate immune system [35, 36]. This observation may be of particular significance when deciding on what kind of PSA should be used for possible therapeutic conjugation, ensuring that the PSA used is least likely to elicit an immune response.

In vertebrates, very few proteins meet the glycoprotein-motif-specific requirements necessary for enzyme-catalyzed polysialylation. In humans, only homopolymers of α-(2,8)-linked 5-*N*-acetylneuraminic acid have been observed. The neural cell adhesion molecule (NCAM) [37–39] and the alpha-subunit of the voltage-sensitive sodium channel [38, 40] were the earliest glycoproteins to be described with *N*-linked PSA expression. While the role of polysialylation in the sodium channel still remains unclear, its role in NCAM function has been implicated in the promotion of plasticity in cell–cell interactions during cell migration, axonal path-finding, branching and fasciculation. Such regulation is achieved through polysialylation of NCAM leading to disruption of its homo- and heterophilic interactions between cell membranes [39, 41].

SynCAM1 is another synaptic cell adhesion molecule that serves as a PSA attachment site on *N*-linked glycan moieties [42]. The cell surface glycoprotein normally forms homo- and heterophilic interactions that have been implicated in organizing excitory synapses between nascent and mature synaptic clefts [43, 44]. Polysialylation obstructs these interactions and has been specifically implicated in mediating the breakdown of cell synapses in polydendrocyte (NG2) cells, which may signal an instructive role in consequential differentiation of these cells into oligodendrocytes or astrocytes, or even neurons [42, 45].

Analysis of human milk has also found PSA associated with the O-linked glycoprotein CD36 [46], a scavenger receptor normally associated with monocyte-derived cell lines [47, 48]. CD36 has been implicated in a number of roles including events leading to platelet activation [49, 50], the removal of apoptotic cells and cell fragments [51], the elimination of oxidized low density lipoproteins [52], and the transport of long chain fatty acids [53]. Despite these various roles, only CD36 integrated in the milk fat globule membrane secreted from the mammary epithelial cells has been confirmed as being polysialylated [46]. It is not yet clear what the purpose of CD36 polysialylation is, but it has been speculated to be of protective and nutritional significance [46]. The presence of polysialic acid on CD36 may serve as a diversion from neonatal infection by bacterial and viral pathogens, which often have receptors that recognize sialic acids [54]. From a nutritional perspective, sialic acid has been associated with the development of the neural system [55, 56].

The neurophilin-2 receptor (NRP-2), which is expressed on the surface of human dendritic cells, is another example of an O-linked polysialylated protein. Removal of PSA from the receptor has been shown to result in an increased ability of dendritic cells to activate T-lymphocytes [57]. Ordinarily, NRP-2 has been associated with cell motility during neuronal and endothelial growth, and tumor cell migration [58, 59]. Unlike previously characterized examples of polysialylated glycoproteins that seem to lose function on being modified, polysialylation of NRP-2 is understood to support and promote its function. The highly positively charged

C-terminus of the chemokine CCL21 is capable of interacting with PSA exhibited by NRP-2, and this in turn provides a scaffold that facilitates interaction between the N-terminus of CCL21 and the chemokine receptor CCR7, which promotes the migratory capacity of dendritic cells and activation of chemotaxis [60, 61]. As discussed above, such additional functions of PSA could enhance its use as a pharmaceutical. In this case, the anti-inflammatory effects of PSA or even sialic acids as shown on immunoglobulins [62] could have benefits such as reduced immunogenicity or adverse reactions.

Intriguingly, the enzymes that catalyze the formation of PSA (SIAT8B [STX] and SIAT8D [PST]) on specific glycoproteins are capable of autopolysialylation which is believed to play a role in the regulation of polysialylation activity [63]. Double knock-outs of these genes in mice have demonstrated axonal growth defects leading to perinatal lethality, and demonstrates particularly the importance of NCAM polysialylation for the appropriate development of neuronal networks and brain function [64]. In a separate study using PST knockout mice chemically treated to exhibit a *status epilepticus* (SE), it was shown that although it was more difficult to elicit an epileptic episode in such mice, the lethality rate increased by 28% when a fit was eventually observed (compared to no deaths observed in non-knockout SE mice). Further, it was shown that the loss of PSA in mutant mice increased anxiety associated behavior, suggesting that PSA may function critically in the development of psychiatric comorbidities in patients with epilepsy [65] an observation that is already recognized in such groups [66]. Taken together, it is clear that the activity of PST is particularly important in ensuring appropriate polysialylation for neural development.

Over-expression of PSA-NCAM has been identified in a number of malignant neuroendocrine tumors, such as Wilms' tumor, neuroblastoma, rhatbdomysarcoma, small cell lung cancer and others [67–72]. It has been suggested that the nature of PSA in destabilizing cell–cell interactions results in the promotion of the growth of tumors and their metastatic activity [70, 71, 73]. In a study using mice, the development of C6 glioma cells transfected with or without polysialyltransferases that catalyze PSA production was investigated. The rates of tumor growth appeared to be similar in both sets of mice; however, in tumors containing the polysialyltransferase gene it was found that there was more polysialylated NCAM. Moreover, these PSA rich cells were observed as invading the *corpus callosum* of the brain more readily than the mock tumor model. This observation led to the conclusion that the attenuation of NCAM-NCAM interactions caused by up-regulated PSA expression was responsible for facilitating aggressive tumor invasion [74]. In a study of patients with neuroendocrine lung cancer a similar correlation between NCAM-PSA expression and the aggressive nature of the tumor and metastatic spread was also noted, and accredited to the reduced adhesion properties between tumor cells expressing NCAM-PSA [70]. Similarly, highly malignant astrocytomas were associated with a greater polysialylation status than less malignant tumors [75]. These "repulsive" or "lubricating" properties of PSA may help explain their usefulness as pharmacokinetic modulators and could even provide advantages yet to be discovered *in vivo*.

There is evidence to suggest that the attenuation of NCAM interactions through polysialylation is more that just the result of a steric hindrance that diminishes molecular interactions. The implications of polysialylation are more far reaching; molecular force measurements have been used to show quantitatively that NCAM polysialylation increases the range and magnitude of intermembrane repulsion leading to the 10–15 nm spacing observed in electron microscopy [76–78]. Essentially an equilibrium between forces of protein attraction and the repulsive steric/ electrostatic forces of PSA dictates the extent of the separation between membranes [77, 78]. Clearly this has implications for the use of polysialic acid in conjugated proteins for therapy, particularly for the length and/or number of PSA chains used. While the negative charge of PSA may serve as a repulsive force to minimize nonspecific membrane interactions and cell uptake (like it does in nature), the chains used should not exert a strong enough repulsive force that may inhibit the therapeutic drug's interaction with its target. A screening study of these forces using a panel of conjugate candidates can readily be performed to assist in selecting the optimum PSA content conjugated to allow effective function. This suggests that PSA modification should be tailored to the particular pharmaceutical agent.

Atomic force microscopy has also demonstrated that PSA chains are capable of forming filament bundles that assemble into networks [6]. The consequence of this is unclear, however the helical coil structure of PSA that arises after a minimum length of residues may be enough to form a tertiary structure that can form associative self-interactions, in much the same way as DNA does, and provide a helical coil epitope for interactions with other proteins. This possibility is reinforced by the observation that CCL21 interactions with PSA become more apparent with longer PSA chains [6]. These conformational factors may also be relevant with respect to the eight-residue minimum seemingly required by most endosialidases for effective cleavage [79, 80] and the fact that anti-NCAM/PSA monoclonal antibodies require a minimum PSA epitope length [81]. Other proteins that have been identified as interacting directly with polysialic acids include members of the neurotrophin family: nerve growth factor (NGF), brain-derived neurotrophic factor (BDNF), neurotrophin-3 (NT-3) and neurotrophin-4 (NT-4). These neurotrophins exhibit activity in the form of noncovalent dimers, but moreover they are capable of interacting directly with polysialic acid to form a neurotrophin-PSA complex. A minimum degree of polymerization (DP) of 12 has been identified for such complexes to form and it has been suggested that PSA acts to sequester the neurotrophins on the neural cell surface and provides a reservoir of these ligands and inhibits their diffusion. Neurotrophins are expressed and function in a time- and space-dependent manner; however, the DP of PSA involved in such binding could affect the ease or difficulty with which the neurotrophin is released or trapped within the vicinity of its biological receptors on the cell surface. In this respect it has been shown that BDNF-polysia complexes putatively lead to increased receptor interactions leading to upregulated growth and/or survival of neuroblastoma cells [82]. This kind of spatiotemporal regulation by PSA may also be the function of PSA binding of the cytokine CCL21 mentioned above.

Another regulatory function of PSA has been described on the activity of histone-1 (H1). As with other histones, H1 is normally associated with mediating DNA packing; however; it has also been identified extracellularly on the surface of several cell types [83–85]. In cerebellar neurons and Schwann cells, H1 has been shown to promote cell migration and neurite outgrowth with positive implications for regeneration *in vivo*. Other members of the histone family have not shown binding to PSA in the same way as H1, demonstrating a unique additional role for this protein [86]. However, the mechanism for H1-PSA regulation of these neurological observations is not yet clear, but it is conceivable that H1 acts as an extracellular molecular scaffold and its interaction with PSA allows modulation of other PSA binding ligands and/or polysialylated glycoproteins on the cell surface.

6.3
PSA Biosynthesis and Biodegradation

Polysialic acid is a naturally occurring polymer also known as neuraminic acid. Carbohydrate chains that include PSA have been identified on the surface of a variety of cells from microbes to vertebrates (see above), and their biosynthesis and conjugation is performed enzymatically through the combined activity of different polysialyltransferase enzymes. These catalyze PSA synthesis at specific glycolipid or glycoprotein precursor groups and can catalyze PSA elongation.

The building blocks of PSA are predominated by three major building units: 5-*N*-glycolylneuraminic acid (Neu5Gc), 5-*N*-acetylneuraminic acid (Neu5Ac or NANA), and 5-deamino3,5-dideoxynuraminic acid (2-keto-3-deoxynonulosonic acid, Kdn) [28]. The latter two are predominant in mammalian PSA, but also in some bacteria such as the neuroinvasive *Escherichia coli* K1, which synthesizes a homopolymer of $\alpha(2,8)$-linked 5-*N*-acetylneuraminic acid (Neu5Ac), a PSA specifically known as colominic acid [87]. In such bacteria, enzymes associated with the biosynthesis of capsular PSA can be found encoded within whole gene clusters. The *kps* gene cluster of *E. coli* K1 is one such group that has been well characterized [88–90]. These genes are strongly conserved among *E. coli* strains that synthesize serologically distinct capsules [88–91]. They encode genes whose products are necessary for PSA synthesis unique to the capsular serotype, and polymer translocation from its cytosolic site of synthesis to the cell surface. The diverse family of enzymes involved in the synthesis of PSA between different strains of bacteria are responsible for the diversity that is distinguished within the sialic acid linkage repeats of the different serotypes. Serogroup B capsular PSA (PSB) found on the surface of *Neissseria meningitidis* B and *E. coli* K1 is a homopolymer of $\alpha(2,8)$-linked 5-*N*-acetylneuraminic acid (colominic acid). Serogroup C capsular polysialic acid (PSC) found on the surface of *Neissseria meningitidis* C is a homopolymer of $\alpha(2,9)$-linked 5-*N*-acetylneuraminic acid, while *E. coli* K92 expresses a heteropolymer of alternate units of $\alpha(2,8)$- $\alpha(2,9)$-linked 5-*N*-acetylneuraminic acid (PSK92). In all three instances the nonreducing end of the acid is anchored *via* a covalently linked phospholipid molecule. The mechanism for chain initiation is not fully understood though it is accepted that polysialyltransferase (*neu* S) is responsible for the addition of sialic acid to endogenous polysialic acid acceptors [92–94]. There is evidence that suggests that initiation and elongation of PSA may take place within a specialized subcellular location; however, it is unclear whether synthesis, translocation and surface attachment of PSA occur independently or as part of closely coupled processes [93, 95].

As discussed above, sialic acids are also ubiquitous components of glycosylated structures in vertebrates. In humans, at least 15 different sialyltransferase (SIAT) enzymes have been identified generating the complexity and specificity observed between the families of classified glycans. The specificities of some of these enzymes have been characterized. Studies have shown that the direct addition of sialic acid to both glycolipids and glycoproteins to form disialic and oligosialic glycans can be performed by SIAT8C (ST8Sia III, α2,3Gal β1,4 GlcNAc α2,8-sialyltransferase), while SIAT8A (ST8Sia I, alpha-N-acetylneuraminate: α2,8-sialyltransferase, GD3 synthase) and SIAT8E (ST8Sia V, α2, 8-polysialytransferase) only utilize glycolipids as acceptors [96–100].

For the specific purpose of sialic acid elongation to form PSA three enzymes have thus far been identified; SIAT8B (ST8Sia II, α2, 8-sialyltransferase, STX) SIAT8C (ST8Sia III, 2,3Gal β1,4 GlcNAc α2,8-sialyltransferase) and SIAT8D (ST8SiaIV, α2, 8-sialyltransferase, PST). All three catalyze the formation of the α2,8 glycosidic linkages between α2,3 α2,6 and α2,8 sialic residues followed by multiple addition of α2,8 sialic acid, but differ in ability to recognize polysialic acid acceptors. For example, it has been shown that SIAT8B, SIAT8C and SIAT8D can add disialyl and oligosialic residues on glycoprotein (α_2-HS-glycoprotein), small oligosaccharides (NeuNAcα2$\rightarrow$3Galβ1$\rightarrow$4Glcβ1$\rightarrow$ceramide and NeuNAcα2$\rightarrow$8NeuNAcα2$\rightarrow$3Galβ1$\rightarrow$4Glc β1$\rightarrow$ceramide) and glycolipid (sialo-paraglobosisde) acceptors, but true polysialylation is not achieved. Further, polysialylation of glycans on NCAM by SIAT8B and SIAT8D is much more efficient than the polysialylation of the glycans alone (though small amounts of polysialylation were observed), suggesting a bias in the evolution of these two enzymes towards the specific polysialylation of NCAM [101]. A study of NCAM structure has identified that the presence of a fibronectin domain as an enzyme docking site is specifically required for polysialylation at adjacently located *N*-linked glycan sites [102]. The necessity of this docking site is further reinforced by the identification of sequences on SIAT8B and SIAT8D that are required for protein-specific polysialylation to take place [103].

The DP on NCAM has been investigated and typical values fall within the 50–150 residue range; however, >370 DP values have also been reported [104, 105]. The *in vivo* regulation of PSA extension has not yet been determined, although the tissue distribution and expressional regulation of the polysialytransferase enzymes (and desialidases) is thought to play a role [63, 106, 107]. Further, a decrease in enzymic activity by autopolysialylation of SIAT8B and SIAT8D has been observed and shown to have an impact on the amount of polysialic acid added to NCAM [63].

Neuraminidases (desialylases) have also been identified that can cleave PSA and provide a natural mechanism for polymer clearance (for example *via* the exposed galactose residues). Their presence may also provide an additional mechanism whereby the length of PSA is regulated, and possibly affect the interactions sialic acid containing glycoproteins have with galactose-recognizing receptors. Neuraminidases have been identified in the cytoplasm, plasma membrane and lysomes, of cells from a variety of tissue types, including the pancreas, skeletal, muscle, kidney, placenta, heart, lung, liver and, to a lesser extent, the brain [108]. They preferentially catalyze the cleavage of terminal α(2-3)-linked and α(2-6)-linked sialic acid residues from mono- or oligosaccharide chains of glycoconjugates, though they are also capable of hydrolyzing α(2-8)-linked sialic acid [108–110]. In the liver, the removal of sialic acid containing glycoconjugates via the hepatic asialoglycoprotein receptor (ASGPR) is preceded by neuraminidase activity. The ASGPR then recognizes the terminal galactose and N-acetylgalactosamine units and mediates the endocytosis of plasma glycoproteins whereby the complex of receptor and ligand is internalized and transported to a sorting organelle where disassociation occurs, the receptor being recycled to the cell membrane [111].

6.4
Pharmacological Effects of PSA

When designing new approaches to drug delivery it is important that their pharmacology is understood. The rate of elucidating the existence of naturally occurring polysialylated proteins and polysialic acid binding proteins has closely paralleled the rate of development of polysialylation as a technology for modulating the pharmacokinetics of therapeutic drugs or proteins. As such, until recently it has not been possible to study the specific effects of exogenous polysialic acid and their conjugates for any possible interactions with other endogenous PSA glycoproteins or PSA binding proteins, and clearly a pharmacological study of any such interactions must be investigated, particularly to determine any long-term effects.

PSA acts as an anti-adhesive between cell–cell/extracellular matrix interactions due to its bulky polyanionic nature, but the fact that PSA can interact specifically with some ligands suggests that these PSA properties will not hinder the activity of any potential conjugates administered for therapy. During an acute toxicity study mice injected with a high single subcutaneous dose of colominic acid displayed no adverse reactions over a 30 day observation period, suggesting good biocompatibility at least in the short term [18].

6.5
PSA Conjugation: Polysialylation for Therapeutic Applications

The use of polysialic acid as a biopolymer alternative to PEGylation was first suggested by Professor Gregory Gregoriadis and his team at the London School of Pharmacy in 1993 [18, 112]. Initial studies using fluorescein as a model drug showed that polysialylation led to the mock drug assuming the half-life of the PSA carrier [112]. To develop this technology further PolyXen® was established in the early 1990s as part of Lipoxen Technologies for the development and production of peptide and drug conjugates with improved stability and *in vivo* pharmacokinetics for therapy (see Figure 6.1 for PSA structures and conjugation). The three major forms of PSA of comparable DP, PSB, PSC and PSK92, have been tested *in vivo* to determine and compare their pharmacokinetics independently. Although it was found that the type of inter-residual linkages that distinguish these types of PSA may influence their clearance rates it was more intriguing to note that the acylation status of their phospholipids moiety also played a significant role in their clearance rates. For example, the clearance rate for acylated PSB (colominic acid) increased from 20 hours to 30 hours when compared with its fully deacetylated counterpart. A possible explanation for this is that the presence of a terminal phospholipids moiety at the reducing end of these polymers creates an ability for these polymers to form micelles and/or aggregate, and such structures may be more readily cleared than their deacetylated counterparts that lose the ability to form them [112]. It would therefore be advantageous to eliminate the phospholipids moieties or at least deacetylate them before using them for any potential conjugation. Specifically, PSB also represents an appropriate choice of PSA to use for conjugation: it is less troublesome to extract from *E. coli* K1 than to extract other PSA types from potentially more virulent microorganisms, and more importantly, as a homopolymer of α2-8 linked sugars it is immunologically identical to that found naturally in human glycoproteins. As such, by itself, it is unlikely to pose any immunogenicity or toxicity for human *in vivo* use [18].

The prolonged residency of polysialylated conjugates increases their potential to reach their target and provide a therapeutic effect before being cleared from circulation, hence extending their duration of action (Figure 6.2). It has also been demonstrated that by varying the length of the PSA chain and/or the number of PSA chains attached to a drug or protein, the half-life of the conjugate can be modulated for optimal therapeutic affect [10, 113, 114].

Lipoxen's PolyXen® uses PSA, a biodegradable and biocompatible human polymer, and with five disclosed non-antibody drug candidates presently in early clinical development it is currently tipped to overtake PEG in leading the stealth platform [115]. This is supported by Baxter's (a top 30 pharmaceutical company and world leader in blood products) integration of Lipoxens PolyXen® technology for possible use with its propriety proteins by entering into an exclusive worldwide development and license agreement to develop improved, longer-acting forms of blood-clotting factors (http://www.lipoxen.com/pipeline/polyxen-product-pipeline.aspx).

For proteins, the therapeutic application of PSA conjugation has been investigated using either reductive amination or thiol specific conjugation. For reductive amination, the nonreducing end of the PSA polymer (Figure 6.1) is activated by mild periodate oxidation and conjugated to primary amines on protein; usually the ε-amino groups of lysine residues and the N-terminal amines. The oxidized PSA is added in a molar excess to the acceptor primarily to drive the reaction to completion, but also by varying the reaction ratio it is possible to manipulate the

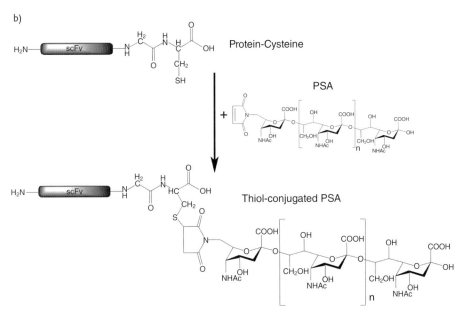

Figure 6.1 Structure of polysialic acid and its chemical attachment to proteins using two methods. (a) Periodate oxidation introduces an aldehyde group to the nonreducing end of PSA. Reductive amination in the presence of sodium cyanoborohydride at a molar excess of PSA (over 2 days at 37 °C) leads to chemical conjugation to amine groups. (b) Maleimide-activated PSA is mixed at a molar excess with mildly reduced protein under milder conditions (2 h at room temperature) leading to thiol-conjugated PSA.

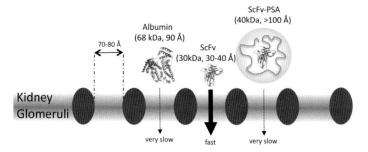

Figure 6.2 The hydrodynamic radius effect of protein polysialylation. Antibody fragments such as scFvs are normally filtered by the kidney, but upon polysialylation, the increase in mass and, more importantly, Stokes radius leads to its exclusion.

conjugation ratio. Invariably, depending on the number of lysine residues available, the reaction leads to heterogeneity in the end product where some molecules could potentially be more polysialylated than others and/or different stoichometric isoforms are made. Subsequent ammonium sulfate precipitation allows conjugated material to be separated from any excess free PSA, while downstream processing such as size-exclusion chromatography or ion-exchange chromatography can allow unconjugated material or even different isoforms to be separated to produce a more homogeneous product [9, 10, 17].

Proteins conjugated in this way include catalase [9], asparaginase [17, 116], insulin [117], and antibody fragments [10, 11]. Where assessed, these conjugates have demonstrated decreased clearance rates *in vivo* and an increase in protein stability relative to native protein [9, 116]. Additionally, in one particular study of immunogenicity using conjugated and unconjugated asparaginase it was demonstrated that only unconjugated material elicited an immune response in mice, while conjugated material, despite having a three- to fourfold longer clearance rate, did not elicit an immune response, suggesting that PSA conjugation can mask antigenic epitopes [116]. Using recombinant antibodies as an example, Fab polysialylation lead to a threefold increase in blood half-life and four- to fivefold increase in blood and tumor bioavailability (Figure 6.3; [10]), with similar observations for scFv polysialylation (Figure 6.3; [11]).

However, an unfavorable consequence of conjugation by reductive animation has been identified with varying detriment against the bioactivity of the conjugated material. Different conjugation ratios that lead to more heavily polysialylated conjugates, and/or conjugation at/near bioactively relevant amino acid residues, has been shown to result in diminished protein function [9–11, 114]. While a tolerable loss of bioactivity was observed with the conjugation of asparaginase, insulin or the Fab antibody fragment H17E2, the process of polysialylation on a scFv antibody fragment (MFE23) was shown to lead to a 20-fold loss of antigen binding (Table 6.1; [10, 11]). Through bioengineering of proteins it is possible to minimize this problem by ensuring lysine residues are not located within bioactively sensitive

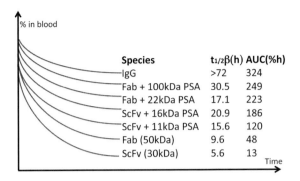

% in blood

Species	$t_{1/2}\beta$(h)	AUC(%h)
IgG	>72	324
Fab + 100kDa PSA	30.5	249
Fab + 22kDa PSA	17.1	223
ScFv + 16kDa PSA	20.9	186
ScFv + 11kDa PSA	15.6	120
Fab (50kDa)	9.6	48
ScFv (30kDa)	5.6	13

Time

Figure 6.3 A schematic comparison showing the relationship between antibody polysialylation and pharmacokinetic enhancement (blood clearance half-life-$t_{1/2}\beta$ and blood exposure-AUC-area under the blood level-time curve).

Table 6.1 Affinity effects upon antibody fragment polysialylation. Site specific-thiol modification *vs* lysine-amine polysialylation.

Species	Kd (nM)	Change in affinity
ScFv	2.6	–
ScFv + PSA (Lys)	53	⇓20.3
ScFv + PSA (Cys)	2.5	None
Fab	45	–
Fab + 22 kDa PSA (Lys)	127	⇓2.8
Fab + 100 kDa PSA (Lys)	122	⇓2.7

domains [118]; however, this approach is not generally applicable to antibodies. An approach based on the bioengineering of C-terminal thiols to recombinant proteins has been thoroughly investigated. Thiol-based polysialylation using maleimide-activated PSA chains has the advantage allowing conjugation to take place on proteins where conjugation to lysine residues causes a loss of bioactivity. This method was successfully applied to the scFv MFE23 resulting in the production of a monopolysialylated homogeneous formulation with fully retained bioactivity and a decreased clearance rate leading to a 30-fold increase in tumor uptake [11]. A similar form of chemistry using maleimide PEG has been applied to the commercially available Cimzia, an anti-TNFα antibody used for the treatment of rheumatoid arthritis and Crohn's disease.

The putative lack of PSA receptors and low immunogenicity provides molecular stealth and adds to the nontoxic properties of PSA. PSA conjugates must therefore be desialylated by neuraminidase-like enzymes before they can be cleared via the hepatic asialoglycoprotein receptor [111]. The biodegradable, nontoxic and highly hydrophilic properties of PSA can therefore be exploited to increase the serum half-life, and consequently action of small peptides and conventional drugs with minimal side-effects. These observations are of particular pharmaceutical interest since polysialylation offers a potentially superior method over current conjugation methods using PEGylation for the therapeutic administration of drugs and peptides, and can be used to improve the performance of other already commercially available therapeutics.

An additional potential advantage of polysialylation is in the prospect of producing recombinant PSA conjugates by utilizing natural biosynthetic pathways. Although much of the chemistry has been refined for chemical polysialylation (and similarly for PEG and other polymers), there are drawbacks to chemical modification approaches. Target proteins can be functionally inactivated, chemically denatured or have a reduced affinity for its ligand/substrate after chemical treatment. Often, downstream processing is needed to remove unreacted components. It would be advantageous to have a cell line producing heterologous proteins in a polysialylated format. Several polysialyltransferase enzymes from different organisms including human [28, 119, 120], xenopus [121] and *E. coli* [29, 30, 122] have now been identified and cloned. However, the glycan-specific recognition requirement of such enzymes for catalysis may require the need to engineer candidate molecules with site-specific motifs where enzymatic conjugation could take place. The advantage of using a recombinant polysialylation system would be the ability to produce PSA conjugates *in vivo* within a system that can potentially continually secrete therapeutic products in a cheap and sustainable manner. Preliminary experiments by Deonarain *et al.* (Chen *et al.*, in preparation) have demonstrated that this can be done with a variety of recombinant antibody fragments using PSA carrier domains from NCAM (as outlined above).

6.6
Summary

Protein polysialylation is rapidly advancing as an alternative to PEGylation and it is only a matter of time before this technology is used to extend the half-life of clinically useful antibodies. The existing data strongly support this concept. The biodegradable and hydrophilic nature of PSA lends itself well to a pharmaco-modulatory role. Observations in nature, such as the lack of receptors for PSA, immuno-regulatory/immuno-dampening function and increased plasticity/lubricating function should all provide clinical benefits. Any concerns that PSA could be inflammatory due to its immuno-privileged location in the brain is somewhat alleviated in that PSA has been found systemically and more importantly PSA conjugates have been well tolerated in clinical trials. Recent data suggesting that PSA conjugates can be made into solid formulations open up the possibility of needle-free administration routes [123] and that nonchemical methods for polysialylation are possible points to a promising outlook for this exciting technology.

References

1 Knop, K., Hoogenboom, R., Fischer, D., and Schubert, U.S. (2010) Poly(ethylene glycol) in drug delivery: pros and cons as well as potential alternatives. *Angew. Chem. Int. Ed. Engl.*, **49** (36), 6288–6308.

2 Bendele, A., Seely, J., Richey, C., Sennello, G., and Shopp, G. (1998) Short communication: renal tubular vacuolation in animals treated with polyethylene-glycol-conjugated proteins. *Toxicol. Sci*, **42** (2), 152–157.

3 Gaberc-Porekar, V., Zore, I., Podobnik, B., and Menart, V. (2008) Obstacles and pitfalls in the PEGylation of therapeutic proteins. *Curr. Opin. Drug Discov. Devel.*, **11** (2), 242–250.

4 Armstrong, J.K., Hempel, G., Koling, S., Chan, L.S., Fisher, T., Meiselman, H.J., and Garratty, G. (2007) Antibody against poly(ethylene glycol) adversely affects PEG-asparaginase therapy in acute lymphoblastic leukemia patients. *Cancer*, **110** (1), 103–111.

5 Riley, S. (2006) Innovation in drug delivery: the future of nanotechnology and non-invasive protein delivery. *Business Insights Report, 2006 (Business Insights Ltd).*

6 Toikka, J., Aalto, J., Hayrinen, J., Pelliniemi, L.J., and Finne, J. (1998) The polysialic acid units of the neural cell adhesion molecule N-CAM form filament bundle networks. *J. Biol. Chem*, **273** (44), 28557–28559.

7 Taylor, M.E., and Drickamer, K. (2006) *Introduction to Glycobiology*, 2nd edn, Oxford University Press, pp. 26–28.

8 Azurmendi, H.F., Vionnet, J., Wrightson, L., Trinh, L.B., Shiloach, J., and Freedberg, D.I. (2007) Extracellular structure of polysialic acid explored by on cell solution NMR. *Proc. Natl. Acad. Sci. U.S.A.*, **104** (28), 11557–11561.

9 Fernandes, A.I., and Gregoriadis, G. (1996) Synthesis, characterization and properties of sialylated catalase. *Biochim. Biophys. Acta*, **1293** (1), 90–96.

10 Constantinou, A., Epenetos, A.A., Hreczuk-Hirst, D., Jain, S., and Deonarain, M.P. (2008) Modulation of antibody pharmacokinetics by chemical polysialylation. *Bioconjug. Chem.*, **19** (3), 643–650.

11 Constantinou, A., Epenetos, A.A., Hreczuk-Hirst, D., Jain, S., Wright, M., Chester, K.A., and Deonarain, M.P. (2009) Site-specific polysialylation of an antitumor single-chain Fv fragment. *Bioconjug. Chem.*, **20** (5), 924–931.

12 Sasahara, K. (1995) Temperature dependence of volume changes on glycine-PEG and L-alanine-PEG in aqueous solution. *Colloid. Polym. Sci.*, **273** (8), 782–786.

13 Antonsen, K.P., and Hoffman, A.S. (1992) Water structure of PEG solutions by differential scanning calorimetry measurement, in *Poly(Ethylene Glycol) Chemistry: Biotechnical and Biomedical Applications* (ed. J.M. Harris), Plenum Press, New York, pp. 15–28.

14 Harris, J.M., and Chess, R.B. (2003) Effect of pegylation on pharmaceuticals. *Nat. Rev. Drug Discov.*, **2** (3), 214–221.

15 Fee, C.J., and Van Alstine, J.M. (2004) Prediction of the viscosity radius and the size exclusion chromatography behavior of PEGylated proteins. *Bioconjug. Chem.*, **15** (6), 1304–1313.

16 Deen, W.M., Bohrer, M.P., and Brenner, B.M. (1979) Macromolecule transport across glomerular capillaries: application of pore theory. *Kidney Int.*, **16** (3), 353–365.

17 Fernandes, A.I., and Gregoriadis, G. (1997) Polysialylated asparaginase: preparation, activity and pharmacokinetics. *Biochim. Biophys. Acta*, **1341** (1), 26–34.

18 Jain, S., Hreczuk-Hirst, D., Laing, P., and Gregoriadis, G. (2004) Polysialylation: the natural way to improve the stability and pharmacokinetics of protein and peptide drugs. *Drug Deliv. Syst. Sci.*, **4** (1), 3–9.

19 Pavlinkova, G., Beresford, G., Booth, B.J., Batra, S.K., and Colcher, D. (1999) Charge-modified single chain antibody constructs of monoclonal antibody

CC49: generation, characterization, pharmacokinetics, and biodistribution analysis. *Nucl. Med. Biol.*, **26** (1), 27–34.

20 Onda, M., Nagata, S., Tsutsumi, Y., Vincent, J.J., Wang, Q., Kreitman, R.J., Lee, B., and Pastan, I. (2001) Lowering the isoelectric point of the Fv portion of recombinant immunotoxins leads to decreased nonspecific animal toxicity without affecting antitumor activity. *Cancer Res.*, **61** (13), 5070–5077.

21 Kim, I., Kobayashi, H., Yoo, T.M., Kim, M.K., Le, N., Han, E.S., Wang, Q.C., Pastan, I., Carrasquillo, J.A., and Paik, C.H. (2002) Lowering of pI by acylation improves the renal uptake of 99mTc-labeled anti-Tac dsFv: effect of different acylating reagents. *Nucl. Med. Biol.*, **29** (8), 795–801.

22 Kim, I.S., Yoo, T.M., Kobayashi, H., Kim, M.K., Le, N., Wang, Q.C., Pastan, I., Carrasquillo, J.A., and Paik, C.H. (1999) Chemical modification to reduce renal uptake of disulfide-bonded variable region fragment of anti-Tac monoclonal antibody labeled with 99mTc. *Bioconjug. Chem.*, **10** (3), 447–453.

23 Lee, H.J., and Pardridge, W.M. (2003) Monoclonal antibody radiopharmaceuticals: cationization, PEGylation, radiometal chelation, pharmacokinetics, and tumor imaging. *Bioconjug. Chem.*, **14** (3), 546–553.

24 Dellian, M., Yuan, F., Trubetskoy, V.S., Torchilin, V.P., and Jain, R.K. (2000) Vascular permeability in a human tumor xenograft: molecular charge dependence. *Br. J. Cancer*, **82** (9), 1513–1518.

25 Behr, T.M., Becker, W.S., Sharkey, R.M., Juweid, M.E., Dunn, R.M., Bair, H.J., Wolf, F.G., and Goldenberg, D.M. (1996) Reduction of renal uptake of monoclonal antibody fragments by amino acid infusion. *J. Nucl. Med.*, **37** (5), 829–833.

26 Woitas, R.P., and Morioka, T. (1996) Influence of isoelectric point on glomerular deposition of antibodies and immune complexes. *Nephron*, **74** (4), 713–719.

27 Behr, T.M., Sharkey, R.M., Juweid, M.E., Blumenthal, R.D., Dunn, R.M., Griffiths, G.L., Bair, H.J., Wolf, F.G.,

Becker, W.S., and Goldenberg, D.M. (1995) Reduction of the renal uptake of radiolabeled monoclonal antibody fragments by cationic amino acids and their derivatives. *Cancer Res.*, **55** (17), 3825–3834.

28 Muhlenhoff, M., Eckhardt, M., and Gerardy-Schahn, R. (1998) Polysialic acid: three-dimensional structure, biosynthesis and function. *Curr. Opin. Struct. Biol.*, **8** (5), 558–564.

29 Bliss, J.M., and Silver, R.P. (1996) Coating the surface: a model for expression of capsular polysialic acid in Escherichia coli K1. *Mol. Microbiol.*, **21** (2), 221–231.

30 Russo, T.A., Wenderoth, S., Carlino, U.B., Merrick, J.M., and Lesse, A.J. (1998) Identification, genomic organization, and analysis of the group III capsular polysaccharide genes kpsD, kpsM, kpsT, and kpsE from an extraintestinal isolate of *Escherichia coli* (CP9, O4/K54/H5). *J. Bacteriol.*, **180** (2), 338–349.

31 Kiss, J.Z., and Rougon, G. (1997) Cell biology of polysialic acid. *Curr. Opin. Neurobiol.*, **7** (5), 640–646.

32 Vogel, U., Weinberger, A., Frank, R., Muller, A., Kohl, J., Atkinson, J.P., and Frosch, M. (1997) Complement factor C3 deposition and serum resistance in isogenic capsule and lipooligosaccharide sialic acid mutants of serogroup B *Neisseria meningitidis. Infect. Immun.*, **65** (10), 4022–4029.

33 Jarvis, G.A. (1995) Recognition and control of neisserial infection by antibody and complement. *Trends Microbiol.*, **3** (5), 198–201.

34 Ferrero, M.A., and Aparicio, L.R. (2010) Biosynthesis and production of polysialic acids in bacteria. *Appl. Microbiol. Biotechnol.*, **86** (6), 1621–1635.

35 Orskov, F., Orskov, I., Sutton, A., Schneerson, R., Lin, W., Egan, W., Hoff, G.E., and Robbins, J.B. (1979) Form variation in Escherichia coli K1: determined by O-acetylation of the capsular polysaccharide. *J. Exp. Med.*, **149** (3), 669–685.

36 Vimr, E.R., and Steenbergen, S.M. (2006) Mobile contingency locus

controlling *Escherichia coli* K1 polysialic acid capsule acetylation. *Mol. Microbiol.*, **60** (4), 828–837.

37 Finne, J. (1982) Occurrence of unique polysialosyl carbohydrate units in glycoproteins of developing brain. *J. Biol. Chem.*, **257** (20), 11966–11970.

38 Zuber, C., Lackie, P.M., Catterall, W.A., and Roth, J. (1992) Polysialic acid is associated with sodium channels and the neural cell adhesion molecule N-CAM in adult rat brain. *J. Biol. Chem.*, **267** (14), 9965–9971.

39 Rutishauser, U., and Landmesser, L. (1996) Polysialic acid in the vertebrate nervous system: a promoter of plasticity in cell-cell interactions. *Trends Neurosci.*, **19** (10), 422–427.

40 James, W.M., and Agnew, W.S. (1987) Multiple oligosaccharide chains in the voltage-sensitive Na channel from electrophorus electricus: evidence for alpha-2,8-linked polysialic acid. *Biochem. Biophys. Res. Commun.*, **148** (2), 817–826.

41 Rutishauser, U. (2008) Polysialic acid in the plasticity of the developing and adult vertebrate nervous system. *Nat. Rev. Neurosci.*, **9** (1), 26–35.

42 Galuska, S.P., Rollenhagen, M., Kaup, M., Eggers, K., Oltmann-Norden, I., Schiff, M., Hartmann, M., Weinhold, B., Hildebrandt, H., Geyer, R., Muhlenhoff, M., and Geyer, H. (2010) Synaptic cell adhesion molecule SynCAM 1 is a target for polysialylation in postnatal mouse brain. *Proc. Natl. Acad. Sci. U.S.A.*, **107** (22), 10250–10255.

43 Biederer, T., Sara, Y., Mozhayeva, M., Atasoy, D., Liu, X., Kavalali, E.T., and Sudhof, T.C. (2002) SynCAM, a synaptic adhesion molecule that drives synapse assembly. *Science*, **297** (5586), 1525–1531.

44 Stagi, M., Fogel, A.I., and Biederer, T. (2010) SynCAM 1 participates in axo-dendritic contact assembly and shapes neuronal growth cones. *Proc. Natl. Acad. Sci. U.S.A.*, **107** (16), 7568–7573.

45 Giza, J., and Biederer, T. (2010) Polysialic acid: a veteran sugar with a new site of action in the brain. *Proc.*

Natl. Acad. Sci. U.S.A., **107** (23), 10335–10336.

46 Yabe, U., Sato, C., Matsuda, T., and Kitajima, K. (2003) Polysialic acid in human milk. CD36 is a new member of mammalian polysialic acid-containing glycoprotein. *J. Biol. Chem.*, **278** (16), 13875–13880.

47 Sampson, M.J., Davies, I.R., Braschi, S., Ivory, K., and Hughes, D.A. (2003) Increased expression of a scavenger receptor (CD36) in monocytes from subjects with Type 2 diabetes. *Atherosclerosis*, **167** (1), 129–134.

48 Harshyne, L.A., Zimmer, M.I., Watkins, S.C., and Barratt-Boyes, S.M. (2003) A role for class A scavenger receptor in dendritic cell nibbling from live cells. *J. Immunol.*, **170** (5), 2302–2309.

49 Tandon, N.N., Kralisz, U., and Jamieson, G.A. (1989) Identification of glycoprotein IV (CD36) as a primary receptor for platelet-collagen adhesion. *J. Biol. Chem.*, **264** (13), 7576–7583.

50 Aiken, M.L., Ginsberg, M.H., Byers-Ward, V., and Plow, E.F. (1990) Effects of OKM5, a monoclonal antibody to glycoprotein IV, on platelet aggregation and thrombospondin surface expression. *Blood*, **76** (12), 2501–2509.

51 Rigotti, A., Acton, S.L., and Krieger, M. (1995) The class B scavenger receptors SR-BI and CD36 are receptors for anionic phospholipids. *J. Biol. Chem.*, **270** (27), 16221–16224.

52 Endemann, G., Stanton, L.W., Madden, K.S., Bryant, C.M., White, R.T., and Protter, A.A. (1993) CD36 is a receptor for oxidized low density lipoprotein. *J. Biol. Chem.*, **268** (16), 11811–11816.

53 Febbraio, M., Abumrad, N.A., Hajjar, D.P., Sharma, K., Cheng, W., Pearce, S.F., and Silverstein, R.L. (1999) A null mutation in murine CD36 reveals an important role in fatty acid and lipoprotein metabolism. *J. Biol. Chem.*, **274** (27), 19055–19062.

54 Karlsson, K.A. (1998) Meaning and therapeutic potential of microbial recognition of host glycoconjugates. *Mol. Microbiol.*, **29** (1), 1–11.

55 Morgan, B.L., and Winick, M. (1980) Effects of environmental stimulation on

brain N-acetylneuraminic acid content and behavior. *J. Nutr.*, **110** (3), 425–432.

56 Carlson, S.E., and House, S.G. (1986) Oral and intraperitoneal administration of N-acetylneuraminic acid: effect on rat cerebral and cerebellar N-acetylneuraminic acid. *J. Nutr.*, **116** (5), 881–886.

57 Curreli, S., Arany, Z., Gerardy-Schahn, R., Mann, D., and Stamatos, N.M. (2007) Polysialylated neuropilin-2 is expressed on the surface of human dendritic cells and modulates dendritic cell–T lymphocyte interactions. *J. Biol. Chem.*, **282** (42), 30346–30356.

58 Geretti, E., Shimizu, A., and Klagsbrun, M. (2008) Neuropilin structure governs VEGF and semaphorin binding and regulates angiogenesis. *Angiogenesis*, **11** (1), 31–39.

59 Pellet-Many, C., Frankel, P., Jia, H., and Zachary, I. (2008) Neuropilins: structure, function and role in disease. *Biochem. J.*, **411** (2), 211–226.

60 Bax, M., van Vliet, S.J., Litjens, M., Garcia-Vallejo, J.J., and van Kooyk, Y. (2009) Interaction of polysialic acid with CCL21 regulates the migratory capacity of human dendritic cells. *PLoS ONE*, **4** (9), e6987.

61 Rey-Gallardo, A., Escribano, C., Delgado-Martin, C., Rodriguez-Fernandez, J.L., Gerardy-Schahn, R., Rutishauser, U., Corbi, A.L., and Vega, M.A. (2010) Polysialylated neuropilin-2 enhances human dendritic cell migration through the basic C-terminal region of CCL21. *Glycobiology*, **20** (9), 1139–1146.

62 Kaneko, Y., Nimmerjahn, F., and Ravetch, J.V. (2006) Anti-inflammatory activity of immunoglobulin G resulting from Fc sialylation. *Science*, **313** (5787), 670–673.

63 Close, B.E., and Colley, K.J. (1998) *In vivo* autopolysialylation and localization of the polysialyltransferases PST and STX. *J. Biol. Chem.*, **273** (51), 34586–34593.

64 Maness, P.F., and Schachner, M. (2007) Neural recognition molecules of the immunoglobulin superfamily: signaling transducers of axon guidance and neuronal migration. *Nat. Neurosci.*, **10** (1), 19–26.

65 Pekcec, A., Weinhold, B., Gerardy-Schahn, R., and Potschka, H. (2010) Polysialic acid affects pathophysiological consequences of status epilepticus. *Neuroreport*, **21** (8), 549–553.

66 Garcia-Morales, I., de la Pena Mayor, P., and Kanner, A.M. (2008) Psychiatric comorbidities in epilepsy: identification and treatment. *Neurologist*, **14** (6 Suppl. 1), S15–S25.

67 Figarella-Branger, D., Dubois, C., Chauvin, P., De Victor, B., Gentet, J.C., and Rougon, G. (1996) Correlation between polysialic-neural cell adhesion molecule levels in CSF and medulloblastoma outcomes. *J. Clin. Oncol.*, **14** (7), 2066–2072.

68 Gluer, S., Schelp, C., von Schweinitz, D., and Gerardy-Schahn, R. (1998) Polysialylated neural cell adhesion molecule in childhood rhabdomyosarcoma. *Pediatr. Res.*, **43** (1), 145–147.

69 Hildebrandt, H., Becker, C., Gluer, S., Rosner, H., Gerardy-Schahn, R., and Rahmann, H. (1998) Polysialic acid on the neural cell adhesion molecule correlates with expression of polysialyltransferases and promotes neuroblastoma cell growth. *Cancer Res.*, **58** (4), 779–784.

70 Lantuejoul, S., Moro, D., Michalides, R.J., Brambilla, C., and Brambilla, E. (1998) Neural cell adhesion molecules (NCAM) and NCAM-PSA expression in neuroendocrine lung tumors. *Am. J. Surg. Pathol.*, **22** (10), 1267–1276.

71 Michalides, R., Kwa, B., Springall, D., van Zandwijk, N., Koopman, J., Hilkens, J., and Mooi, W. (1994) NCAM and lung cancer. *Int. J. Cancer Suppl.*, **8**, 34–37.

72 Mayanil, C.S., George, D., Mania-Farnell, B., Bremer, C.L., McLone, D.G., and Bremer, E.G. (2000) Overexpression of murine Pax3 increases NCAM polysialylation in a human medulloblastoma cell line. *J. Biol. Chem.*, **275** (30), 23259–23266.

73 Fukuda, M. (1996) Possible roles of tumor-associated carbohydrate antigens. *Cancer Res.*, **56** (10), 2237–2244.

74 Suzuki, M., Nakayama, J., Suzuki, A., Angata, K., Chen, S., Sakai, K., Hagihara, K., Yamaguchi, Y., and Fukuda, M. (2005) Polysialic acid facilitates tumor invasion by glioma cells. *Glycobiology*, **15** (9), 887–894.

75 Petridis, A.K., Wedderkopp, H., Hugo, H.H., and Maximilian Mehdorn, H. (2009) Polysialic acid overexpression in malignant astrocytomas. *Acta Neurochir. (Wien)*, **151** (6), 601–603. discussion 603–604.

76 Yang, P., Yin, X., and Rutishauser, U. (1992) Intercellular space is affected by the polysialic acid content of NCAM. *J. Cell Biol.*, **116** (6), 1487–1496.

77 Johnson, C.P., Fujimoto, I., Rutishauser, U., and Leckband, D.E. (2005) Direct evidence that neural cell adhesion molecule (NCAM) polysialylation increases intermembrane repulsion and abrogates adhesion. *J. Biol. Chem*, **280** (1), 137–145.

78 Rutishauser, U. (1996) Polysialic acid and the regulation of cell interactions. *Curr. Opin. Cell Biol.*, **8** (5), 679–684.

79 Finne, J., and Makela, P.H. (1985) Cleavage of the polysialosyl units of brain glycoproteins by a bacteriophage endosialidase. Involvement of a long oligosaccharide segment in molecular interactions of polysialic acid. *J. Biol. Chem.*, **260** (2), 1265–1270.

80 Pelkonen, S., Pelkonen, J., and Finne, J. (1989) Common cleavage pattern of polysialic acid by bacteriophage endosialidases of different properties and origins. *J. Virol.*, **63** (10), 4409–4416.

81 Chung, W.W., Lagenaur, C.F., Yan, Y.M., and Lund, J.S. (1991) Developmental expression of neural cell adhesion molecules in the mouse neocortex and olfactory bulb. *J. Comp. Neurol.*, **314** (2), 290–305.

82 Kanato, Y., Kitajima, K., and Sato, C. (2008) Direct binding of polysialic acid to a brain-derived neurotrophic factor depends on the degree of polymerization. *Glycobiology*, **18** (12), 1044–1053.

83 Watson, K., Edwards, R.J., Shaunak, S., Parmelee, D.C., Sarraf, C., Gooderham, N.J., and Davies, D.S. (1995) Extranuclear location of histones in activated human peripheral blood lymphocytes and cultured T-cells. *Biochem. Pharmacol.*, **50** (3), 299–309.

84 Brix, K., Summa, W., Lottspeich, F., and Herzog, V. (1998) Extracellularly occurring histone H1 mediates the binding of thyroglobulin to the cell surface of mouse macrophages. *J. Clin. Invest.*, **102** (2), 283–293.

85 Henriquez, J.P., Casar, J.C., Fuentealba, L., Carey, D.J., and Brandan, E. (2002) Extracellular matrix histone H1 binds to perlecan, is present in regenerating skeletal muscle and stimulates myoblast proliferation. *J. Cell Sci.*, **115** (Pt 10), 2041–2051.

86 Mishra, B., von der Ohe, M., Schulze, C., Bian, S., Makhina, T., Loers, G., Kleene, R., and Schachner, M. (2010) Functional role of the interaction between polysialic acid and extracellular histone H1. *J. Neurosci.*, **30** (37), 12400–12413.

87 Robbins, J.B., McCracken, G.H. Jr, Gotschlich, E.C., Orskov, F., Orskov, I., and Hanson, L.A. (1974) *Escherichia coli* K1 capsular polysaccharide associated with neonatal meningitis. *N. Engl. J. Med.*, **290** (22), 1216–1220.

88 Rick, P., and Silver, R. (1996) Endobacterial common antigen and capsular polysaccharides, in *Escherichia coli and Salmonella: Cellular and Molecular Biology* (ed. F.C. Neidhardt), American Society for Microbiology, Washington, DC, pp. 104–122.

89 Roberts, I.S. (1995) Bacterial polysaccharides in sickness and in health. The 1995 Fleming Lecture. *Microbiology*, **141** (Pt 9), 2023–2031.

90 Vimr, E., Steenbergen, S., and Cieslewicz, M. (1995) Biosynthesis of the polysialic acid capsule in *Escherichia coli* K1. *J. Ind. Microbiol.*, **15** (4), 352–360.

91 Vimr, E.R. (1992) Selective synthesis and labeling of the polysialic acid capsule in *Escherichia coli* K1 strains with mutations in nanA and neuB. *J. Bacteriol.*, **174** (19), 6191–6197.

92 Troy, F.A. 2nd (1992) Polysialylation: from bacteria to brains. *Glycobiology*, **2** (1), 5–23.

93 Steenbergen, S.M., and Vimr, E.R. (2008) Biosynthesis of the *Escherichia coli* K1 group 2 polysialic acid capsule occurs within a protected cytoplasmic compartment. *Mol. Microbiol.*, **68** (5), 1252–1267.

94 Troy, F.A. 2nd (1995) Sialobiology and the polysialic acid glycotope: occurrence, structure, function, synthesis and glycopathy, in *Biology of the Sialic Acids* (ed. A. Rosenberg), Plenium Press, New York, pp. 95–144.

95 Bliss, J.M., and Silver, R.P. (1997) Evidence that KpsT, the ATP-binding component of an ATP-binding cassette transporter, is exposed to the periplasm and associates with polymer during translocation of the polysialic acid capsule of Escherichia coli K1. *J. Bacteriol.*, **179** (4), 1400–1403.

96 Nara, K., Watanabe, Y., Maruyama, K., Kasahara, K., Nagai, Y., and Sanai, Y. (1994) Expression cloning of a CMP-NeuAc:NeuAc alpha 2-3Gal beta 1-4Glc beta 1-1'Cer alpha 2,8-sialyltransferase (GD3 synthase) from human melanoma cells. *Proc. Natl. Acad. Sci. U.S.A.*, **91** (17), 7952–7956.

97 Sasaki, K., Kurata, K., Kojima, N., Kurosawa, N., Ohta, S., Hanai, N., Tsuji, S., and Nishi, T. (1994) Expression cloning of a GM3-specific alpha-2,8-sialyltransferase (GD3 synthase). *J. Biol. Chem.*, **269** (22), 15950–15956.

98 Haraguchi, M., Yamashiro, S., Yamamoto, A., Furukawa, K., Takamiya, K., Lloyd, K.O., and Shiku, H. (1994) Isolation of GD3 synthase gene by expression cloning of GM3 alpha-2,8-sialyltransferase cDNA using anti-GD2 monoclonal antibody. *Proc. Natl. Acad. Sci. U.S.A.*, **91** (22), 10455–10459.

99 Nakayama, J., Fukuda, M.N., Hirabayashi, Y., Kanamori, A., Sasaki, K., Nishi, T., and Fukuda, M. (1996) Expression cloning of a human GT3 synthase. GD3 AND GT3 are synthesized by a single enzyme. *J. Biol. Chem.*, **271** (7), 3684–3691.

100 Kono, M., Yoshida, Y., Kojima, N., and Tsuji, S. (1996) Molecular cloning and expression of a fifth type of alpha2,8-sialyltransferase (ST8Sia V). Its substrate specificity is similar to that of SAT-V/ III, which synthesize GD1c, GT1a, GQ1b and GT3. *J. Biol. Chem.*, **271** (46), 29366–29371.

101 Angata, K., Suzuki, M., McAuliffe, J., Ding, Y., Hindsgaul, O., and Fukuda, M. (2000) Differential biosynthesis of polysialic acid on neural cell adhesion molecule (NCAM) and oligosaccharide acceptors by three distinct alpha 2,8-sialyltransferases, ST8Sia IV (PST), ST8Sia II (STX), and ST8Sia III. *J. Biol. Chem.*, **275** (24), 18594–18601.

102 Nelson, R.W., Bates, P.A., and Rutishauser, U. (1995) Protein determinants for specific polysialylation of the neural cell adhesion molecule. *J. Biol. Chem.*, **270** (29), 17171–17179.

103 Foley, D.A., Swartzentruber, K.G., and Colley, K.J. (2009) Identification of sequences in the polysialyltransferases ST8Sia II and ST8Sia IV that are required for the protein-specific polysialylation of the neural cell adhesion molecule, NCAM. *J. Biol. Chem.*, **284** (23), 15505–15516.

104 Inoue, S., and Inoue, Y. (2001) Developmental profile of neural cell adhesion molecule glycoforms with a varying degree of polymerization of polysialic acid chains. *J. Biol. Chem.*, **276** (34), 31863–31870.

105 Nakata, D., and Troy, F.A. 2nd (2005) Degree of polymerization (DP) of polysialic acid (polySia) on neural cell adhesion molecules (N-CAMS): development and application of a new strategy to accurately determine the DP of polySia chains on N-CAMS. *J. Biol. Chem.*, **280** (46), 38305–38316.

106 Lackie, P.M., Zuber, C., and Roth, J. (1994) Polysialic acid of the neural cell adhesion molecule (N-CAM) is widely expressed during organogenesis in mesodermal and endodermal derivatives. *Differentiation*, **57** (2), 119–131.

107 Seidenfaden, R., Gerardy-Schahn, R., and Hildebrandt, H. (2000) Control of NCAM polysialylation by the differential expression of polysialyltransferases

ST8SiaII and ST8SiaIV. *Eur. J. Cell Biol.*, **79** (10), 680–688.

108 Monti, E., Preti, A., Venerando, B., and Borsani, G. (2002) Recent development in mammalian sialidase molecular biology. *Neurochem. Res.*, **27** (7–8), 649–663.

109 Saito, M.Y., and Yu, R.K. (1995) Biochemistry and role of sialic acids, in *Biology of the Sialic Acids* (ed. A. Rosenberg), Plenum Press, New York, pp. 261–313.

110 Schauer, R., Kelm, S., Reuter, G., and Roggentin, A. (1995) Biochemistry and role of sialic acids, in *Biology of the Sialic Acids* (ed. A. Rosenberg), Plenum Press, New York, pp. 7–67.

111 Stockert, R.J. (1995) The asialoglycoprotein receptor: relationships between structure, function, and expression. *Physiol. Rev.*, **75** (3), 591–609.

112 Gregoriadis, G., McCormack, B., Wang, Z., and Lifely, R. (1993) Polysialic acids: potential in drug delivery. *FEBS Lett.*, **315** (3), 271–276.

113 Gregoriadis, G., Fernandes, A., McCormack, B., Mital, M., and Zhang, X. (1999) Polysialic acids: potential role in therapeutic constructs. *Biotechnol. Genet. Eng. Rev.*, **16**, 203–215.

114 Gregoriadis, G., Fernandes, A., Mital, M., and McCormack, B. (2000) Polysialic acids: potential in improving the stability and pharmacokinetics of proteins and other therapeutics. *Cell. Mol. Life Sci.*, **57** (13–14), 1964–1969.

115 Fraser-Moodie, I. (2008) Delivery mechanisms for large molecule drugs: successes and failures of leading technologies and key drivers for market success. *Business Insights Report, 2006 (Business Insights Ltd)*.

116 Fernandes, A.I., and Gregoriadis, G. (2001) The effect of polysialylation on the immunogenicity and antigenicity of asparaginase: implication in its pharmacokinetics. *Int. J. Pharm.*, **217** (1–2), 215–224.

117 Jain, S., Hreczuk-Hirst, D.H., McCormack, B., Mital, M., Epenetos, A., Laing, P., and Gregoriadis, G. (2003) Polysialylated insulin: synthesis, characterization and biological activity *in vivo. Biochim. Biophys. Acta*, **1622** (1), 42–49.

118 Adams, G.P., Shaller, C.C., Chappell, L.L., Wu, C., Horak, E.M., Simmons, H.H., Litwin, S., Marks, J.D., Weiner, L.M., and Brechbiel, M.W. (2000) Delivery of the alpha-emitting radioisotope bismuth-213 to solid tumors via single-chain Fv and diabody molecules. *Nucl. Med. Biol.*, **27** (4), 339–346.

119 Nakayama, J., Fukuda, M.N., Fredette, B., Ranscht, B., and Fukuda, M. (1995) Expression cloning of a human polysialyltransferase that forms the polysialylated neural cell adhesion molecule present in embryonic brain. *Proc. Natl. Acad. Sci. U.S.A.*, **92** (15), 7031–7035.

120 Jacobs, C.L., Goon, S., Yarema, K.J., Hinderlich, S., Hang, H.C., Chai, D.H., and Bertozzi, C.R. (2001) Substrate specificity of the sialic acid biosynthetic pathway. *Biochemistry*, **40** (43), 12864–12874.

121 Kudo, M., Takayama, E., Tashiro, K., Fukamachi, H., Nakata, T., Tadakuma, T., Kitajima, K., Inoue, Y., and Shiokawa, K. (1998) Cloning and expression of an alpha-2,8-polysialyltransferase (STX) from *Xenopus laevis. Glycobiology*, **8** (8), 771–777.

122 Cho, J.W., and Troy, F.A. 2nd (1994) Polysialic acid engineering: synthesis of polysialylated neoglycosphingolipids by using the polysialyltransferase from neuroinvasive *Escherichia coli* K1. *Proc. Natl. Acad. Sci. U.S.A.*, **91** (24), 11427–11431.

123 Zhang, R., Jain, S., Rowland, M., Hussain, N., Agarwal, M., and Gregoriadis, G. (2010) Development and testing of solid dose formulations containing polysialic acid insulin conjugate: next generation of long-acting insulin. *J. Diabetes Sci. Technol.*, **4** (3), 532–539.

7
Half-Life Extension through HESylation®

Thomas Hey, Helmut Knoller, and Peter Vorstheim

7.1
Introduction

Drug delivery technologies are of increasing importance for the optimization of pharmaceutical products. Particularly in the area of therapeutic proteins technologies are needed to cope with the challenges of elimination, immunogenicity, dose, stability and effectiveness. Recently there has been a tendency to turn away from the optimization of therapeutic proteins already on the market, towards innovative research approaches. Especially focused on by researchers are developments towards high dosage applications or chronic administration. Therefore plasma half-life prolongation technologies with new properties are demanded. Toxicological concerns are growing with the increasing amount of new requirements for administered polymer regarding production, formulation or the route of administration of polymer-bound therapeutic proteins, which arise due to viscosity problems accompanied by highly concentrated solutions. Since the PEGylation technique has been well known for several decades, and several drugs are on the market, the limitations of this technology are coming under critical review [1]. One of the main disadvantages of the polymer used is its lack of biodegradability, which makes it difficult for patients to eliminate the polymer from the body, especially if the molecular weight of polyethylene glycol exceeds the renal threshold. The increased need for safety is noticeable and new toxicological concerns regarding the safety of polymers like PEG are under extensive discussion. Recently, Webster *et al.* reviewed the metabolism and toxicity of PEG and PEG conjugates [2]. Furthermore, Armstrong *et al.* reported a markedly higher occurrence of IgM and IgG anti-PEG antibodies in up to 25% of healthy donors [3]. Inspired by the success of PEGylated drugs, the excellent safety profile in combination with the fine tuning potential by small chemical modifications of the hydroxyethyl starch (HES) polymer has led to a growing interest in improving the capabilities of proteins by attaching HES molecules.

Therapeutic Proteins: Strategies to Modulate Their Plasma Half-Lives, First Edition. Edited by Roland Kontermann.
© 2012 Wiley-VCH Verlag GmbH & Co. KGaA. Published 2012 by Wiley-VCH Verlag GmbH & Co. KGaA.

7.2
Hydroxyethyl Starch (HES)

7.2.1
Production and Characteristics

Hydroxyethyl starch is a modified natural polymer derived from amylopectin. The starch most commonly used for the preparation of HES is a waxy species of maize but other plants like potatoes are also used as raw materials. In waxy maize starch, the predominant polysaccharide is amylopectin (>95%) which consists mainly of α-1,4 linked D-glucose units with approximately 6% branching via α-1,6 linkages (Figure 7.1). In the first step of the HES manufacturing process (see Figure 7.2), the high molecular weight amylopectin molecules are broken down to the appropriate molecular weight by mechanical stress and acidic hydrolysis. The second step is hydroxyethylation of the resulting fragments. The physical properties of the starch fragments are altered by attaching hydroxethyl groups in two ways, so that the polymer can be used for human application: firstly, starch is poorly water-soluble; by introducing the hydroxyethyl groups the molecules become expanded which leads to an increase of the water-binding capacity and a decrease of viscosity, and secondly an unsubstituted amylopectin molecule would be immediately. Cleaved in the blood by plasmatic α-amylase and excreted via the renal pathway. The hydroxyethyl groups are capable of delaying the latter process in an adjustable manner.

Finally, ultrafiltration eliminates small HES fractions and process-related impurities. Furthermore, by applying different membranes in the ultrafiltration step the molecular weight and the polydispersity of the product can be adjusted. The

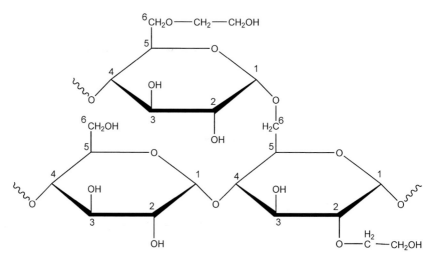

Figure 7.1   Detail of the HES structure.

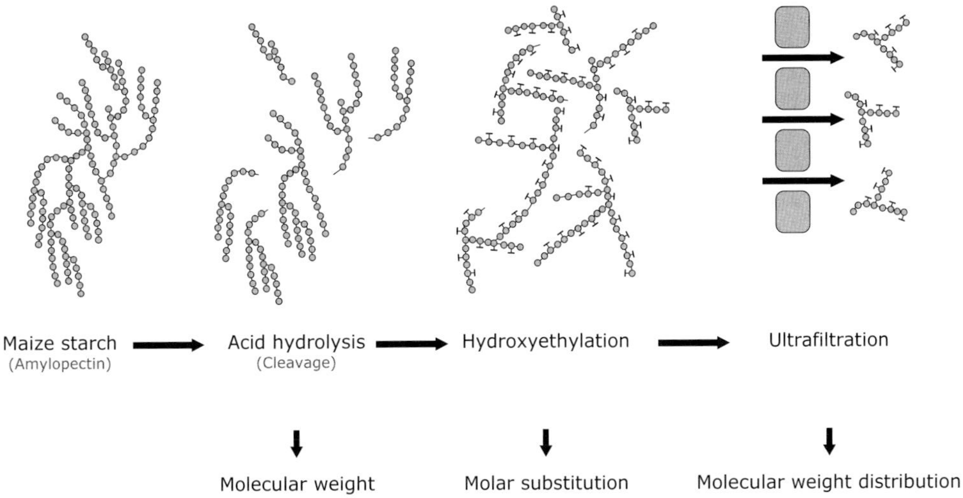

Maize starch ⟶ Acid hydrolysis ⟶ Hydroxyethylation ⟶ Ultrafiltration
(Amylopectin)   (Cleavage)

↓   ↓   ↓

Molecular weight   Molar substitution   Molecular weight distribution

Figure 7.2 HES manufacturing process.

different specifications of the resulting HES polymer are inherently stable, globular and highly water-soluble polysaccharides.

7.2.2
HES Parameters

7.2.2.1 Mean Molecular Weight

HES solutions are polydisperse compositions, wherein each molecule differs from the next with respect to the degree of polymerization, the number and pattern of branching sites and the hydroxyethyl substitution pattern. HES solutions, therefore, are mixtures of HES compounds with different molecular weights. Consequently, a particular HES solution is determined by its average molecular weight by statistical means. In this context, the number average molecular weight, M_n, is calculated as the arithmetic mean depending on the number of molecules and is defined by the following equation:

$$\bar{M}_n = \frac{\sum_i n_i \cdot M_i}{\sum_i n_i}$$

where n_i is the number of molecules of species i of molar mass M_i. Alternatively, the weight average molecular weight, M_w, represents a way for defining the molecular size of the HES molecules. M_w is defined by the following equation:

$$\bar{M}_w = \frac{\sum_i n_i \cdot M_i^2}{\sum_i n_i M_i}$$

where n_i is the number of molecules of species i of molar mass M_i. The M_w relates to the weight as determined according to a SEC-MALLS (size exclusion chromatography-multi-angle laser light scattering) method described in the European Pharmacopeia hydroxyethyl starch monograph (2010) [4].

7.2.2.2 Molar Substitution (MS)

The molar substitution (MS) is calculated as the average number of hydroxyethyl groups per glucose moiety in the polymer. MS is determined by gas chromatography after total hydrolysis of the HES molecule [4].

7.2.2.3 Other Parameters

Besides the two main dimensions for characterizing HES, molecular weight and molar substitution, the polydispersity and the substitution pattern are further parameters that can be used. The polydispersity is a measure of the distribution of molecular mass in a given polymer sample and it is denoted as:

$$P = \frac{\bar{M}_w}{\bar{M}_n}$$

The substitution pattern is generally described as the so called C_2/C_6 ratio, the number of hydroxyethyl groups located at C_2 over the number of hydroxyethyl groups located at C_6.

7.3
Clinical Use of HES

HES is one of the most frequently used plasma volume substitutes [5]. The first HES product Hespan® was launched in the US market in the 1970s. Since then, numerous types of HES solutions have been introduced worldwide, which differ in their pharmacological properties and, even today, a variety of different HES specifications are still marketed. In Table 7.1 the most prominent HES preparations are listed with their respective pharmacologically important parameters.

Many different HES specifications are still available on the market with a mean molecular weight which varies from 70 kDa up to 670 kDa, and – more importantly regarding the pharmacokinetic behavior – with molar substitutions in the range from 0.40 up to 0.75.

The high safety margin for the polymer HES is remarkable and is reflected in the maximal intravenous daily administration doses in the range from 90 g up to more than 200 g referring to an adult weighing 75 kg. For example Neff *et al.* [6]

Table 7.1 Characteristics of HES preparations [7].

	Mean molecular weight, kDa	Molar substitution	C_2/C_6 ratio	Maximum daily dose, $g\,kg^{-1}$
HES 670/7.5	670	0.75	4.5 : 1	1.2
HES 600/0.7	600	0.70	5 : 1	1.2
HES 450/0.7	480	0.70	5 : 1	1.2
HES 200/0.62	200	0.62	9 : 1	1.2
HES 70/0.5	70	0.50	3 : 1	1.2
HES 200/0.5	200	0.50	5 : 1	2.0
HES 130/0.42	130	0.42	6 : 1	3.0
HES 130/0.4	130	0.40	9 : 1	3.0

report that patients have received up to 66 L of a 6% HES 130/0.4 solution within 28 days, without compromising renal function, coagulation or causing bleeding complications. The unquestioned safety and biocompatibility of HES was the starting point for the development of attaching HES to various classes of pharmaceutical ingredients with the aim to improve parameters like:

- plasma half life
- stability
- solubility
- immunogenicity
- toxicity and/or
- efficacy.

7.4
HES Metabolism and Toxicology

7.4.1
Metabolic Pathways

Two metabolic pathways are described, both with significant importance for the degradation of HES. The main metabolic path way for a short-term degradation of HES is catalyzed by the plasmatic α-amylase. α-Amylase cleaves the α-1,4 glycosidic bond of polycarbohydrates and degrades unsubstituted starch in the plasma within minutes to small oligomers down to maltose followed by renal elimination. By attaching hydroxyethyl groups, the degradation process is slowed down by steric hindrance and restricted access of the α-amylase to its site of action, the α-1,4 glycosidic bond. Consequently, the degradability of HES decreases with increasing degree of hydroxyethylation (i.e., increasing MS) (Figure 7.3).

The second pathway plays a major role in the intracellular degradation of HES molecules and is presumably catalyzed by the α-glucosidase enzyme [8]. When

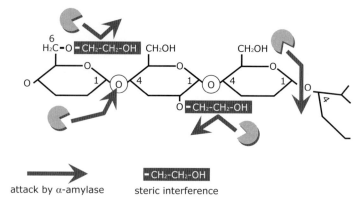

attack by α-amylase   steric interference

Figure 7.3 HES metabolism via plasmatic α-amylase.

administered to mammals macromolecules are efficiently incorporated via phago-cytosis into blood-derived or tissue-dependent macrophages. The amount of incorporated HES is directly dependent on MS values. Ständer *et al.* [9] investi-gated the HES storage over a long time period, up to seven years after infusion. HES storage takes place in several different organs like the liver, kidney, lung, skeletal muscle, lymph node and skin. With the exception of renal tubular cells, liver parenchyma cells and cutaneous basal cells, all the other cells exclusively belong to the reticuloendothelial system. Various groups [10–12] have investigated the question of how the functionality of macrophages is influenced by the incor-poration and storage of HES. The results of most of the studies indicate that HES has only a minor or no influence on the functionality of the macrophages. This supports the hypothesis that HES is a nearly inert immunological molecule, probably due to its high similarity to well known molecules of human origin, such as glycogen. This hypothesis was furthermore supported by clinical data generated in numerous clinical studies over recent decades. Laxenaire [13] inves-tigated the anaphylactoid reaction of HES compared with other blood volume substitutes in a prospective multicenter clinical trial with 19 593 patients. The rate of anaphylactoid reactions is approximately 0.058% of infusions, which is considerably lower than the incidence after the administration of dextran. Diet-erich *et al.* [14] investigated HES reactive antibodies in the blood sera of patients who had received HES. The incidence of antibody formation was 1 in 1004 patients. The investigators measured low titers of IgM antibodies but the patient showed no adverse reactions despite being exposed to HES several times. However, it is certain that the storage of HES is transient and decreases over time in all tissues with different kinetics. The storage in liver and lung is only short term. Lymphatic organs like the spleen are intermediate and the skin is the tissue with the longest storage detectable for up to 54 months. The kinetics of the HES elimination is individually different and is potentially dependent on the enzymatic set up of the respective organism, but nevertheless the HES tissue storage is a considerably important element of the *in vivo* degradation and elimination of HES.

7.5
HESylation® – Conjugation of Hydroxyethyl Starch to Drug Substances

7.5.1
The Origin of HES Protein Coupling

One of the first descriptions of HES protein conjugates dates back to the 1970s when Richter and de Belder synthesized HES albumin conjugates applying the chemistry and expertise for the modification of polymeric carbohydrates established at the Swedish company, Pharmacia, for chromatography materials based on dextran or sepharose (e.g., cyanogen bromide activation) [15]. In contrast to the early polyethylene glycol (PEG) conjugation experiments conducted by Abuchowski with blood proteins like albumin [16] directed towards a reduction in immunogenicity, Richter and de Belder used the HES albumin conjugates for the immunization of rabbits in order to obtain antisera for the diagnostics of the carbohydrate polymer that had already been in use as a plasma expander for several years.

At the same time, researchers at the companies Fresenius and Hematech were working on blood substitutes – a hot topic in the 1970s that also triggered PEGylation technology. They cross-linked hemoglobin with HES using either a random coupling approach with a cyanogen bromide activation of HES or alternatively an amination of aldehyde groups generated by periodate oxidation [17, 18].

But while the PEGylation technology was further developed during the next decade, subsequently leading to the approval of the first PEG-modified biotherapeutic Adagen® in 1990, only a few groups continued the work on HES conjugates – mainly with hemoglobin as target [19–22]. The success of PEG and the scientific challenge of designing artificial blood was responsible for waking up the "sleeping beauty" HESylation at the end of the 1990s, when again researchers at Fresenius developed a chemistry for the site-specific modification of HES with target substances [23].

HES–hemoglobin conjugates prepared earlier, often suffered from toxicity and manufacturing problems since the random coupling approach using a multivalent HES (numerous BrCN-activated or aldehyde sites) in combination with a multivalent target (multiple lysine residues and a reactive N-terminus in the case of proteins) was often accompanied by a polymerization of the reaction partners. The size and stoichiometry of the reaction products was uncontrollable; and the efficacy and tolerance of such preparations in animal experiments was low, presumably due to the appearance of heavily crosslinked aggregate-like high-molecular weight compounds.

7.5.2
**Going from Multivalent to Site-Specific Functionalization of HES by Selective
Oxidation of the Reducing End**

Sommermeyer and Eichner solved the problems described above by using a HES molecule carrying only one activated site [23]. As hydroxyethyl starch is a directional molecule, a property inherited from the synthesis pathway of its precursor molecule amylopectin, it carries a single reducing end group in the stem region, while the aldehyde functionalities of all terminal glucoses in the branched regions are involved in glycosidic bond formation.

It was possible to perform an oxidation reaction under mild conditions with hypoiodide in a slightly alkaline medium that selectively and quantitatively converted the glucose or hydroxyethylated glucose at the reducing end into an aldonic acid. The unique carboxy group was then used directly for the reaction of amino groups of the target molecule. In this "linkerless" approach, the HES aldonic acid molecule was converted into its lactone by intensive drying procedures and the target molecule was added under conditions where water was excluded as far as possible. The reaction was performed in a water-free, polar solvent such as dimethyl sulfoxide to achieve high yields – excessive water would destroy the lactone before ring opening by an amino group of the target substance could occur. The procedure worked well for hemoglobin, but the number of suitable target molecules is limited, since many proteins are denatured in the above mentioned solvents.

An improvement overcoming these restrictions was the *in situ* activation of the carboxy group as an active ester, for example, by reaction with disuccinimidyl carbonate or EDC, followed by the addition of the amino-compound in an aqueous solution. These activated esters are also prone to hydrolysis, but their stability was sufficient to yield significant amounts of HES conjugates with serum albumin or nucleic acid molecules [24, 25]. Although NHS esters have a good selectivity for amino groups, their high reactivity can lead to unwanted side-reactions with hydroxy groups or thiols. The resulting esters or thioesters are expected to be unstable in aqueous media but nonetheless, due to the high number of hydroxyl groups in HES, crosslinking events between HES molecules cannot be totally excluded. As a consequence the isolation of these acylation reagents as a defined material becomes almost impossible.

As activated HES in a ready-to-use and stable storage form was essential for the development of the technology and HES reagents with specificities other than amine reactivity were demanded, the linkerless approaches described above were complemented by coupling methods via linkers. Suitable linker substances for the conversion of the oxidized reducing end of the HES were developed that are designed to be attached by a reaction with the lactonized HES acid in dry solvents and introduce another reactive group or serve as an intermediate for the generation of the desired reactivity. Examples are aliphatic diamines, such as diaminobutan or hexamethylendiamin yielding amino-HES which were then reacted with another bifunctional linker in an acylation or reductive amination reaction. With the linker strategy, highly reactive HES derivatives for several important conjugation methods such as thiol-modification, aldehyde- or amine-reactive coupling became accessible (Figure 7.4) [26].

7.5.3
HES Derivatives Based on Non-Oxidized HES

A further generation of HES derivatives uses the same reactive groups but it is no longer based on regioselective oxidation of HES at its reducing end. The latter presents a hemiacetal structure and only poorly acts as an aldehyde – only ~1% of the glucose ring is present in the reactive open-chain conformation. Nonetheless, an efficient attachment of certain linker structures by an amination reaction was possible under certain conditions [27, 28]. One set of linkers employed the high reactivity of hydroxylamine compounds with carbonyl groups yielding oxime linkages – this reaction, also included in the category "native chemical ligation", is a common chemistry used for example, in glycopeptide synthesis or commercially available PEG polymers bearing an aminooxy group. The oxime linkage is fairly stable under conditions compatible with protein stability. Respective conjugates prepared by the reaction of aminooxy-PEG with a protein carrying a keto-functionalized artificial amino acid are already in clinical testing (ARX-201, www.ambrx.com).

The "classical reductive amination" approach, the modification of the reducing end of HES with an amino compound, followed by a reduction of the resulting Schiff base to a secondary amine, lead to substantial amounts of site-selectively modified polymer as well when certain reaction parameters were applied. Abundant functional groups of HES molecules and the low reactivity of HES renders its site-specific functionalization – as already mentioned above – much more difficult than the synthesis of for example, the respective PEG derivatives. This can be nicely elucidated using the example of amine-reactive polymers suitable for reductive amination to the N-terminus of a protein: The widely used PEG propionaldehyde is easily prepared by reaction between PEG alkoxide and 3-chloro-propionaldehyde (mostly in its acetal protected form). The synthesis of a similar compound based on HES involves a reductive amination with 3-amino propionaldehyde-diethylacetal – the protection of the acetal is essential in this case, since otherwise stable concatemers of the linker would be formed by reaction between the amino and the free aldehyde. The reactive aldehyde group on the final polymer derivative is usually liberated by acid hydrolysis of the acetal [29]. Here the chemical nature of the HES causes a severe problem, since starch itself contains numerous acetal groups – exactly one per glucose subunit – forming the α-1,4 or α-1,6 glycosidic bonds. Consequently hydrolysis conditions efficient in aldehyde deprotection will in general cause cleavage and thus also damage the polymer moiety. Deprotection conditions for each of the linker structures had to be optimized by systematic screening but finally enabled the synthesis of these HES reagents for the N-terminal modification of proteins in high yield [30].

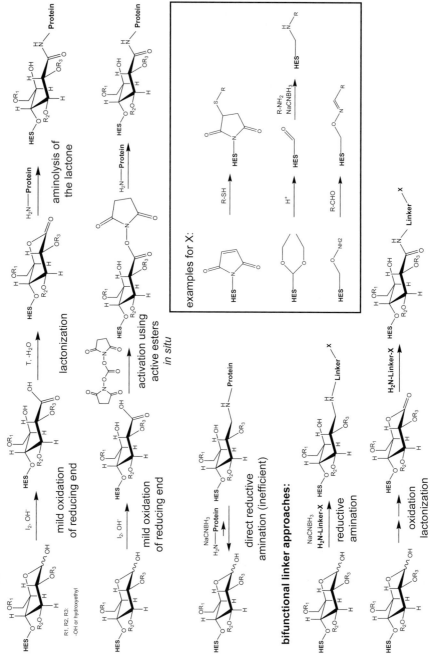

Figure 7.4 Chemistry of regioselective modification of the reducing end of HES.

The regioselective modification of HES with a reactive linker at its unique reducing end is the preferred way to obtain conjugates with a defined stoichiometry which is advantageous from a regulatory point of view. Nonetheless the random conjugation strategy to multiple functionalized sites of HES was pursued in industry and academia [31], for example, by approaches employing a periodate oxidation step to generate numerous aldehyde groups on the polymer that were either reacted with a heterobifunctional linker to introduce various functionalities suitable for the modification of thiol-containing peptides or aldehyde groups for conjugation to proteins in a reversible manner – via Schiff bases without a reduction step [32, 33].

7.6
HES–Protein Conjugates – Two Case Studies

7.6.1
Erythropoietin Polymer Conjugates

7.6.1.1 Erythropoietin Products on the Market
The glycoprotein hormone erythropoietin (EPO) is in broad use for the treatment of anemia in the context of cancer and kidney diseases with a total market volume of ~9B USD. Since the approval of the first recombinant version Epogen® in 1989, various attempts have been made to improve the protein, which suffers from fast renal clearance due to its size of ~30 kDa. The introduction of two additional *N*-glycosylation sites increased the *in vivo* half-life and enhanced the efficacy of the resulting molecule (up to threefold) compared with Epogen making darbepoietin (Aranesp®), the gold standard of erythropoiesis-stimulating agents (ESAs) for many years. Despite – or due to – the success of Aranesp many other compounds capable of stimulating red blood cell production are under development, applying for example, half-life prolongation technologies such as PEGylation or using artificial activating peptide ligands for the EPO receptor which are easier to produce than a human protein with a glycosylation pattern that is critical for its biological function [34].

The large market and the commercial availability of the starting material and already improved variants that could serve as benchmarks made EPO – besides other therapeutic proteins like G-CSF or interferon α – to an ideal target molecule for the proof of principle or validation of a half-life extension technology.

7.6.1.2 Chemistry of Polymer-Modified Erythropoietin
HESylation of EPO was performed following two different conjugation strategies: one targeted the glycosylation sites for the attachment of the polymer, while the second approach used a reductive amination reaction allowing the preferential modification of the protein's N-terminal α-amino group.

The degree of sialic acid modification has a strong influence on the half-life of glycoproteins since molecules exposing free galactose residues on their glycan structures are rapidly cleared from the bloodstream by asialoglycoprotein receptors located in the liver. This effect is investigated for EPO in detail and prevents the practical use of a protein with incomplete sialylation due to for example, production under nonideal culture conditions or expression systems.

To overcome the liver clearance mechanism, the glycan structures of such EPO molecules were selectively shielded with HES. This was achieved either by a selective oxidation of the few existing sialic acid residues under mild conditions with periodate or alternatively in an enzymatic approach using galactose-oxidase as a catalyst for the formation of carbonyl functions that were subsequently reacted with an aldehyde reactive HES bearing a hydrazone or aminooxy linker. Alternatively, an HES molecule of comparable size derivatized with a propionaldehyde as reactive group was conjugated to an untreated EPO under slightly acidic conditions. This was done in order to direct the polymer to the N-terminal amino function that is expected to be more reactive than the ε-amino groups of the lysine residues due to a lower pK_a value. The N-terminally modified protein was enriched up to 80% of the total conjugate with the particular HES size and linker structure used as determined by proteolytic peptide mapping. The remaining 20% of modification were mainly localized at positions K45 and K52, as well as at K152. This pattern is in line with the distribution of coupling sites described for the approved PEGylated version of EPO (Mircera®) that is synthesized by an amine-specific acylation reaction with a 30 kDa PEG carrying a succinimidyl butanoic acid linker. The respective lysine residues in EPO seem to be activated compared with "normal" lysines which can be a result of the local environment for example, involvement in surface charge networks.

7.6.1.3 *In Vitro* Activity of Polymer-Modified Erythropoietin Variants

All isolated conjugates showed significant residual activity in an *in vitro* assay using the cell line UT-7 with the cell number/viability as a measure of the EPO-dependent stimulation of proliferation (Figure 7.5). Activities of the two types of HES EPO conjugates were measured relative to an unmodified EPO standard, hyperglycosylated EPO (Aranesp) and Mircera (30 kDa PEG-EPO conjugate). Interestingly, the dose–response curve shape obtained for the improved versions of PEG differed significantly from the EPO calibration standard. The absence of a clear plateau for high concentrations of the test substances prevented the calculation of the potency as an EC_{50} value. This finding, most prominent for the polymer-modified proteins, might reflect altered binding properties of the modified EPO molecules at the EPO receptor. A decreased affinity for the EPO receptor could also be responsible for a lower clearance rate by cellular uptake and thus further prolong the circulation half-life of the conjugate *in vivo* [35].

The unmodified EPO had the highest *in vitro* activity as expected, followed by Aranesp with an estimated EC_{50} value in the range of $1 \, ng \, mL^{-1}$ and still an almost sigmoidal curve shape. The HESylation of the naturally occurring glyco structures following activation by periodate further reduced the *in vitro* potency, but the effect was less pronounced than in the case of the polymer-conjugates derived from amine-modification. Mircera, as well as the N-terminally HESylated EPO showed

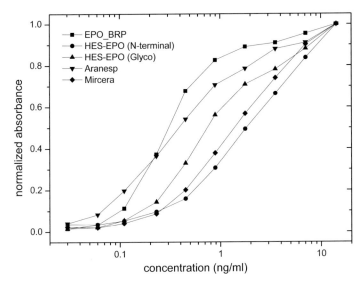

Figure 7.5 *In vitro* activity of EPO and EPO-polymer conjugates. The proliferative response of UT-7 cells after treatment with various erythropoietin compounds was measured in a WST-1 viability assay. Each dose–response curve was normalized against the maximum value for better comparison. EPO BRP (3rd European Pharmacopeia Biological Reference) Preparation of erythropoietin was used as a control substance representing the unmodified protein.

the greatest reduction in activity, reaching only 15–20% of the EPO reference. A closer look on the 3D structures of EPO/EPO-R fragment complexes reveals a potential explanation – the secondary coupling sites K45, K52 and K152 are deeply buried in the complex structure making critical interactions with the receptor. The N-terminal residues seem to be accessible but they are still located in the proximity of the main binding sites, while the region carrying the glyco structures is far away from them [36].

7.6.1.4 *In Vivo* Activity of Polymer-Modified Erythropoietin Variants

The effect of the different modifications on the *in vivo* efficacy of the respective EPO compound was tested in a simple mouse model using the change of hematocrit after subcutaneous (s.c.) injection of the test substance as readout for the erythropoietic activity. Figure 7.6 shows that unmodified EPO elicits only a weak and temporary response due to its fast renal excretion. The published threefold increase in half-life for the hyperglycosylated variant results also in an improved efficacy with a sustained increase of the hematocrit peaking at six days, that is reverted to the starting values two weeks after injection [37]. The glyco-HESylated protein reaches about the same peak level as Aranesp but seems to maintain a somewhat higher level between days 8 and 15. Mircera and the respective HES EPO conjugate targeting amino groups – despite having the lowest *in vitro* bioactivity – show the highest AUC of all compounds (~2–2.5fold of Aranesp). The

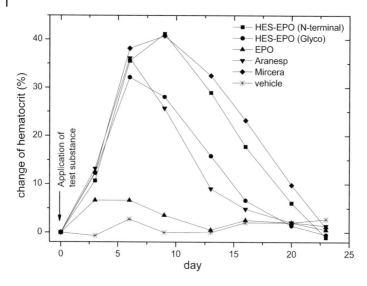

Figure 7.6 *In vivo* activity of EPO and EPO-polymer conjugates. BALB/C mice were dosed with 50 μg kg⁻¹ of the respective erythropoietin test substance subcutaneously. The erythropoietic effect was followed by measurement of the hematocrit.

maximum hematocrit values for these two compounds are detected 8 days post injection and their decline is significantly slower compared with the other test substances with the normal level being reached after >20 days. The dosing scheme applied might even result in an underestimation of the long-acting polymer-modified EPOs: the hematocrit values for Mircera and the amino-HESylated EPO went up to 70% or more potentially triggering feedback mechanisms in the animals and leading to an active downregulation to prevent a life-threatening polycythemia-like condition.

In summary, EPO HESylated at either the *N*-glycan structures or the N-terminus showed a significant residual *in vitro* activity (approx. 20–40% of the EPO standard) in an EPO-dependent cell proliferation assay. In an *in vivo* situation looking at the efficacy marker hematocrit, this impaired activity is overcompensated by the increased half-life of the HESylated compounds leading to an improved AUC compared with the current standard of care Aranesp. An HES EPO conjugate having coupling sites also found in Mircera showed an *in vitro* and *in vivo* bioactivity profile well comparable to the PEGylated version.

7.6.2
Polymer-Modified Interferon α Variants

7.6.2.1 PEGylated Interferon α Products on the Market
Recombinant interferon α is in clinical use for the treatment of HCV for about 25 years marketed under the brand names Roferon® (IFNα 2a, Roche) and Intron®A (IFNα 2b, Schering Plough). The 19 kDa cytokine shows a fast renal clearance with

a half-life of only 2–6 h in man, so a thrice-weekly dose is necessary to obtain acceptable effectiveness against the virus. However, even this frequent administration schedule in many cases leads to an increasing viral load, since the high protein doses needed for virus suppression are reached only at the first day after injection and the virus rapidly propagates itself in the following days, though other antiviral agents such as Ribavirin® are present. Furthermore, the interferon therapy is typically accompanied by fever and fatigue and even more severe adverse reactions such as leukopenia, caused by interferon's antiproliferative effect also on bone-marrow cells. A long-acting form capable of reducing the high peak concentrations directly after injection, while maintaining a sufficiently high drug level on the subsequent days, was expected to show an improved therapeutic effectiveness with reduced adverse events.

This led to the development of PEGylated versions of the first generation biologics – PegIntron® and PEGasys®, both conjugates prepared using an acylation reaction with succinimido-activated polymers. PegIntron (Merck/Schering-Plough) is made with a 12 kDa PEG succinimidyl carbonate resulting in a mixture of different positional isomers and unstable linkages (e.g., with His-34). Due to the use of a small PEG and the release of unmodified IFNα by hydrolysis of the protein-polymer, the bioactivity of the starting material is preserved to ~28%, but this chemical instability also necessitates the formulation as a lyophilized powder. PEGasys on the other hand is a liquid formulation in a prefilled syringe since it is prepared from a next-generation branched PEG NHS ester. This chemistry in combination with a larger 2×20 kDa polymer leads to a more selective conjugation to only four predominant lysine residues of the protein. The bioactivity is strongly impaired by the polymer modification (7% residual activity) but the improved pharmacokinetic features of the conjugate overcompensate this low activity making PEGasys (in combination with Ribavirin) the current standard of care for HCV treatment [38].

7.6.2.2 HESylation of rhIFNα-2b

The availability of the unmodified proteins, as well as the polymer-modified benchmark in pharmaceutical quality was an ideal prerequisite for comparative studies using HESylated versions of IFNα in proof-of-concept studies for the HESylation technology platform. In contrast to the PEG conjugates mentioned above, a reductive amination reaction with an amine-reactive HES derivative was performed in order to obtain better product homogeneity by regioselective conjugation to the N-terminus of the protein. The first optimization experiments revealed a high reactivity of IFNα with the aldehyde-containing HES even at low pH values of about 4.0 with a single digit molar excess leading to complete conversion of protein into the conjugate under certain reaction conditions. Unfortunately, these quantitative reactions were accompanied by the formation of a di-modified IFNα species explained by the presence of a second coupling site at Lys-121 identified by peptide mapping. Further optimization of the reaction conditions aiming for a conjugate with a 1:1 stoichiometry showed that the di-HESylation process became relevant only at coupling yields > 80%. In addition, the di-HESylation could be reduced by the use of larger HES molecules preventing the reaction with Lys-121 due to steric hindrance by the presumably faster occurring HES modification at the more reactive N-terminus.

Although HES is a semi-synthetic natural polymer showing a polydispersity in its mass distribution that is slightly higher than that of modern PEGs, the resulting HES conjugate can in general be well separated from the unmodified protein by ion exchange chromatography techniques. For the analytical characterization of HES conjugates or reaction mixtures, in addition other chromatographic techniques such as SE-HPLC or RP-HPLC are also used. The separation principle behind these methods is the shielding of either charged amino acids or hydrophobic surface patches of the protein by the hydrophilic polymer resulting in weakened interactions with the column and earlier elution compared with the unmodified protein.

The conjugate isolated by the above-mentioned standard chromatographic techniques is then analyzed for the number of HES coupling sites by methods such as SEC or field flow fractionation (aFFFF) combined with multiple detection including MALLS The position of the modification can be located by proteolytic or chemical cleavage followed by RP-HPLC/MS analysis of the fragments.

An example of the conjugate analysis by aFFFF with multiple detection can be seen in Figure 7.7. The components of the sample are separated by size based on their different mobility in a special flow chamber and are then detected by UV absorption, refractive index (RI) monitors and light-scattering. While the RI and scattering signal is used to calculate the absolute conjugate mass, the UV signal at a protein specific wavelength yields information on the contribution of the

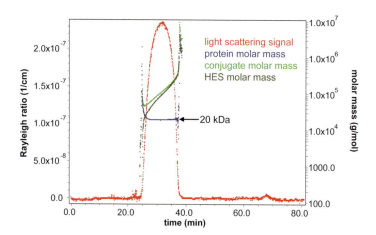

Figure 7.7 Analysis of conjugate size and stoichiometry for HES-IFNα. The conjugate size was measured by asymmetric flow field flow fractionation with triple detection (UV, RI, MALLS). The chromatogram shows the light scattering signal (90°) for the conjugate peak in red, together with the molecular weights calculated for the conjugate (brown) and its components HES (green) and protein (blue). The calculated protein mass in the conjugate is in-line with the theoretical value for IFNα-2b (19 265 Da).

protein to the mass, if its extinction coefficient is known or has been determined with the same instrument. The protein mass in the conjugate remains constant over the whole peak, while the conjugate mass – and consequently the modifier mass – increases from left to right due to the separation by mass reflecting the use of a polydisperse polymer. The quantitative analysis yields average values for the protein, polymeric modifier in the conjugate, as well as the total conjugate mass. The conjugate stoichiometry is calculated based on the molecular weights of its educts.

For the current example, such a calculation yields an HES-to-protein ratio of ≤1.05:1, that is, only a negligible portion of the conjugate seems to be di-HESylated, explainable by the presence of at least one highly reactive lysine residue as discussed above.

The assignment of the putative second modification site to a particular amino acid residue in IFNα by peptide mapping was not possible due to the low amount of di-HESylated product. The only peptide fragment in a tryptic digest showing a clear difference between the unmodified protein control and the conjugate sample was T1 – the cleavage product including the N-terminal amino-group as the reaction site for a reductive amination. This peak is no longer visible in the HES conjugate, instead there is a very broad additional signal putatively representing the HES-modified T1-peptide. The polydisperse HES – also showing some affinity to the column under these conditions – causes this changed elution behavior. The disappearance or shifting of a peak serves as an indirect proof for the complete modification of the N-terminus (Figure 7.8).

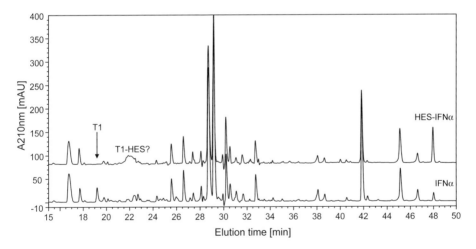

Figure 7.8 Identification of the polymer coupling site for HES-IFNα. The HES-IFNα conjugate (upper chromatogram) and the unmodified IFNα were subjected to a tryptic peptide mapping with subsequent RP-HPLC-MS analysis. The chromatograms are nearly superimposable – except for one peak present in IFNα that is missing in the conjugate (N-terminal peptide T1; 19.5 min). The HES modification of the peptide presumably leads to peak broadening and a shift in its elution time (~22 min).

7.6.2.3 *In Vitro* Activity of HESylated rhIFNα-2b

The reductive amination approach allowing such a regioselective conjugation was used for the preparation of HES-IFNα conjugates with a size-matrix of aldehyde functionalized HES derivatives. The effect of the polymer size on the conjugate bioactivity was investigated using a CPE (cytopathic effect) assay, a pharmacopeia method for *in vitro* activity measurement of interferons, based on the protective effect of the cytokine on a cell culture against virus infection. The commercially available first generation product Intron A served as a control in this assay and the potency of the test substances that were also based on Interferon α-2b was normalized against Intron A activity set to 100%. Figure 7.9 reveals the correlation between conjugate size and reduction of *in vitro* bioactivity most likely caused by interference of the cytokine-receptor interaction by the bulky polymer. Interestingly, the reduction in bioactivity shows a linear dependency for increasing the conjugate mass from 50 to 110 kDa, while the *in vitro* bioactivity of even larger conjugates remains nearly constant in the range of 2–3% relative to Intron A. In a similar experiment conducted with different PEG sizes the bioactivity decreased with the PEG size in the expected manner, although the use of a 60 kDa PEG for the modification of IFNα-2a yielded a conjugate with a significantly lower activity compared with the PEG40 kDa sample [39]. This slight difference between the two polymers can be explained by the different coupling chemistries and topologies. While the flexible and long-ranging PEG was randomly conjugated via NHS chemistry, the more rigid HES molecule was selectively attached to the N-terminus.

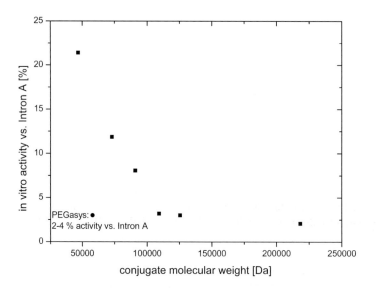

Figure 7.9 *In vitro* activity of HES-IFNα conjugates having variable size. The bioactivity of various IFNα compounds was determined in a CPE-Assay as protective effect on MDBK cells against infection with VSV. Intron®A was used as unmodified IFNα-2b standard.

When a critical size of the HES molecule regarding conjugate bioactivity is reached, a further increase in polymer size seems not to cause a stronger inhibition of the *in vitro* activity as may be the case for PEG-IFNα, since in the case of HES the mass increase occurs in three dimensions instead of two as it is the case for a branched PEG.

Interestingly, in our assay the PEGasys control repeatedly exhibited a bioactivity of only 2% relative to Intron A which is significantly lower than the 7% published by Bailon *et al.* [38]. The reasons for this finding are unclear but a higher bioactivity of HES-protein conjugates having a hydrodynamic radius comparable to PEGasys is explained by the more localized attachment of the "rigid" HES to the N-terminus. Here it is expected to show less interference with the receptor binding region located at positions 35–40 and 123–140. In PEGasys the major conjugation sites for PEG are located within or in close proximity to these sequence areas, explaining the stronger reduction in bioactivity.

7.6.2.4 Pharmacokinetics of HES-IFNα Compared with PEGasys

Despite the low *in vitro* activity, PEGasys has become the standard of care in HCV therapy due to its excellent *in vivo* efficacy. The determining factor here is the PK profile and the maintenance of drug levels sufficiently high to inhibit virus replication over a longer period. In patients, PEGasys showed an (age-dependent) elimination half-life of 60–110 h after s.c. administration allowing for a once weekly dose of 180 µg drug substance.

PEGasys and an HES-IFNα conjugate of comparable size were investigated regarding their PK profile in a comparative study conducted in rabbits. The protein part of the conjugates was radioactively labeled with a standard radioiodination protocol targeting the tyrosine residues as ELISA assays with polymer-modified proteins are often impaired by poor signal recovery due to shielding effects by the polymer against the detection antibodies and the sensitivity of the CPE assay is limited due to the low *in vitro* bioactivities of the conjugates.

Figure 7.10 shows an intensity vs. time curve typical of a s.c. administration for both test substances. The HES-IFNα conjugate is more rapidly mobilized after injection and the C_{max} value is reached significantly earlier (36 h) compared with PEGasys (72 h), which is known to have an extensive absorption half-life of ~50 h in man. The faster t_{max} value can presumably be attributed to the high hydrophilicity of HES and might be of special interest regarding a faster initial reduction of the virus load by the HES conjugate compared with PEGasys. The C_{max} values and AUCs are roughly comparable for both substances, while the last data points of the PK experiment point towards a longer elimination half-life (102 vs. 51 h) and thus more sustained therapeutic action of the HESylated IFNα.

7.6.2.5 Viscosity: HES Compared with PEG

The rheological properties of interferon polymer conjugates were considered to be an interesting feature differentiating HES from PEG, though without practical relevance in the case of highly potent cytokine administered in µg-doses. The viscosity of concentrated conjugates (~2% protein content) prepared by reductive

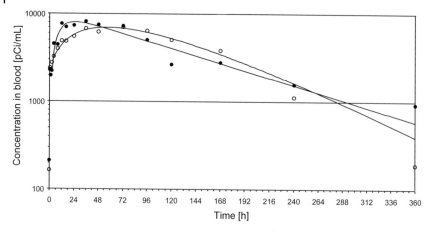

Figure 7.10 Pharmacokinetics of HES-IFNα vs. PEGasys® in rabbits. The test substances (• HES-IFNα; ○ PEGasys) radiolabeled with iodine-125 using a chloramine-T approach were administered s.c. at a dose of $8\,\mu Ci\,kg^{-1}$ (25 μg protein kg^{-1}) to New Zealand White rabbits and the radioactivity in blood samples was followed over the time-points as indicated. Fitting of the data using a noncompartmental model yielded an AUC of ~120 000 pCi*h mL^{-1} for both substances. T½ was 51 h for PEGasys and 102 h for HES-IFNα; C_{max} was 7250 pCi mL^{-1} reached after a t_{max} of 72 h in case of PEGasys and 8130 pCi mL^{-1} at 36 h post-injection time for HES-IFNα.

amination with either PEG- or HES-aldehyde were analyzed in a rheometer setup. The solution viscosity of the HES-IFNα conjugate was approximately two- to threefold lower compared with its counterpart prepared with a 40 kDa branched PEG resulting in similar pharmacokinetics as discussed above (Figure 7.11). The reason for the decreased viscosity is again rooted in the different topology of the polymers: while PEG in aqueous solution adopts an extended coil structure, the rigidity and local mass concentration of the HES is higher due to the lower amount of flexible bonds in the branched structure. As a consequence, the Mark–Houwink coefficient used for the calculation of the intrinsic viscosity already differs by a factor of two.

7.7
Summary and Conclusion

HESylation uses hydroxyethyl starch polymers to modify drug substances predominantly in the area of biopharmaceuticals. As a semisynthetic polymer HES parameters like MS and MW can be tailored for each specific coupling application. The excellent safety profile of HES resulting from its biocompatibility and biodegradability is supported by the long term pharmaceutical use as blood volume expander. In proven safety aspects, HES is positioned uniquely compared with other polymers currently in use for protein modification. Besides its degradability,

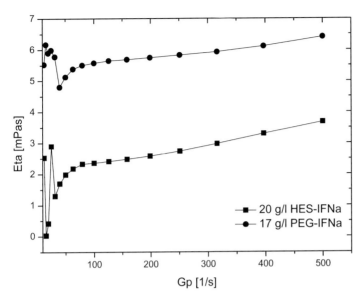

Figure 7.11 Solution viscosity of N-terminally modified HES-IFNα vs. PEG-IFNα. Interferon α modified at the N-terminus with either branched PEG40k or an HES molecule resulting in a comparable or better PK profile (see Figure 7.10) was concentrated up to ~20 mg mL^{-1} protein content and the solution viscosity was measured rheometrically.

further product specific advantages of HES often become apparent in head to head comparison with PEG polymers, the most important being:

- Lower immunogenicity avoiding neutralizing antibodies;
- Less interference with binding sites/active sites and thus higher *in vitro* activity;
- Higher *in vivo* recovery due to the hydrophilic nature of the polymer; and
- Lower viscosity, enabling high concentration-formulations or smaller needle sizes;
- Clinically proven safety profile with known degradation pathways.

Because of these interesting properties, HESylation attracts a high level of attention in the area of drug delivery for new pharmaceuticals, especially when dosing at high concentration is expected in a chronic disease setting.

References

1 Veronese, F.M., and Pasut, G. (2005) PEGylation, successful approach to drug delivery. *Drug Discov. Today*, **10**, 1451–1458.
2 Webster, R., Elliott, V., Park, B.K., Walker, D., Hankin, M., and Taupin, P. (2009) PEG and PEG conjugates toxicity: towards an understanding of the toxicity of PEG and its relevance to PEGylated biologicals, in *Pegylated Protein Drugs: Basic Science and Clinical Applications* (ed. Veronese, F.M.), Birkhäuser Verlag, Basel, pp. 127–146.

3 Armstrong, J.K., Hempel, G., Koling, S., Chan, L.S., Fisher, T., Meiselman, H.J., and Garratty, G. (2007) Antibody against poly(ethylene glycol) adversely affects PEG-asparaginase therapy in acute lymphoblastic leukemia patients. *Cancer*, **110**, 103–111.

4 European Pharmacopeia (2010) Hydroxyethylstarch monograph 1785.

5 Finfer, S., Liu, B., Taylor, C., Bellomo, R., Billot, L., Cook, D., Du, B., McArthur, C., and Myburgh, J. (2010) Resuscitation fluid use in critically ill adults: an international cross-sectional study in 391 intensive care units. *Crit. Care*, **14**, R185.

6 Neff, T.A., Doelberg, M., Jungheinrich, C., Sauerland, A., Spahn, D.R., and Stocker, R. (2003) Repetitive large-dose infusion of the novel hydroxyethyl starch 130/0.4 in patients with severe head injury. *Anesth. Analg.*, **96**, 1453–1459. table.

7 Westphal, M., James, M.F., Kozek-Langenecker, S., Stocker, R., Guidet, B., and Van, A.H. (2009) Hydroxyethyl starches: different products–different effects. *Anesthesiology*, **111**, 187–202.

8 Kiehl, P., Metze, D., Kresse, H., Reimann, S., Kraft, D., and Kapp, A. (1998) Decreased activity of acid alpha-glucosidase in a patient with persistent periocular swelling after infusions of hydroxyethyl starch. *Br. J. Dermatol.*, **138**, 672–677.

9 Ständer, S., Szepfalusi, Z., Bohle, B., Ständer, H., Kraft, D., Luger, T.A., and Metze, D. (2001) Differential storage of hydroxyethyl starch (HES) in the skin: an immunoelectron-microscopical long-term study. *Cell Tissue Res.*, **304**, 261–269.

10 Jaeger, K., Jüttner, B., Heine, J., Ruschulte, H., Scheinichen, D., and Piepenbrock, S. (2000) Effects of hydroxyethyl starch and modified fluid gelatine on phagocytic activity of human neutrophils and monocytes–results of a randomized, prospective clinical study. *Infus. Ther. Transfus. Med.*, **27**, 256–260.

11 Nohe, B., Burchard, M., Zanke, C., Eichner, M., Krump-Konvalinkova, V., Kirkpatrick, C.J., and Dieterich, H.J. (2002) Endothelial accumulation of hydroxyethyl starch and functional consequences on leukocyte-endothelial interactions. *Eur. Surg. Res.*, **34**, 364–372.

12 Szepfalusi, Z., Parth, E., Jurecka, W., Luger, T.A., and Kraft, D. (1993) Human monocytes and keratinocytes in culture ingest hydroxyethylstarch. *Arch. Dermatol. Res.*, **285**, 144–150.

13 Laxenaire, M.C., Charpentier, C., and Feldman, L. (1994) [Anaphylactoid reactions to colloid plasma substitutes: incidence, risk factors, mechanisms. A French multicenter prospective study]. *Ann. Fr. Anesth. Reanim.*, **13**, 301–310.

14 Dieterich, H.J., Kraft, D., Sirtl, C., Laubenthal, H., Schimetta, W., Polz, W., Gerlach, E., and Peter, K. (1998) Hydroxyethyl starch antibodies in humans: incidence and clinical relevance. *Anesth. Analg.*, **86**, 1123–1126.

15 Richter, A.W., and de Belder, A.N. (1976) Antibodies against hydroxyethylstarch produced in rabbits by immunization with a protein-hydroxyetylstarch conjugate. *Int. Arch. Allergy Appl. Immunol.*, **52**, 307–314.

16 Abuchowski, A., van Es, T., Palczuk, N.C., and Davis, F.F. (1977) Alteration of immunological properties of bovine serum albumin by covalent attachment of polyethylene glycol. *J. Biol. Chem.*, **252**, 3578–3581.

17 Beez, Michael and Rothe, Wolfgang (1976) Kolloidales Volumenersatzmittel aus Hydroxyaethylstaerke und Haemoglobin. Dr. Eduard Fresenius, Chemischpharmazetische Industrie KG. DE Patent 2616086.1., filed 13.04.1976 and issued 03.04.1986.

18 Wong, Jeffrey Tze-Fei (1976) Blood substitute based on hemoglobin. Hematech Inc. US Patent 4064118, filed 08.10.1976 and issued 20.12.1977.

19 Baldwin, J.E., Gill, B., Whitten, J.P., and Taegtmeyer, H. (1981) Synthesis of polymer-bound hemoglobin samples. *Tetrahedron*, **37**, 1723–1726.

20 Cerny, L.C., Stasiw, D.M., Cerny, E.L., Baldwin, J.E., and Gill, B. (1982) A hydroxyethyl starch-hemoglobin polymer as a blood substitute. *Clin. Hemorheol.*, **2**, 355–365.

21 Lee, C.R., McKenzie, C.A., Webster, K.D., and Whaley, R. (1991) Pegademase bovine: replacement therapy for severe

combined immunodeficiency disease. *DICP : The Annals of Pharmacotherapy*, **25**, 1092–1095.

22 Maout, E., Grandgeorge, M., and Dellacherie, E. (1993) Hydroxyethylstarch conjugated to human hemoglobin for use in blood transfusion: comparison with Dextran conjugates, in *Carbohydrates and Carbohydrate Polymers* (ed. Yalpani, M.), ATL Press, Shrewsbury, pp. 132–140.

23 Sommermeyer, Klaus and Eichner, Wolfram (1997) Haemoglobin-Hydroxyethyl Starch conjugates as oxygen carriers. Fresenius AG. EP Patent 0912197, filed 07.07.1997 and issued 05.12.2001.

24 Sommermeyer, Klaus (2005) Method for producing conjugates of polysaccharides and polynucleotides. Noxxon Pharma AG, Supramol Parenteral Colloid GmbH. Patent application WO2005/074993, filed 08.02.2005.

25 Sommermeyer, Klaus, Eichner, Wolfram, Frie, Sven, Jungheinrich, Cornelius, Scharpf, Roland, Lutterbeck, Katharina, Hemberger, Jürgen, and Orlando, Michele (2002) Conjugates of hydroxyalkyl starch and an active agent. Fresenius Kabi Deutschland GmbH. US Patent 7816516, filed 15.03.2002 and issued 19.10.2010.

26 Conradt, Harald S., Grabenhorst, Eckart, Nimtz, Manfred, Zander, Norbert, Frank, Ronald, and Eichner, Wolfram (2003). HASylated polypeptides, especially HASylated erythropoietin. Fresenius Kabi Deutschland GmbH. EP Patent 1398322, filed 11.09.2003 and issued 19.04.2006.

27 Zander, Norbert (2003) Method of producing hydroxyalkyl starch derivatives. Fresenius Kabi Deutschland GmbH. EP Patent 1398327, filed 11.09.2003 and issued 27.08.2008.

28 Zander, Norbert, Eichner, Wolfram and Conradt, Harald (2003) Hydroxyalkyl starch derivatives. Fresenius Kabi Deutschland GmbH. EP Patent 1398328, filed 11.09.2003 and issued 18.11.2009.

29 Bentley, Michael David, and Harris, J. Milton (1998) Poly(ethylene glycol) aldehyde hydrates and related polymers and applications in modifying amines. Shearwater Corp. US Patent 5,990,237, filed 20.05.1998 and issued 23.11.1999.

30 Hacket, Frank, Hey, Thomas, Hauschild, Franziska, Knoller, Helmut, Schimmel, Martin, and Sommermeyer, Klaus (2009). Hydroxyalkyl starch derivatives and process for their preparation. Fresenius Kabi Deutschland GmbH. Patent application WO2009/077154.

31 Besheer, A., Hertel, T.C., Kressler, J., Mader, K., and Pietzsch, M. (2009) Enzymatically catalyzed HES conjugation using microbial transglutaminase: Proof of feasibility. *J. Pharm. Sci.*, **98**, 4420–4428.

32 Frank, Hans-Georg, and Haberl, Udo (2006) Supravalent Compounds. Aplagen Deutschland GmbH. Patent application PCT/EP2006/006097, filed 23.06.2006. 23-6-2006.

33 Adamson, Gord, Bell, David, and Brookes, Steven (2009). Polysaccharide-protein conjugates reversibly coupled via imine bonds. Therapure Biopharma Inc. Patent application PCT/CA2009/000885, filed 26.06.2009.

34 McDougall Iain, C. (2006) Recent advances in erythropoietic-agents in renal anemia. *Semin. Nephrol.*, **26**, 313–318.

35 Jarsch, M., Brandt, M., Lanzendorfer, M., and Haselbeck, A. (2008) Comparative erythropoietin receptor binding kinetics of C.E.R.A. and epoetin-beta determined by surface plasmon resonance and competition binding assay. *Pharmacology*, **81**, 63–69.

36 Syed, R.S., Reid, S.W., Li, C., Cheetham, J.C., Aoki, K.H., Liu, B., Zhan, H., Osslund, T.D., Chirino, A.J., Zhang, J., Finer-Moore, J., Elliott, S., Sitney, K., Katz, B.A., Matthews, D.J., Wendoloski, J.J., Egrie, J., and Stroud, R.M. (1998) Efficiency of signalling through cytokine receptors depends critically on receptor orientation. *Nature*, **395**, 511–516.

37 Egrie, J.C., and Browne, J.K. (2001) Development and characterization of novel erythropoiesis stimulating protein (NESP). *Br. J. Cancer*, **84** (Suppl. 1), 3–10. 3–10.

38 Bailon, P., Palleroni, A., Schaffer, C.A., Spence, C.L., Fung, W.J., Porter, J.E., Ehrlich, G.K., Pan, W., Xu, Z.X., Modi,

M.W., Farid, A., Berthold, W., and
Graves, M. (2001) Rational design of a
potent, long-lasting form of interferon: a
40 kDa branched polyethylene glycol-
conjugated interferon alpha-2a for the
treatment of hepatitis C. *Bioconjug.
Chem.*, **12**, 195–202.

39 Bailon, P., and Won, C.Y. (2009)
PEG-modified biopharmaceuticals. *Expert
Opin. Drug Deliv.*, **6**, 1–16.

Part Three
Half-Life Modulation Involving Recycling by
the Neonatal Fc Receptor

8
The Biology of the Neonatal Fc Receptor (FcRn)

Jonghan Kim

8.1
Homeostasis of Albumin and Immunoglobulin

Plasma proteins are required for a variety of vital functions; for example, albumin provides colloidal osmotic pressure in plasma and buffering capacity in the blood. Immunoglobulins guard against foreign molecules, and transferrin transports iron for hemoglobin synthesis. The steady-state concentrations of plasma proteins are maintained by the delicate balance between production and catabolism, while recycling and tissue distribution also contribute to the magnitudes of the concentrations (Figure 8.1). Albumin and immunoglobulin G (IgG) are the two most abundant plasma proteins, covering about 70% of total proteins in plasma. Furthermore, half-lives of both proteins are also much longer than other proteins in humans (Figure 8.2).

Although albumin and IgG are structurally disparate and functionally unrelated, a large body of evidence has indicated that the two proteins share at least one common property: an inverse relationship between their concentrations and half-lives in serum, while other plasma proteins demonstrate either a positive or no relationship. To account for the inverse association of concentration–survival time of IgG, about a half century ago Brambell proposed a saturable Fc receptor-mediated salvage process [5].

The rapid evolution of molecular and genetic information about these proteins and their receptors has uncovered a comprehensive understanding of the mechanisms involved in the homeostasis of albumin and IgG. Today we recognize that the turnover of both proteins is tightly regulated by the neonatal Fc receptor (FcRn), which is a heterodimer consisting of a nonclassical major histocompatibility complex class I (MHC-I) protein and β2-microglobulin (B2M). IgG and albumin bind the FcRn heterodimer independently of each other and are rescued from degradation by an intracellular recycling mechanism that functions throughout life [1, 6–10]. Thus, both proteins demonstrate longer half-lives and higher plasma concentrations than other plasma proteins. Patients and animals with FcRn-deficiency display hypercatabolism of both proteins, leading to hypoalbuminemia and hypogammaglobulinemia [11, 12]. FcRn also transports IgG from mother to young

Therapeutic Proteins: Strategies to Modulate Their Plasma Half-Lives, First Edition. Edited by Roland Kontermann.
© 2012 Wiley-VCH Verlag GmbH & Co. KGaA. Published 2012 by Wiley-VCH Verlag GmbH & Co. KGaA.

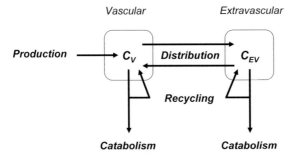

Figure 8.1 Plasma protein turnover. Plasma proteins are produced by biosynthesis (liver, immune cells, etc) and secreted into plasma or vascular space where they are transported to and distributed reversibly in the extravascular space. Proteins are continuously pinocytosed into cells and catabolized via lysosomal degradation. The steady-state concentrations of plasma proteins in the vascular (C_v) and extravascular spaces (C_{EV}) are maintained by two major turnover processes, production and catabolism, while recycling and distribution processes also contribute to the magnitudes of the concentrations in each space.

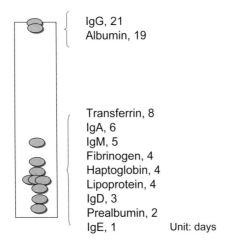

IgG, 21
Albumin, 19

Transferrin, 8
IgA, 6
IgM, 5
Fibrinogen, 4
Haptoglobin, 4
Lipoprotein, 4
IgD, 3
Prealbumin, 2
IgE, 1   Unit: days

Figure 8.2 Half-lives of plasma proteins in humans. It has long been known that IgG and albumin, the two major plasma proteins, exhibit remarkably prolonged lifespans (about three weeks of half-lives), whereas other plasma proteins are relatively short-lived, half-lives of most being <1 week [1–4]. FcRn accounts for the unusually long half-lives of the two proteins.

during the perinatal period. An interesting evolutionary perspective is that FcRn may have been chosen to be a protective receptor to reduce a higher burden of production of the two major plasma proteins, thus providing bioenergetic advantages.

In order to understand the turnover pathway of the FcRn–protein complex and its regulation, this chapter presents the physiology of albumin and IgG homeostasis. Its major focus is on the biochemistry of FcRn and how its function contributes to the homeostasis of both ligands. This chapter also summarizes new and emerging information about the bioengineering of therapeutic proteins that takes advantage of the FcRn biology. Finally, the potential benefits of treatment to ameliorate the autoimmune condition promoted by autoantibodies are considered.

8.2
Neonatal Fc Receptor Biochemistry

The neonatal Fc receptor (FcRn) is a heterodimeric transmembrane glycoprotein consisting of a heavy α-chain that is structurally related to MHC class I molecules and a B2M that is noncovalently associated with the α-chain [13]. The expression of both units is required for FcRn functions since mice deficient in either FcRn α-chain or B2M gene display severe hypercatabolic hypoproteinemia [1, 6–10]. The absence of B2M results in receptor degradation in the endoplasmic reticulum, suggesting that B2M is important for the correct folding and cell surface expression of FcRn [14, 15].

FcRn binds to IgG in a unique manner, in contrast to the binding between IgG and other classical Fcγ receptors. First, unlike other classical MHC-I molecules, the peptide binding cleft of FcRn is empty, disabling antigen presentation and the subsequent immune responses [16, 17]. Instead, FcRn binds to IgG at the junction between the CH2 and CH3 domains. Secondly, FcRn–IgG binding is pH-dependent. FcRn binds to IgG with a very high affinity (nanomolar K_D) at an acidic condition, but with negligible binding affinity at neutral or physiological pH. At pH 6, FcRn becomes more thermally stable and shows slower IgG dissociation kinetics than at pH 8, with no considerable conformational change in structure between the two pH values. Instead, the pH dependence of FcRn–IgG binding results from chemical properties of IgG and FcRn; at an acidic pH, the histidine residues (H310 and H435) in the CH2-CH3 hinge region of the Fc molecule are predominantly protonated, which allows more favorable interactions with anionic pockets on FcRn than at pH 8 [18–20]. *In vivo* results, based on site-directed mutagenesis and pharmacokinetic analysis, also suggested that these histidine residues play a central role in the systemic maintenance of IgG homeostasis in mice [21]. In addition, hydrophobic interaction between FcRn and the Fc region is stabilized at an acidic pH, contributing to FcRn–IgG binding [22].

Concerning binding stoichiometry, one or two molecules of FcRn can bind one IgG molecule. Under equilibrium conditions, two molecules of FcRn bind one IgG molecule (2:1 stoichiometry), whereas alterations of the carbohydrate moieties on FcRn can result in 1:1 binding stoichiometry under nonequilibrium conditions [23]. While equimolar binding affinity of FcRn–IgG is low, FcRn dimerization, required for IgG binding with high affinity [24], is stabilized at an acidic pH by the formation of a histidine-mediated salt bridge and a side chain rearrangement [18]. Wild-type Fc molecule of IgG has two symmetrical binding sites to FcRn, and heterodimeric Fc molecule having only one FcRn binding site undergoes more lysosomal degradation compared with its wild-type Fc containing two FcRn binding sites, suggesting that ligand bivalency plays a critical role in IgG homeostasis [25].

Albumin binds to FcRn in a similar fashion. The binding affinity of albumin to FcRn reduces dramatically (~200-fold) from acidic to neutral pH through rapid association and dissociation kinetics [26]. IgG and albumin bind to FcRn independently and noncooperatively. Thus, the two proteins bind at different sites of FcRn without interactions and with high binding affinities in acidic intracellular endosomes, but with negligible binding affinity at physiologic pH, allowing continuous rounds of recycling between intracellular and extracellular fluids (Figure 8.3). Unlike FcRn–IgG binding, albumin binds FcRn with a 1:1 stoichiometry, and a large positive change in entropy upon binding suggests a hydrophobic interaction. The domain III (DIII) of albumin is responsible for FcRn binding and the long serum half-life [26]. Pharmacokinetic analysis in mice using mutant DIII scaffold indicates that H464 appears to be most crucial for FcRn binding, followed by H510 and H535 [27]. *In vitro* mutagenesis and binding studies showed that the conserved H166 residue of the human FcRn heavy chain, located opposite to the FcRn–IgG interaction site, plays a critical role in the pH-dependent FcRn–albumin interaction [26, 28].

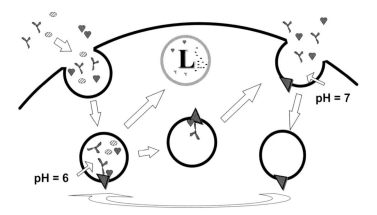

Figure 8.3 FcRn-mediated recycling and transport of IgG and albumin. Following pinocytosis, plasma proteins move to acidic endosomal compartments where IgG (Y-shape) and albumin (heart-shape) bind FcRn with high affinity at distinct binding sites. Unbound plasma proteins (hatched circles) or excess ligands, when FcRn is saturated, are destined for degradation in lysosomes (L), while bound complex is diverted to the cell membrane where IgG and albumin dissociate from FcRn at physiologic pH, providing longer survival of the two proteins. Unbound receptor is either recycled back to endosome for another round of recycling or degraded in the lysosome. For simplicity, exact stoichiometry between ligands and the receptor is omitted. The same type of intracellular trafficking occurs for FcRn-mediated transport from one to the other side of the cell (not shown).

FcRn–ligand binding is species-specific, which has been in part attributed to the poor pharmacokinetics in humans of many therapeutic monoclonal antibodies raised in mice. For example, mouse FcRn binds both mouse and human IgG, while the binding affinity is low between human FcRn and mouse IgG [29]. On the other hand, mouse FcRn has low binding affinity to human albumin but human FcRn binds mouse albumin as well as human albumin [30]. Nonconserved amino acid residues found in FcRn appear to be responsible for these cross-species differences [30, 31]. Transgenic mice expressing human FcRn as well as human IgG may provide a useful animal model for the development of FcRn-related therapeutic proteins during the preclinical studies [22].

8.3
FcRn Function: Recycling

In mice homozygous for a targeted disruption of the B2M gene, IgG is cleared more rapidly from the systemic circulation compared with wild-type siblings [7, 8] while the clearance of chicken IgY or mouse IgA is unchanged [6, 9]. Studies using the α-chain and B2M double knockout mouse strains showed that IgG survives longer in the α-chain$^{-/-}$ B2M$^{+/+}$ mouse strain than in α-chain$^{+/+}$ B2M$^{-/-}$ and α-chain$^{-/-}$ B2M$^{-/-}$ strains with no difference in IgG half-lives between the latter two strains. Furthermore, steady-state endogenous plasma concentrations of both IgG and albumin are higher in the α-chain-deficient strain than in either the B2M-deficient or double-deficient strain. Thus, it appears to suggest that a B2M-associated effect other than FcRn prolongs the survival of both IgG and albumin [32], but further study is necessary to rule out other possibilities such as leaky gene transcription in the α-chain-deficient strain.

Sites of FcRn expression and functions are summarized in Table 8.1. Vascular endothelial cells are the main sites for both degradation and recycling of plasma proteins. While FcRn is expressed in the endothelium of small arterioles and capillaries rather than in larger vessels, it is found within vesicular structures but not on the membrane [33]. *In vitro* study, using IgG molecules having different binding affinities to FcRn in human endothelial cells, showed that unbound IgG accumulates in the lysosome [34]. FcRn is also highly expressed in bone marrow-derived cells and professional antigen presenting cells in different tissues. *In vivo* bone marrow chimera experiments showed that FcRn-deficiency significantly shortened IgG half-life in mice, indicating that bone marrow-derived phagocytic cells play an important role in IgG homeostasis [35].

FcRn deficiency occurs, although rarely, in humans. Two siblings from a consanguineous marriage were markedly deficient in both albumin and IgG due to their rapid degradation as reflected by short half-lives of both proteins [12]. They presented a similar phenotype to the ones observed in FcRn-deficient or B2M-deficient mice [6]. This manifestation implied a lack of FcRn function. Gene sequencing revealed that a single nucleotide transversion in the B2M gene led to mutation of alanine to proline at the midpoint of the signal sequence. The familial

Table 8.1 Sites of FcRn expression and functions of FcRn.

Cells/Tissues/Organs	Known or proposed functions
Vascular endothelial cells	Recycling of IgG and albumin from degradation
	Vascular–extravascular transport of IgG and albumin
Bone marrow-derived cells	Recycling of IgG and albumin from degradation
Intestine	Absorption of IgG
Placenta	Transmission of maternal IgG to fetus
Lung	Absorption of IgG
Kidney	Elimination of IgG from podocytes
Liver	Recycling of IgG and albumin from degradation
Brain	Efflux of antigen-IgG complex from the brain to blood
Eye	IgG elimination across the blood–retinal barrier into the blood system
Mucosal immune cells, dendritic cells	Bidirectional transport of IgG and IgG–antigen complex to facilitate antigen presentation

hypercatabolic hypoproteinemia presented by the two siblings supports the idea that FcRn binds IgG and albumin, salvages both from a degradative fate, and maintains their physiologic concentrations across the mammals [11].

FcRn does not modulate production of albumin or IgG, as assessed by a direct measurement using biosynthetically labeled albumin in plasma [36] or inferential kinetic analysis using IgG catabolism [37], respectively. Notably, the production rates for both albumin (FcRn ligand) and transferrin (having no binding affinity to FcRn) increase by approximately 20% in FcRn-deficient mice compared with normal mice, suggesting a likely compensatory mechanism for the lowered plasma oncotic pressure that is caused by hypoalbuminemia in FcRn-deficient mice. Using the steady-state condition, it was concluded that the amount of albumin saved from degradation by FcRn-mediated recycling is the same as that produced by the liver [36].

A mechanism-based kinetic model was developed to estimate and quantify FcRn-mediated recycling of both ligands in humans using easily "accessible" plasma concentrations rather than unmeasurable endosomal concentrations [38]. The analysis based on FcRn saturation kinetics showed that the FcRn-mediated fractional recycling rates of IgG and albumin are 142% and 44% of their fractional catabolic rates, respectively, signifying that FcRn-mediated recycling is the primary determinant for the high endogenous concentrations of these proteins. The kinetic model also explained familial hypercatabolic hypoproteinemia caused by complete FcRn deficiency. The model also postulated that the hypercatabolic IgG deficiency of myotonic dystrophy, characterized by a shorter IgG half-life but normal albumin level, could be associated with a normal number of FcRn with lowered affinity for IgG but normal affinity for albumin as the binding sites for both ligands are distinctively separated [38].

8.4
FcRn Function: Transport

FcRn protein is detected in adult human intestine, mostly in the epithelial cells, indicating FcRn-mediated transport of intestinal IgG beyond neonatal life in humans [39], whereas rodent intestine expresses FcRn during the neonatal period but not in adult [7, 40]. In both humans and rodents, FcRn–IgG binding is favored at slightly acidic pH in the proximal intestine and the bound complex undergoes transcytosis via vesicular transport across the gut epithelial cells, after which IgG is released from FcRn at physiologic pH.

FcRn also transports IgG from mother to young during gestation through the placenta, providing fetuses and newborns with immunity [10]. Using the yolk sac placenta of mouse fetuses (day 19–20) of the three FcRn genotypes resulting from matings of FcRn heterozygous parents, the IgG concentrations in FcRn-deficient fetuses were negligible ($1.5\,\mu g\,ml^{-1}$) compared with wild-type fetuses ($336\,\mu g\,ml^{-1}$), confirming that FcRn is required for IgG transport from mother to fetus. Interestingly, heterozygous fetuses have IgG concentrations about a half that of wild-type fetuses, indicating the need for a complete expression for its full function. In the yolk sac placenta, FcRn is only found in the endoderm from which IgG appears to exit by FcRn after constitutive uptake of IgG. It is notable that albumin concentrations are also lower by 32% in FcRn-deficient fetuses than in wild-type fetuses, which could be explained by two FcRn-mediated processes, albumin recycling from degradation and its placenta transport in the conceptus [10]. Further study is necessary to distinguish these two poorly understood processes. A similar pattern is observed in humans; placental IgG transfer requires two cellular layers, syncytiotrophoblast through which FcRn transports IgG and villus endothelium where an FcRn-independent process mediates IgG transport [41–44].

FcRn is expressed in the lung epithelium [45] where it transports IgG across the lung epithelial cells in humans and mice [46]. Through this process, Fc fusion proteins such as Fc-conjugated erythropoietin can be absorbed from the lung to the systemic circulation. Furthermore, FcRn-mediated absorption is enhanced when Fc–erythropoietin was deposited in the upper and central airways of the lung in monkeys, where FcRn expression is pronounced [47]. While alveolar epithelium in itself has a capability of protein transport, which is also favored by large surface area, FcRn-mediated transport across the lung can be exploited for rapid systemic delivery and improved bioavailability of therapeutic proteins and peptides.

In the kidney, FcRn is expressed on the podocytes and the brush border of the proximal tubular epithelium [48] and appears to be involved in clearance of IgG which otherwise would have been accumulated in the glomerular basement membrane, as seen in FcRn-deficient mice [49]. Saturation of clearance mechanism potentiated the pathogenicity of nephrotoxic sera, supporting the idea that podo-cytes play an active role in removing proteins and FcRn facilitates elimination of IgG [49]. In contrast, kidney transplantation experiments revealed that FcRn in the kidney appears to reclaim albumin, as suggested by the progressive develop-ment of hypoalbuminemia in wild-type mice transplanted with FcRn-deficient kidneys, while serum albumin levels returned to normal in FcRn-deficient mice transplanted with wild-type kidneys [50]. It is unclear how these two FcRn ligands are handled in a different manner in the kidney.

While blood–brain barrier (BBB) restricts IgG transport from the blood to brain, IgG following intracerebral injection is rapidly exported to the blood via reverse transcytosis across the BBB. FcRn expression in the brain capillary endothelium and choroid plexus epithelium raised a question whether brain FcRn might be involved in IgG transport [51]. Pharmacokinetic analysis of IgG distribution in the brain after intravenous administration of radiolabeled IgG antibody revealed that there is no significant difference between brain-to-plasma concentration ratios determined from control and knockout animals and thus suggests that FcRn does not contribute significantly to IgG distribution across the BBB in mice [52]. However, FcRn appears to modify the pharmacokinetics of pathogenic protein as evidenced by removal of Aβ peptide, the Alzheimer's pathogenic amyloid beta peptide of aging brain [53]. FcRn-deficient mice or normal mice treated with FcRn inhibitor showed reduced clearance of endogenous Aβ from the brain following administration of anti-Aβ antibody. FcRn thus appears to play a key role in rapid removal of immune complex of Aβ from the brain to blood across the BBB [53].

FcRn is also present in the eye, another immunologically-privileged tissue [54], where it may influence IgG elimination across the blood–retinal barrier into the blood, based on an observation that intravitreally-administered IgG penetrates across the blood–retinal barrier in normal mice but not in FcRn-deficient mice [55].

8.5
FcRn Function: Mucosal Immune

FcRn is found in immune cells such as human monocytes, macrophages, and dendritic cells [56]. Immunocytochemical data using the polarized human intesti-nal T84 cell line has shown that FcRn is involved in bidirectional IgG transport across epithelial barriers, implying an important effect of FcRn on IgG function in immune surveillance at mucosal surfaces [57]. In the context of FcRn-related immune functions, it has been proposed that FcRn secretes IgG into the intestinal lumen where IgG binds bacterial pathogens. Then, it would retrieve immune complex from the lumen such that adaptive immune responses can be initiated in regional organized lymphoid structure, providing host defense and mucosal immune functions [22, 54, 58]. In dendritic cells, FcRn appears to promote antigen presentation associated with IgG [59]. FcRn-deficient mice immunized with flagel-lin developed less severe colitis when challenged with dextran sodium sulfate than wild-type mice [60]. However, more colitis was observed in FcRn-deficient mice given wild-type bone marrow than in wild-type mice given FcRn-deficient bone marrow. Interestingly, in the latter comparison serum anti-flagellin IgG levels were similar between the two groups, indicating an equal contribution of hematopoietic and nonhematopoeitic cells to FcRn-mediated IgG protection [60].

8.6
Therapeutic Implications of FcRn

Since FcRn controls homeostasis of IgG and albumin, the two major plasma proteins, and both proteins carry out essential physiologic functions, several approaches have been taken to improve pharmacokinetics of therapeutic proteins and gain clinical benefits (Table 8.2). While more extensive, detailed information is available in the later chapters, some therapeutic implications and applications are briefly reviewed here.

Intravenous immunoglobulin (IVIG) ameliorates immune thrombocytopenia, an autoimmune disease characterized by platelet destruction, in rats. This can be explained in part by more rapid clearance of pathogenic antiplatelet antibody due to FcRn saturation [61], although other immunomodulatory effects are also involved including anti-idiotypic neutralization of antiplatelet antibodies, engagement of inhibitory FcγRIIB receptor, and antagonism of FcγR receptor activation. FcRn-deficient mice were partially or completely protected from developing

Table 8.2 Therapeutic implications and applications of FcRn[a].

Strategy	Effect
Intravenous immunoglobulin (IVIG)	Increased catabolism of pathogenic IgG by FcRn saturation
Monoclonal antibody against FcRn	Increased IgG catabolism by blocking FcRn–IgG binding
Antibody with increased binding affinity for FcRn	Increased IgG catabolism by outcompeting endogenous IgG for FcRn binding
Engineered therapeutic antibody with increased binding affinity for FcRn	Prolonged survival of therapeutic antibody
Conjugation to FcRn ligands (Fc molecule, albumin)	Increased epithelial absorption and placental delivery of proteins or peptides with prolonged survival
Antibody against pathogenic antigen in the brain (amyloid beta peptide)	FcRn-mediated efflux of IgG–antigen to reduce antigen level in the brain

a) Adapted from Roopenian and Akilesh [22].

autoimmune arthritis, while IVIG ameliorated arthritis in wild-type mice [62], indicating that inhibition of FcRn can be a therapeutic strategy for pathogenic antibody-mediated autoimmune diseases.

Proteins and peptides that have poor bioavailability or exhibit rapid clearance can be conjugated to FcRn ligands (IgG, Fc molecule, and albumin) to improve their pharmacokinetics owing to FcRn-mediated recycling and transport, and thereby providing greater therapeutic effects. Also, mutant IgG or Fc molecules binding to FcRn with even higher affinity at acidic pH than their wild-type counterpart but unchanged or lower affinity at neutral pH improve their *in vivo* half-lives in serum [63]. By contrast, a mutant IgG antibody that inhibits FcRn–IgG interactions can modulate the endogenous IgG concentrations; mice administered with this FcRn blocker display a rapid decrease of endogenous IgG level, providing a therapeutic tool to eliminate unwanted antibody such as pathogenic antibody and IgG-toxin complexes [64].

8.7
Conclusions

The neonatal Fc receptor (FcRn), an MHC-I-related protein bound to B2M light chain, salvages IgG and albumin from an intracellular degradative fate and recycles them out of cells, prolonging the half-lives of both. These proteins bind FcRn at acidic pH but not at physiologic pH. The two ligands bind noncooperatively and independently to FcRn by distinctive mechanisms. For regulation of humoral immunity, FcRn governs IgG concentrations throughout the lifespan; transporting IgG through the placenta in the prenatal stage and across the gut of the neonate as well as protecting IgG from degradation throughout life. FcRn recycles and conserves albumin to maintain colloidal osmotic pressure of the plasma and to transport many essential nutrients and drugs. FcRn thereby reduces the workload of the liver, which otherwise must be twice the normal size to compensate for catabolic loss of albumin in the absence of FcRn in mice. As well, an immune system four times the normal would be necessary without FcRn. The large capacity of FcRn controlling the homeostasis of the two most abundant plasma proteins allows pharmacological interventions to enhance clinical benefits by improved pharmacokinetics of therapeutic proteins and peptides.

References

1 Anderson, C.L., *et al.* (2006) Perspective–FcRn transports albumin: relevance to immunology and medicine. *Trends Immunol.*, **27**, 343–348.

2 Schultze, H.E., and Heremans, J.F. (1966) *Molecular Biology of Human Proteins: With Special Reference to Plasma Proteins. Vol. 1. Nature and Metabolism of Extracellular Proteins*, Elsevier.

3 Mariani, G., and Strober, W. (1990) Immunoglobulin metabolism, in *Fc Receptors and the Action of Antibodies* (ed. H. Metzger), American Society for Microbiology, pp. 94–180.

4 Peters, T. Jr (1996) *All about Albumin: Biochemistry, Genetics, and Medical Applications*, Academic Press.

5 Brambell, F.W., *et al.* (1964) A theoretical model of gamma-globulin catabolism. *Nature*, **203**, 1352–1354.

6 Chaudhury, C., *et al.* (2003) The major histocompatibility complex-related Fc receptor for IgG (FcRn) binds albumin and prolongs its lifespan. *J. Exp. Med.*, **197**, 315–322.

7 Ghetie, V., *et al.* (1996) Abnormally short serum half-lives of IgG in beta 2-microglobulin-deficient mice. *Eur. J. Immunol.*, **26**, 690–696.

8 Israel, E.J., *et al.* (1996) Increased clearance of IgG in mice that lack beta 2-microglobulin: possible protective role of FcRn. *Immunology*, **89**, 573–578.

9 Junghans, R.P., and Anderson, C.L. (1996) The protection receptor for IgG catabolism is the beta2-microglobulin-containing neonatal intestinal transport receptor. *Proc. Natl. Acad. Sci. U.S.A.*, **93**, 5512–5516.

10 Kim, J., *et al.* (2009) FcRn in the yolk sac endoderm of mouse is required for IgG transport to fetus. *J. Immunol.*, **182**, 2583–2589.

11 Wani, M.A., *et al.* (2006) Familial hypercatabolic hypoproteinemia caused by deficiency of the neonatal Fc receptor, FcRn, due to a mutant beta2-microglobulin gene. *Proc. Natl. Acad. Sci. U.S.A.*, **103**, 5084–5089.

12 Waldmann, T.A., and Terry, W.D. (1990) Familial hypercatabolic hypoproteinemia. A disorder of endogenous catabolism of albumin and immunoglobulin. *J. Clin. Invest.*, **86**, 2093–2098.

13 Simister, N.E., and Mostov, K.E. (1989) An Fc receptor structurally related to MHC class I antigens. *Nature*, **337**, 184–187.

14 Zhu, X., *et al.* (2002) The heavy chain of neonatal Fc receptor for IgG is sequestered in endoplasmic reticulum by forming oligomers in the absence of beta2-microglobulin association. *Biochem. J.*, **367**, 703–714.

15 Praetor, A., and Hunziker, W. (2002) Beta(2)-Microglobulin is important for cell surface expression and pH-dependent IgG binding of human FcRn. *J. Cell. Sci.*, **115**, 2389–2397.

16 Burmeister, W.P., *et al.* (1994) Crystal structure at 2.2 A resolution of the MHC-related neonatal Fc receptor. *Nature*, **372**, 336–343.

17 Burmeister, W.P., *et al.* (1994) Crystal structure of the complex of rat neonatal Fc receptor with Fc. *Nature*, **372**, 379–383.

18 Vaughn, D.E., and Bjorkman, P.J. (1998) Structural basis of pH-dependent antibody binding by the neonatal Fc receptor. *Structure*, **6**, 63–73.

19 Raghavan, M., *et al.* (1995) Analysis of the pH dependence of the neonatal Fc receptor/immunoglobulin G interaction using antibody and receptor variants. *Biochemistry*, **34**, 14649–14657.

20 Kim, J.K., *et al.* (1994) Localization of the site of the murine IgG1 molecule that is involved in binding to the murine intestinal Fc receptor. *Eur. J. Immunol.*, **24**, 2429–2434.

21 Kim, J.K., *et al.* (1999) Mapping the site on human IgG for binding of the MHC class I-related receptor, FcRn. *Eur. J. Immunol.*, **29**, 2819–2825.

22 Roopenian, D.C., and Akilesh, S. (2007) FcRn: the neonatal Fc receptor comes of age. *Nat. Rev. Immunol.*, **7**, 715–725.

23 Sanchez, L.M., *et al.* (1999) Stoichiometry of the interaction between the major histocompatibility complex-related Fc receptor and its Fc ligand. *Biochemistry*, **38**, 9471–9476.

24 Raghavan, M., *et al.* (1995) Effects of receptor dimerization on the interaction between the class I major histocompatibility complex-related Fc receptor and IgG. *Proc. Natl. Acad. Sci. U.S.A.*, **92**, 11200–11204.

25 Tesar, D.B., *et al.* (2006) Ligand valency affects transcytosis, recycling and intracellular trafficking mediated by the neonatal Fc receptor. *Traffic*, **7**, 1127–1142.

26 Chaudhury, C., *et al.* (2006) Albumin binding to FcRn: distinct from the FcRn–IgG interaction. *Biochemistry*, **45**, 4983–4990.

27 Kenanova, V.E., *et al.* (2010) Tuning the serum persistence of human serum albumin domain III: diabody fusion

proteins. *Protein Eng. Des. Sel.*, **23**, 789–798.

28 Andersen, J.T., *et al.* (2006) The conserved histidine 166 residue of the human neonatal Fc receptor heavy chain is critical for the pH-dependent binding to albumin. *Eur. J. Immunol.*, **36**, 3044–3051.

29 Ober, R.J., *et al.* (2001) Differences in promiscuity for antibody-FcRn interactions across species: implications for therapeutic antibodies. *Int. Immunol.*, **13**, 1551–1559.

30 Andersen, J.T., *et al.* (2010) Cross-species binding analyses of mouse and human neonatal Fc receptor show dramatic differences in immunoglobulin G and albumin binding. *J. Biol. Chem.*, **285**, 4826–4836.

31 Ward, E.S., and Ober, R.J. (2009) Chapter 4: Multitasking by exploitation of intracellular transport functions the many faces of FcRn. *Adv. Immunol.*, **103**, 77–115.

32 Kim, J., *et al.* (2008) Beta 2-microglobulin deficient mice catabolize IgG more rapidly than FcRn- alpha-chain deficient mice. *Exp. Biol. Med. (Maywood)*, **233**, 603–609.

33 Borvak, J., *et al.* (1998) Functional expression of the MHC class I-related receptor, FcRn, in endothelial cells of mice. *Int. Immunol.*, **10**, 1289–1298.

34 Ward, E.S., *et al.* (2003) Evidence to support the cellular mechanism involved in serum IgG homeostasis in humans. *Int. Immunol.*, **15**, 187–195.

35 Akilesh, S., *et al.* (2007) Neonatal FcR expression in bone marrow-derived cells functions to protect serum IgG from catabolism. *J. Immunol.*, **179**, 4580–4588.

36 Kim, J., *et al.* (2006) Albumin turnover: FcRn-mediated recycling saves as much albumin from degradation as the liver produces. *Am. J. Physiol. Gastrointest. Liver Physiol.*, **290**, G352–G360.

37 Junghans, R.P. (1997) IgG biosynthesis: no "immunoregulatory feedback". *Blood*, **90**, 3815–3818.

38 Kim, J., *et al.* (2007) Kinetics of FcRn-mediated recycling of IgG and albumin in human: pathophysiology and therapeutic implications using a simplified mechanism-based model. *Clin. Immunol.*, **122**, 146–155.

39 Israel, E.J., *et al.* (1997) Expression of the neonatal Fc receptor, FcRn, on human intestinal epithelial cells. *Immunology*, **92**, 69–74.

40 Yoshida, M., *et al.* (2006) IgG transport across mucosal barriers by neonatal Fc receptor for IgG and mucosal immunity. *Springer Semin. Immunopathol.*, **28**, 397–403.

41 Simister, N.E. (2003) Placental transport of immunoglobulin G. *Vaccine*, **21**, 3365–3369.

42 Takizawa, T., *et al.* (2005) A novel Fc gamma R-defined, IgG-containing organelle in placental endothelium. *J. Immunol.*, **175**, 2331–2339.

43 Leach, J.L., *et al.* (1996) Isolation from human placenta of the IgG transporter, FcRn, and localization to the syncytiotrophoblast: implications for maternal-fetal antibody transport. *J. Immunol.*, **157**, 3317–3322.

44 Simister, N.E., *et al.* (1996) An IgG-transporting Fc receptor expressed in the syncytiotrophoblast of human placenta. *Eur. J. Immunol.*, **26**, 1527–1531.

45 Kim, K.J., *et al.* (2004) Net absorption of IgG via FcRn-mediated transcytosis across rat alveolar epithelial cell monolayers. *Am. J. Physiol. Lung Cell Mol. Physiol.*, **287**, L616–L622.

46 Spiekermann, G.M., *et al.* (2002) Receptor-mediated immunoglobulin G transport across mucosal barriers in adult life: functional expression of FcRn in the mammalian lung. *J. Exp. Med.*, **196**, 303–310.

47 Bitonti, A.J., *et al.* (2004) Pulmonary delivery of an erythropoietin Fc fusion protein in non-human primates through an immunoglobulin transport pathway. *Proc. Natl. Acad. Sci. U.S.A.*, **101**, 9763–9768.

48 Haymann, J.P., *et al.* (2000) Characterization and localization of the neonatal Fc receptor in adult human kidney. *J. Am. Soc. Nephrol.*, **11**, 632–639.

49 Akilesh, S., *et al.* (2008) Podocytes use FcRn to clear IgG from the glomerular basement membrane. *Proc. Natl. Acad. Sci. U.S.A.*, **105**, 967–972.

50 Sarav, M., *et al.* (2009) Renal FcRn reclaims albumin but facilitates elimination of IgG. *J. Am. Soc. Nephrol.*, **20**, 1941–1952.

51 Schlachetzki, F., *et al.* (2002) Expression of the neonatal Fc receptor (FcRn) at the blood-brain barrier. *J. Neurochem.*, **81**, 203–206.

52 Garg, A., and Balthasar, J.P. (2009) Investigation of the influence of FcRn on the distribution of IgG to the brain. *AAPS J.*, **11**, 553–557.

53 Deane, R., *et al.* (2005) IgG-assisted age-dependent clearance of Alzheimer's amyloid beta peptide by the blood-brain barrier neonatal Fc receptor. *J. Neurosci.* **25**, 11495–11503.

54 Baker, K., *et al.* (2009) Immune and non-immune functions of the (not so) neonatal Fc receptor, FcRn. *Semin. Immunopathol.*, **31**, 223–236.

55 Kim, H., *et al.* (2009) FcRn receptor-mediated pharmacokinetics of therapeutic IgG in the eye. *Mol. Vis.*, **15**, 2803–2812.

56 Zhu, X., *et al.* (2001) MHC class I-related neonatal Fc receptor for IgG is functionally expressed in monocytes, intestinal macrophages, and dendritic cells. *J. Immunol.*, **166**, 3266–3276.

57 Dickinson, B.L., *et al.* (1999) Bidirectional FcRn-dependent IgG transport in a polarized human intestinal epithelial cell line. *J. Clin. Invest.*, **104**, 903–911.

58 Yoshida, M., *et al.* (2006) Neonatal Fc receptor for IgG regulates mucosal immune responses to luminal bacteria. *J. Clin. Invest.*, **116**, 2142–2151.

59 Qiao, S.W., *et al.* (2008) Dependence of antibody-mediated presentation of antigen on FcRn. *Proc. Natl. Acad. Sci. U.S.A.*, **105**, 9337–9342.

60 Kobayashi, K., *et al.* (2009) An FcRn-dependent role for anti-flagellin immunoglobulin G in pathogenesis of colitis in mice. *Gastroenterology*, **137**, 1746–1756.

61 Hansen, R.J., and Balthasar, J.P. (2002) Effects of intravenous immunoglobulin on platelet count and antiplatelet antibody disposition in a rat model of immune thrombocytopenia. *Blood*, **100**, 2087–2093.

62 Akilesh, S., *et al.* (2004) The MHC class I-like Fc receptor promotes humorally mediated autoimmune disease. *J. Clin. Invest.*, **113**, 1328–1333.

63 Hinton, P.R., *et al.* (2004) Engineered human IgG antibodies with longer serum half-lives in primates. *J. Biol. Chem.*, **279**, 6213–6216.

64 Vaccaro, C., *et al.* (2005) Engineering the Fc region of immunoglobulin G to modulate *in vivo* antibody levels. *Nat. Biotechnol.*, **23**, 1283–1288.

9
Half-Life Extension by Fusion to the Fc Region

Jalal A. Jazayeri and Graeme J. Carroll

9.1
Introduction

Monomeric therapeutic proteins/cytokines are cleared rapidly from the body because they are small and thus have a relatively short serum half-life *in vivo* (5–50 min). This is due to several factors, which include rapid clearance by the kidneys, degradation by proteolytic enzymes and peripheral tissues, and receptor-mediated endocytosis. The main clinical implication of a short serum half-life is the need for frequent administration of the therapeutic proteins either by self-injection or by infusion at a specialized clinic or hospital. In view of these therapeutic limitations, the design of protein-based biological agents with an increased serum half- life is an important priority.

Various techniques have been used to construct a wide range of therapeutic proteins with increased serum half-life. These include PEGylation, which has widespread applications, and more recently various antibody fragments have been used to generate antibody fusion proteins [1, 2]. In particular, the genetic fusion of the constant region of immunoglobulin G (IgG), the Fc region, to a therapeutic protein of interest, such as an extracellular domain of a receptor, ligand, enzyme, hormone, or peptides, has become increasingly popular in recent years and has emerged as an important alternative to monoclonal antibodies (mAbs) in drug development.

The Fc-chimeric proteins have some of the characteristics of antibodies, with several benefits. These include improved serum stability and half-life due to the homo-dimeric nature of Fc fusion proteins, which increases their molecular weight and in turn constrains clearance via the kidneys. The extent of such pharmacokinetic (PK) improvements depends on the IgG isotype used. In some instances the Fc region improves the biophysical characteristics of fusion proteins as well, such as the solubility and stability of the protein [3]. In addition, the expression level of Fc-dimeric proteins often leads to higher yields in cultures. The fusion proteins can be purified using protein-A affinity chromatography, simplifying the downstream manufacturing process [4].

Therapeutic Proteins: Strategies to Modulate Their Plasma Half-Lives, First Edition. Edited by Roland Kontermann.
© 2012 Wiley-VCH Verlag GmbH & Co. KGaA. Published 2012 by Wiley-VCH Verlag GmbH & Co. KGaA.

In this chapter, we focus on recent advances in the construction and applications of the Fc-based proteins as biotherapeutics. To elucidate the Fc fusion technology, an introduction to the biology of the IgG1-Fc region and its receptor is provided, followed by a discussion of the methodologies applied for the expression and purification of Fc-based fusion proteins. The chapter also addresses design considerations for the construction of fusion proteins and introduces some plasmid-based expression vectors now available commercially. Several examples of Fc fusion constructs, which are either in clinical use or being evaluated in therapeutic trials, are also provided (Tables 9.1 and 9.2).

9.2
Immunoglobulin G

Immunoglobulin G (IgG) contains four polypeptide chains, two light chains and two heavy chains. Each chain contains an NH_2-terminal antigen-binding variable

Table 9.1 Examples of clinical applications of Fc and Fab antibody fragments.

IgG subclass	Product	Target	Trade name	Applications	Company	Ref
IgG1	TNFR55-Fc	TNF-α,	Lenercept	RA, PsA	Hoffmann-La Roche	[5]
IgG1	TNFR75-Fc	TNF-α,	Etanercept (Enbrel)	RA, Ank Spond, PsA	Amgen and Wyeth	[5]
IgG1	CD4-Fc	HIV gp120	PRO542	HIV infection	Genentech Inc.	[6]
IgG1	LFA3-Fc	CD2	Alefacept, LFA3TIP	Psoriasis	Biogen (Cambridge, MA)	[7]
IgG1	CTLA-4-Fc	B7	Abatacept (Orencia)	RA, PsA	Bristol-Myers Squibb	[8]
IgG1	sIL-4R-Fc	IL-4	Nuvance (altrakincept)	Asthma Phase II	Amgen/ Immunex	[9]
IgG1	Chimeric mAb	IL-2Rα	Basiliximab (Simulect)	Organ transplantation	Novartis	
IgG1	Humanized mAb	IL-2Rα	Daclizumab (Zenapax)	Organ transplantation	Hoffman-La Roche	[10]
	Humanized mAb	Anti-IgE-Fc	Xolair	Allergy	Genentech Inc. Tanox/Novartis	
IgG1	Humanized VEGF-Fc	Cancer cells	Aflibercept	Reduce vascularization of tumors	Sanofi-Aventis; Regeneron Pharmaceuticals	[11]

Table 9.2 Examples of Fc-fused cytokines/proteins.

Protein	Applications	Ref
ADAMTS-Fc	Constructed to overexpress and facilitate purification of proteins in insect cells for biophysical analysis of proteins	[12]
ATR-Fc	Anthrax toxin receptor (ATR) Fc construct – used for the prevention of anthrax infection	[13]
ALK1-Fc	Inhibits multiple mediators of angiogenesis and suppresses tumor growth. Activin receptor-like kinase-1 (ALK1) is a type I, endothelial cell-specific member of the transforming growth factor-β superfamily.	[14]
B7.1-Fc	Constructed and assessed for its anti-tumor potential when used alone or in combination with regulatory T cells	[15]
BP1700-Fc	Gamma1 (700) chimeric gene. Used for protection against gram-negative infection (*E. coli*) in high risk individuals	[16]
BR3-Fc	A soluble BAFF antagonist. Shown to block BAFF-mediated proliferation of cynomolgus monkey B cells	[17]
CD99-Fc	Use for the recognition of mycoplasma hyorhinis	[18]
CD22-Fc	Used to demonstrate that sialylated ligands for CD22 are expressed on sinusoidal endothelial cells of murine bone marrow	[19]
CD68-Fc	A promising new tool for preventing macrophage/foam cell formation. It offers a novel therapeutic strategy for patients with acute coronary syndrome by modulating the generation of vulnerable plaques.	[20]
CD47-Fc	CD47 antagonist – inhibits the production of pro-inflammatory cytokines in human dendritic cells (DCs). A murine CD47-Fc construct is shown to alleviate inflammation in a mouse model of rheumatoid arthritis.	[21]
CNTO 530	A 58 kD antibody Fc domain fusion protein, created using Centocor's MIMETIBODY platform. Stimulates erythropoiesis. Erythropoietin (EPO) mimetic peptides (EMPs) are novel peptides that mimic the actions of erythropoietin, fused to the Fc fragment of human IgG4. Produced sustained increases in red blood cell parameters in rats and rabbits, the serum half-life of CNTO 530 was 2 days in rabbits and 3 days in rats.	[22]
CEA-Fc	Carcinoembryonic antigen-IgG-Fc fusion protein – used for the activation and detection of anti-CEA designer T cells	[23]
CR2-Fc	C3-binding region of complement receptor type 2 linked to a complement-activating human IgG1 Fc domain, targets and amplifies complement deposition on HIV virions and enhances the efficiency of HIV lysis	[24]

Continued

Table 9.2 Continued

Protein	Applications	Ref
DNase-Fc	Inhaled into the airways where it degrades DNA to lower molecular weight fragments, thus reducing the viscoelasticity of sputum and improving lung function in cystic fibrosis patients	[4]
ELC-Fc	CKR-7 (ELC-Fc) is recommended for detection of the chemokine receptor family members (CKR-7) of mouse and human origin by immunofluorescence.	[25]
Epo-Fc	Effective in treating hematopoietic disorders such as anemia in chronic renal failure or due to blood loss	[26]
EGF-Fc	Recombinant epidermal growth factor receptor (EGFR) fused to Fc. Used for the construction of artificial extracellular matrices	[27]
FLSC R/T-Fc	Targets CCR5, the major co-receptor for HIV-1 during primary infection	[28]
FcRn-Fc	Used as a tool to analyze not only FcRn–IgG interactions but also other complex protein–protein interactions	[29]
FSH-Fc	Follicle stimulating hormone fusion protein for treatment of both male and female infertility	[30]
FIX-Fc	Under development for the treatment of hemophilia B (Syntonix, USA and Biovitrum AB, Sweden)	[31]
Factor IX-Fc:	Used for treatment of hemophilia B, contains a single Factor IX molecule attached to the Fc region of IgG	[32]
FVIII: Fc	Under development for the treatment of hemophilia A (Syntonix, USA and Biovitrum AB, Sweden)	[33]
GPV1-Fc	Glycoprotein V1, a major platelet collagen receptor, Fc construct controls the onset of pathological arterial thrombosis.	[34]
GLP-1-Fc	Intramuscular gene transfer of the plasmid in db/db mice demonstrated that expression of the GLP-1/Fc peptide normalizes glucose tolerance by enhancing insulin secretion and suppressing glucagon release.	[35]
hGHR-Fc	Human growth hormone-Fc fusion proteins under investigation for transcytosis induction	[36]
IL-2-Fc	Binds with high affinity to IL-2R and is capable of recruiting host Ab-dependent cell-mediated cytotoxicity and CDC activities, with a serum half-life of 25 h due to the Fc fusion	[37]
IL-15-Fc	IL-15-Fc mutant/Fcγ2a (CRB-15), targets the IL-15 receptor and has been shown to block delayed-type hypersensitivity	[38]

Table 9.2 Continued

Protein	Applications	Ref
IL-4-Fc	IL-4 promotes growth of both T and B-lymphocytes. IL-4-Fc inhibits the production of TNF, IL-1 and IL-6 by macrophages.	[39]
IL-10-Fc	Potential anti-inflammatory agent in the treatment of septic shock and in T-cell mediated autoimmune diseases	[40]
IL-18bp-Fc	Inhibits inflammatory bowel disease (IBD) and significantly inhibits the intestinal inflammation induced by DSS	[41]
IFNα-Fc	Human IFN-α2b linked to the Fc as a novel potential therapeutic for the treatment of hepatitis C virus infection	[42]
INFβ-Fc	Potential applications include the treatment of multiple sclerosis.	[42]
IP-10:Fc	Inducible 10 (IP-10) fused to mouse IgG2a. Binds specifically to CXCR3 and directs CDC and ADCC to CXCR3+ cells	[43]
IL-17R-Fc	Inhibits T cell proliferation and prolongs allograft survival in mice. Also blocks IL-17 in an animal model of arthritis	[44]
rILT3-Fc	Recombinant Ig-like transcript (ILT) 3. Potential to suppress the immune response in autoimmunity or transplantation	[45]
LIF05-Fc	A mutant form of human leukemia inhibitory factor (hLIF) with reduced binding affinity for LIF receptor gp130	[46]
MH35-BD-Fc	A mutant hybrid form of hLIF and murine LIF (mLIF), with reduced gp130 binding but increased LIF-R binding affinity	[46]
MOG-Fc	Composed of the extracellular Ig-like domain of human myelin oligodendrocyte glycoprotein (MOG) fused to Fc. Has been shown to reduce the number of circulating MOG-reactive B cells in an anti-MOG Ig heavy chain knock-in mouse	[47]
OX40L-Fc	Immunotherapeutic which enhances local tissue responses to and the killing of *Leishmania donovani*.	[48]
PD-L1:Fc	Prevents the development of transplant arteriosclerosis post-CD154 mAb therapy in a murine cardiac allograft model	[49]
PT-Fc	PT-Fc is designed on the basis of sequences derived from TNF-α heavy chain. It inhibits TNF-α induced apoptosis.	[50]
scFv-Fc	Specifically recognizes the epitope sequence of prion protein in ELISA. Used as a tool for the detection of mammalian prion	[51]
Syndecan-Fc	Exhibits a broad range of antiviral activity against primary HIV-1 isolates	[52]

Continued

Table 9.2 Continued

Protein	Applications	Ref
TrkB-Fc	Carrier for brain-derived neurotrophic factor and thus enhances the delivery or penetration of this polypeptide into the brain	[53]
TNFR-Fc	Used in the evaluation of a pseudotyped adeno-associated virus as a means for the delivery and distribution of soluble proteins	[54]
Tim-3:Fc	T-cell Ig- and mucin-domain-containing molecule 3. Used *in vitro* and *in vivo* to study development of Th1 responses	[55]
TACI:Fc	Total Anterior Circulation Infarct (TACI) scavenging B cell growth factor (BAFF). Alleviates ovalbumin-induced bronchial asthma	[56]
VCP-Fc	Vaccinia virus complement control protein used in xenograft transplantation methods	[57]
VEGF-Fc (KH902)	The extracellular domains of vascular endothelial growth factor (VEGF) receptors 1 and 2 and the Fc portion of IgG1. It has a high affinity for VEGF and inhibits the proliferation of human umbilical vein endothelial cells (HUVECs) induced by VEGF.	[11]

domain (Fab region), comprising VL (110 amino acids) and CL (110 amino acids), and a COOH-terminal constant heavy domain (Fc region), comprising CH2 and CH3, and VH (330–440 amino acids) and CH1 (110 amino acids) (Figure 9.1a). The two chains are held together by disulfide bonds in the hinge region. The Fc domain (the heavy chain) is the non-antigen binding part of the IgG molecule and has a flexible hinge region in between CH2 and CH1 forming a dimeric structure. The type of heavy chain determines the immunoglobulin isotype and defines the class of immunoglobulin.

The Fv portion of an antibody consists of the variable region of a heavy (VH) and a light chain (VL), and is the smallest fragment that maintains the binding specificity and affinity of the whole antibody.

The five major classes are IgM (μ), IgD (δ), IgG (γ), IgE (ϵ) and IgA (α). They share more than 95% homology in the amino acid sequences of the Fc regions, but show major differences in the amino acid composition and structure of the hinge region [58]. The secreted immunoglobulins, notably IgA, IgM, and IgG, provide a first line of defense against microorganisms. IgG is the most common of the five different types of antibodies found in the blood or other body fluids in humans and in most vertebrates.

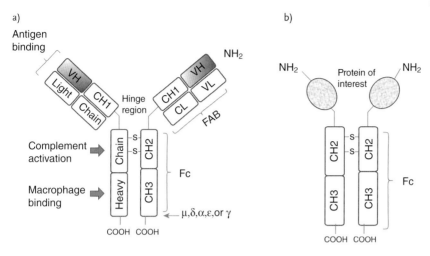

Figure 9.1 Schematic diagram representing (a) IgG general structure and (b) Fc fusion protein structure.

9.2.1
The Fc Region

The Fc (fragment, crystallizable) domain, which can be liberated from IgG by digestion with papain, is the main factor responsible for the half-life and distribution of the IgG in body fluids. It also interacts with cell surface receptors such as Fc receptors and complement proteins (Table 9.3).

The Fc regions of all IgG classes are somewhat similar and classified into five types. IgG which has several subclasses ($\gamma1$, $\gamma2$, $\gamma3$, and $\gamma4$) is the most abundant immunoglobulin in the serum (~70–80%). The Fc region of IgG has a half-life of about 22 days, compared with other immunoglobulin isotypes (2.5–6 days) and a much longer survival than Fab fragments [67, 68]. This is due to its strong binding affinity for Fc receptor (FcR). The IgG in complex with FcR is recycled to the cell surface where at physiological pH in the blood, it is released into the circulation. As a result, the Fc domain of IgG in particular, has been used to construct dimeric fusion molecules with many proteins, cytokines, and growth factors.

9.2.2
The Fc Receptor

The FcR is a protein found on the cell surface of many cells, which can deliver signals when activated. The different types of FcRs are classified according to the type of antibody they bind. A Latin letter is used to classify the type of antibody. The Fc receptors that bind IgG are called *Fc-gamma receptors* (FcγR), those that bind IgA, *Fc-alpha receptors* (FcαR) and those that bind IgE, *Fc-epsilon receptors*

Table 9.3 Fc receptors.

Receptor	Ligand (antibody)	Tissue/cell distribution	Function	Ref
High affinity (Kd $>10^{-9}$ M)				
FcεR1	IgE	Mast cells, Eosinophils	Tetrameric in structure, comprised of 45-kD α and 30-kD β chains and a homodimer of two disulfide linked 10-kD γ chains; it releases inflammatory mediators from cytoplasmic granules.	[59]
FcγRI (CD64)	IgG1 and IgG3	Macrophages, Neutrophils, Eosinophils	Contributes substantially to severity of arthritis, hypersensitivity responses, and protection from bacterial infection. 40 kD protein involved in phagocytosis, cell activation and induction of microbe killing. Three distinct forms in humans: FcγRIA (CD64A), FcγRIB (CD64B), and FcγRIC (CD64C)	[60]
Fcα/μR	IgA and IgM	B cells, Macrophages	Expressed on the majority of B-lymphocytes and macrophages in the spleen. The Fcα/μR mediates endocytosis of *Staphylococcus aureus*/ anti-*S. aureus* IgM antibody immune complexes by B-lymphocytes.	[61]
Low affinity (Kd $>10^{-5}$ M)				
FcγRIIA (CD32)	IgG	Macrophages, Neutrophils,	Soluble FcγRIIa inhibits rheumatoid factor binding to immune complexes (ICs).	[62]
FcγRIIB1 (CD32)	IgG	B Cells	Potent transducer of signals that block antigen-induced B cell activation. Contributes to enhanced calcium response in B cells from patients with systemic Lupus Erythematosus	[63]
FcγRIIB2 (CD32)	IgG	Macrophages	Controls antibody production by down-regulating cell activation, when co-clustered with B cell antigen receptors (BCR) *in vivo*, human B cells express FcγRIIb1 and FcγRIIb2, differing only in a 19 amino acid long insert in the cytoplasmic tail of the former.	[64]

Table 9.3 Continued

Receptor	Ligand (antibody)	Tissue/cell distribution	Function	Ref
FcγRIIIA (CD16a)	IgG	Macrophages	It is a transmembranous isoform of the receptor expressed on the surface of natural killer (NK) cells and macrophages, plays a significant role in the clearance of immune complexes, antibody- dependent cellular cytotoxicity of K/NK cells, phagocytosis and antigen presentation.	[65]
FcεRII (CD23)	IgE	B cells, Eosinophils	Found on mature B cells, activated macrophages, eosinophils, follicular dendritic cells and platelets	[66]
FcαR1 (CD89)	IgA	Monocytes, macrophages	FcαR1 triggering antibody-dependent cell-mediated cytotoxicity (ADCC). It plays important roles in the clearance of IgA in the circulation. Due to its specific effector functions, IgA has therapeutic potential for mucosal protection against viruses and bacteria.	[66]
FcRn	IgG	Macrophages, Neutrophils	Transfers IgG from mother to fetus across the placenta. Protects IgG from degradation	[66]

(FcεR) (Table 9.3). Immunoglobulin G (IgG) molecules bind three classes of FcR specific for the IgG class of antibody, namely FcγRI, FcγRII and FcγRIII, all of which are expressed by a variety of cells such as B-lymphocytes, monocytes and macrophages [66]. Fc receptors bind to antibodies that are attached to infected cells or invading pathogens. The nature of the response depends on the cell type. Every antibody class has an FcR of its own, which, depending on the cell surface receptor expression profile, triggers different cellular responses, such as phagocytosis, antibody-dependent cellular cytotoxicity (ADCC), antigen presentation, and cytokine production. For example, the Fc portion of Rituximab (a chimeric mono-clonal antibody against the protein CD20, found on the surface of B cells) mediates ADCC and complement-dependent cytotoxicity (CDC) [69].

Other Fc receptors include FcγR1 (CD64) and FcγRII (CD32), which bind to IgG1; FcαR1 (CD89) binds to IgA1; FcεR1 and FcεRII bind to IgE; FcμR binds IgM, and FcδR binds IgD [66, 70]. Each class of receptor has a unique ligand-binding chain (α-chain), which dimerizes with the FcR-γ-chain. When activated, FcR-γ generates signals within cells through a central activation motif known as immunoreceptor tyrosine-based activation motifs (ITAM).[1] In contrast to B cell receptors (BCRs) and T cell receptors (TCRs), FcRs are not involved in antigen recognition. In an antigen–antibody complex they only recognize the Fc portion of antibodies. Upon binding to FcR, antibodies give antigen-specificity to many types of cells, which do not have antigen recognition on their own.

The FcR generally exists in two forms; membrane bound and as a soluble molecule (sFcR). Examples of the former include receptors found on cells such as monocytes, macrophages, neutrophils and eosinophils (FcγR1; CD64) and FcγRII; CD32), natural killer cells (FcγRIII; CD16) and B cells (FcεRII; CD23). Fc receptors are classified as low and high affinity. High affinity (Kd > 10^{-9} M), referred to as FcR1 (with subclasses such as FcγRI, FcγRII, FcγRIII and FcδRIII) and low affinity (Kd > 10^{-5} M) such as FcRII. The diversity in the structure of the FcR allows the same ligand to induce a variety of biological responses.

Mutation studies have revealed that there are two regions of the CH2 domain, which are critical for FcγR and complement C1q binding. These two regions also have unique sequences in two subclasses, notably IgG2 and IgG4. For example Shields shows that substitution into human IgG2 (residues 233–236) and IgG4 (residues 327, 330 and 331) greatly reduces ADCC and CDC activation [72].

9.2.3
Fc-Mediated Antibody Functions and Their Optimization

Because the Fc region is the portion of the IgG molecule that interacts with neonatal Fc receptor (FcRn),[2] it has been targeted for the creation of mutations leading to increased half–life and better pharmacokinetic properties [73]. For example the amino acid sequences of the Fc region have been modified and glycosylated to allow tuning of effector functions, which include ADCC and CDC.

Also IgG2 mutants with increased binding affinity to human FcRn, at pH 6.0, have been generated by introduction of mutations at FcRn binding region positions 434, 433, 256, 254 and 252 [74]. These molecules are reported to have a serum half-life about twofold longer than the wild-type antibody. In addition, Hinton *et al.* [75] using molecular modeling, have generated several IgG mutants, at positions 250 and 428, generating T250Q and M428L respectively with three- and sevenfold increases in the IgG affinity to the FcRn and a twofold increase in its serum half-life compared with the wild type.

---

1) Immunoreceptor tyrosine-based activation motifs (ITAM), are short sequences with two consensus sequence (Tyr-X-X-Leu/Ile) repeats, separated by six to eight amino acids. They represent a unique module linking antigen and Fc-receptors to their signaling cascades. The complex of FcR without ITAM does not trigger cell activation [71].

2) Neonatal Fc receptor, the FcRn, for neonatal rat intestine, the tissue from which it was first cloned, mediates both immunoglobulin G (IgG) transport, providing immunity from mother to fetus, and IgG protection. In the acidic environment of endosomes the affinity of IgG for FcRn increases. FcRn interacts with the Fc region of IgG in a pH dependent fashion. While in complex with FcRn, IgG escapes intracellular degradation and enters the blood stream.

Improvements in the half-lives of Fc fusion proteins have also been achieved by changing the isotype of the human heavy chain C region from IgG1 (γ1) or IgG3 (γ3), which are high affinity binders, to those with reduced binding to FcR, for example, IgG4 (γ4) which is a low affinity binder [75, 76].

Examples of this approach include those described by Gillies *et al.* [77] who improved the half-lives of two different immunocytokines (fusion proteins between whole antibodies and cytokines) by changing the isotype of the human heavy chain C region from IgG1 or IgG3 to those with reduced binding to FcR, such as IgG4. They improved the serum half-life of an IL-2-Ab fusion protein by reducing its interaction with Fc receptor. This was achieved by changing the H-chain isotype of the human heavy chain C region from IgG1 to IgG4, which has low binding affinity for FcR. Such IgG4-based protein constructs, have been shown to have equivalent or better anti-tumor activity in mouse tumor models such as severe combined immunodeficiency mouse xenograft models. They then modified the sequences so as to increase the half-life while retaining the ability to activate Fc receptor-mediated effector functions.

Extending the serum half-life of proteins is not always a desired outcome. This is the case for example in antibody-mediated diseases, where a reduced IgG level is desired to improve safety or, in the case of some imaging applications, where a reduced serum half-life (from weeks to hours or minutes) is preferred. This is achieved by using small antibodies that lack a functional Fc region to prevent recycling by FcRn. Vaccaro *et al.* [78] have engineered the Fc region of human IgG to bind to FcRn with much higher affinity and with reduced pH dependence. They showed that such binding inhibits FcRn–IgG interactions and induces a rapid decrease of IgG levels in mice. Examples of such constructs include the integrin-specific Fab abciximab (ReoPro; Eli Lilly), an antithrombotic which works by preventing the clumping of blood platelets, and the VEGF-specific Fab ranibizumab (Lucentis; Genentech) for the treatment of neovascular (wet) age-related macular degeneration (AMD).

It may sometimes be advantageous to adjust the pharmacokinetic properties to achieve half-lives that are intermediate. This can be achieved by creating mutations at the Fc region that weaken the interaction with FcRn and decrease antibody half-life [79]. In fact, IgG molecules with specific mutations that prevent FcR binding have been generated and used in the treatment of autoimmune diseases and transplant rejection. Such constructs are capable of retaining the antigen specificity of the antibody, but are unable to initiate any undesired immune-cell reactions, such as neutrophil degranulation.

9.3
Strategies to Increase Cytokine Serum Stability and Half-Life

Several Fc-based strategies have been used to enhance the therapeutic efficacy of proteins/cytokines and to reduce their frequency of administration. These include:

9.3.1
Fc-Fusion Dimeric

This is basically the fusion of the protein of interest to the Fc region of an antibody molecule such as immunoglobulin G1 (IgG1). In this strategy the antibody combining region (Fab) of the antibody molecule is replaced by the protein of choice, while the Fc constant region and the hinge regions are retained (Figure 9.1b). The Fc region, in addition to providing immune effector functions, has conformational flexibility as well. Retention of the flexible hinge region ensures that proper folding of the construct is achieved.

Use of the Fc region has been shown to enhance the serum half-life of the protein by several fold *in vivo* and to be highly efficacious in the treatment of immuno-inflammatory diseases (42). Early examples of such Fc constructs have been reported by Peppel *et al.* [80] and include the tumor necrosis factor or TNF antagonists, in which the TNF receptors (p55-kD and p75kD) were fused to Fc using multistep PCR. These constructs were later developed as lenercept and etanercept (Enbrel™) respectively with a view to blocking the proinflammatory activity of tumor necrosis factor [5]. Etanercept is the first example of a soluble Fc fusion protein to be used as a therapeutic drug [81]. It neutralizes both membrane bound and the soluble form of TNF-alpha. Etanercept was approved by the FDA in the USA in 1998 for the treatment of rheumatoid arthritis and subsequently for the treatment of other forms of arthritis such as ankylosing spondylitis and psoriatic arthritis. The Fc component of etanercept contains the CH3 domain and the CH2 domain and hinge region, but not the CH1 domain (43).

Another example is alefacept, a dimeric fusion protein consisting of the extracellular CD2-binding portion of the human leukocyte function antigen-3 (LFA-3) linked to Fc. Alefacept or LFA-3-Fc, reduces disease expression in patients with chronic plaque psoriasis. It binds to CD2 on memory-effector T cells, inhibiting their activation and also reducing the number of these cells by Fc-dependent cytotoxicity.

There are many similar constructs, among them IFNα-Fc, which is used in the treatment of hepatitis B and C (Table 9.4). In its native form, IFNα must be administered frequently due to its short half-life, however once fused with the Fc portion of IgG1 its serum half-life becomes 20-fold greater [84].

9.3.1.1 **Protein Domains Fused to Fc**
Many proteins consist of several structural domains, which vary in length (20–500 amino acids) to form multifunctional proteins. Generally two protein domains can be fused with the Fc fragment. To construct a fusion protein, the Fc region should be inserted into the protein of interest at a point which has minimum impact on its function. These include the cytoplasmic and extracellular domains to which the Fc constant fragment can be fused to either N- or C-terminus. In the case of type 1 and type II transmembrane proteins (single pass molecules), only the extracellular domain of cell receptor proteins, rather than the transmembrane and

Table 9.4 Vectors for the construction of Fc fusion proteins.

Vector	Application	Ref
pFUSE-Fc	For the construction of Fc fusion proteins. Available with several isotypes with distinct properties. Fuses a sequence encoding a given protein of interest to the Fc region of an immunoglobulin. This region comprises the CH2 and CH3 domains of the IgG heavy chain and the hinge region.	InvivoGen
Fc-X (pdCs-Fc-X)	General expression system that enhances the production and secretion of proteins in mammalian cells. The protein of interest is expressed as a fusion to a signal peptide and the Fc fragment of immunoglobulin as the N-terminal fusion partner.	[82]
pPIC9-Fc	A vector system based on pPIC9- Fc, in which the hinge, CH2 and CH3 domains (Fc fragment) of mouse IgG1 and His-tag were cloned into the *Pichia* expression vector pPIC9.	[83]

cytoplasmic domain, is inserted *in frame* into the cDNA encoding Fc region of the protein of interest. Examples of such constructs include the CD47: Fc which is a soluble recombinant CD47-Fc fusion protein that is made up of the extracellular domain of native or mutated human CD47 or viral CD47 linked to the Fc portion of human IgG. On the other hand for secreted proteins, which typically have an N-terminal hydrophobic signal sequence (3–60 amino acids long peptide) that facilitates protein translocation into the endoplasmic reticulum (ER), the IgG domain should be inserted at the C-terminus of the fusion protein, since the N-terminus secretory signals are usually removed from the mature protein.

9.3.2
Fc-Fusion Monomeric

There are disadvantages associated with increasing the molecular weight of proteins by fusion with Fc fragments of IgG in its dimeric form. This is because there is an inverse relationship between the size of the molecule and the rate at which it can diffuse through mucus. Thus large dimeric Fc fusion constructs potentially have a lower rate of diffusion. Consequently second-generation Fc fusion molecules were developed which were monomeric with respect to the therapeutic protein and dimeric with respect to the Fc region (Figure 9.2a). Such constructs have been shown to result in an even greater extension of the serum half-life, with improved pharmacokinetic parameters for protein therapeutics, and have been extensively reviewed [42].

Several monomeric Fc–cytokine constructs have been reported. These include an erythropoietin–Fc fusion molecule (Epo-Fc) which was constructed to reduce the size and charge of the dimeric molecule and was shown to have about a twofold increase in binding affinity to FcRn compared with Epo-Fc in its dimeric form [85]. Another example is the factor IX (FIX) protein, used for treatment of Hemophilia B, which has been attached to the Fc region creating rFIX-Fc, The half-life of rFIX-Fc was determined to be approximately three- to fourfold longer than that of rFIX in all species tested [32]. Other similar monomeric constructs include IFNα-Fc and IFNβ-Fc both of which have been shown to have increased efficacy compared with their dimeric counterparts [42]. More examples of the Fc fusion-based drugs on the market are shown in Tables 9.3 and 9.4.

9.3.3
Fc-Peptide Fusion Protein (Peptibody)

In addition to the application of Fc fusion proteins as antagonists, many peptides, being potent agonists, are also good therapeutic targets. For example Romiplostim (NPlate®), which is a peptide mimic of thrombopoietin (TPO) fused to human Fc has been developed by Amgen for use as an antihemorrhagic agent. It activates intracellular transcriptional pathways leading to increased platelet production via the TPO receptor. Nplate is used for the treatment of thrombocytopenia in patients who do not respond to corticosteroids, immunoglobulins or splenectomy [86, 87].

9.3.4
Other Antibody-Engineered Constructs

These include whole antibody fusions which are constructed by fusing the full length of an antibody molecule, at either the carboxyl or the amino-terminus of each H chain, to non-IgG ligands (Figure 9.2d,e). These constructs can be used as specific delivery vehicles to selectively target a metastatic tumor and deliver an immunostimulatory molecule like a cytokine or growth factor. Examples of this approach include (i) construction of B7.her2.IgG3 in which the T-cell stimulatory ligand B7.1 is fused to an anti-HER2/neu IgG3 at the N-terminus of the heavy chain [88]. This construct has been shown to be an effective therapeutic agent targeting metastatic breast cancer through the HER/neu antigen.

9.3.4.1 Fab Fusions
Fab fragments encompass one constant and one variable domain of each of the heavy and the light chains respectively (VH, CH1 and CL, VL) fragments. The molecular weight of the heterodimer is usually around 50 kDa. Fab fragments can be prepared by papain digestions of whole antibodies. Fab fusion protein immunoligands are molecules in which the variable region is retained and the fusion partner (cytokine or growth factor) replaces the Fc domain (Figure 9.2c). Examples

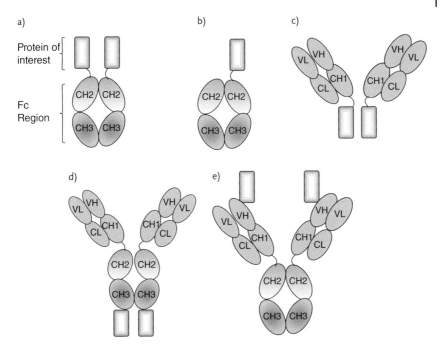

Figure 9.2 Antibody fusion proteins. Schematic diagram depicting the protein of interest (rectangles): (a) as a dimer fused to the Fc part of the antibody with antigen-binding regions replaced; (b) as a monomer fused to the Fc part of the antibody with one antigen-binding region replaced; (c) as a dimer fused to a bivalent Fab (pepsin fragment) [F (ab')₂]; (d) as a dimer fused to the whole antibody at the carboxyl-terminus; and (e) as a dimer fused to the whole antibody at the amino-terminus.

include Fab-IL-8 in which the 2–72 amino acid form of human IL-8 was fused to a Fab fragment with specificity for the human epidermal growth factor receptor (hEGF-R) [89].

9.3.4.2 Antibody without the Fc Region (Diabodies)

Diabodies, similar in size to Fab fragments, are small bivalent antibody fragments comprising two polypeptide chains and two Fv fragments. Such constructs, which are without the Fc fragment, lack the undesirable side effects associated with the Fc domain. In addition, due to their smaller size they can penetrate tumor cells more efficiently than the whole antibody. Furthermore, they can be coupled to drugs, thereby facilitating drug delivery to tumor targets [90]. The main disadvantage of this approach is the rapid clearance of these smaller molecules. However, since they lack the Fc receptor and hence the capacity to recruit natural immune functions, they can be designed to have effector domains that target specific tumors. (e.g., anti-CDs). Applications include diabody-based molecules with affinity for HER2/neu, which can be used in the radioimmunotherapy (RAIT) of breast cancer and vaccination against the HER-2/neu oncogenic protein [91, 92].

9.4
Methods to Construct Fc-Fusion Dimeric Proteins

9.4.1
Polymerase Chain Reaction (PCR) Approach

Most commonly the protein of interest is covalently linked to the Fc domains CH2 through CH3 but not CH1. The Fc domain is inserted at either the N- or the C-terminus of the protein depending on which end is free. This is achieved by two-step polymerase chain reaction (PCR). In step 1 the cDNA fragment of the protein of interest and the Fc fragments encoding the hinge region, CH2 and CH3 domains of the IgG heavy chain, are separately amplified with appropriate restriction sites incorporated into them. In the second step the two fragments are mixed together (with forward and reverse primers corresponding to cDNA of the protein of interest and the Fc fragment respectively) and amplification is carried out. Fc forward and reverse primers we have successfully used include, 5′ CTG GTT CCGCAT GGA TCCGAG CCC AGA TCT TGT GAC 3′; and, 5′ GCG AAT TCT CAT TTA CCC GGA GAC AG 3′ respectively. A thrombin cleavage site is also incorporated into the amplified region (underlined).

9.4.2
Fc -Plasmid Vectors

Plasmid vectors available for the construction of Fc fusion proteins include:

i) **pFUSE-Fc (InvivoGen):** This is an expression plasmid with two promoters, EF1 prom/HTLV 5′UTR driving the Fc fusion and cytomegalovirus (CMV) enh/FerL prom driving the selectable marker Zeocin™. This region comprises the CH2 and CH3 domains of the IgG heavy chain and the hinge region. The hinge serves as a flexible spacer between the two parts of the Fc fusion protein, allowing each part of the molecule to function independently. Several variations of the plasmid are constructed; notably pFUSE-Fc1 containing a native signal sequence and pFUSE-Fc2 containing an IL-2 signal sequence for the generation of Fc fusion constructs derived from proteins that are not naturally secreted (Table 9.1). Advantages of such recombinant constructs include ease of detection in the supernatant of the transfected cells by SDS-PAGE and Western blot analysis and ease of purification by single-step protein A or protein G affinity chromatography. In addition the Fc regions can be obtained from various species. Human Fc: IgG1, IgG2, IgG3, IgG4; Mouse Fc: IgG1, IgG2a, IgG2b, IgG3; Rabbit Fc: IgG and Rat Fc: IgG2b (Invivogen). In addition to increased half-life, the Fc regions can be engineered to have increased or reduced effector functions (ADCC and CDC).

ii) ***Fc-X fusion vector pdCs-Fc-x*** made by EMD Lexigen Pharmaceuticals (Billerica, MA, USA) [82] . This vector has over 30 subtypes. The expression in pdCs utilizes the enhancer/promoter of the human CMV and the SV40 polyadenylation signal. The vector contains the mutant dihydrofolate reductase gene as a selection marker. For mammalian expression systems constitutive promoters such as CMV and SV40 are most commonly used in contrast to the *E. coli* expression system in which most expression vectors have inducible promoters such as IPTG. The protein of interest is expressed as a fusion to the N-terminus region of the Fc fragment of IgG. It is secreted into the culture medium due to the inclusion of a signal peptide (a short 3–60 amino acid long peptide). Up to $100\,mg\,mL^{-1}$ is produced in permanent cell lines and in conditioned media [82].

iii) ***pPIC9-Fc*** is another vector in which the hinge, the Fc fragment of mouse IgG and His-tag are cloned into the *Pichia* expression vector pPIC9. The vector has been used for the production of single chain Fv (ScFv-Fc) [83]. In this construct the scFv fragment was introduced into pPIC9-Fc, which can bind glutathione-S-transferase (GST) from *Schistosoma japonicum,* to yield the expression cassette pPIC9-ScFv-Fc [83].

9.4.3
Design Considerations

9.4.3.1 Choice of Linkers
Linkers are used to join the Fc fragment and the protein of interest together and to allow the whole construct to be separated from an affinity chromatography column in the purification steps. They are also required to adopt an extended protein conformation and allow structural flexibility. Several linkers can be used to join the Fc fragment with the protein of interest. These include the serine protease, thrombin, and serine endopeptidase, Factor Xa. Thrombin is a linker in the form of protease cleavage and it recognizes the sequence Leu-Val-Pro-Arg-Gly-Ser, commonly referred to as thrombin cleavage site (TCS), and cuts the peptide bond between Arg and Gly. Factor Xa is often used to remove the common histidine tag, from expressed proteins. It cleaves after the last amino acid in the sequence Ile-Glu-Gly-Arg.

The purpose of such linkers is to selectively cleave the fusion constructs, for applications where the Fc/His tags are not required and are only used for protein purification purposes. After purification of the fusion protein, the Fc/His domains can be cleaved off while still immobilized on an affinity chromatography purification column. Chiquito *et al.* [93] have developed a program called LINKER which can generate a set of linker sequences capable of adopting extended conformations as determined by X-ray crystallography and NMR. The program searches a loop library derived from the Brookhaven Protein Data Bank.

9.4.3.2 **Codon Optimization**

Upon cloning and construction of the Fc fusion proteins and before proceeding to large-scale production, in order to maximize protein expression in a correct and biologically active form, it is common practice to "optimize" the DNA sequence-encoding gene for the protein of interest. In particular, highly repeated sequences, long sequences, low GC content and sequences with large numbers of parameters need to be evaluated. Codon optimization involves modification of a variety of critical factors related to the different stages of protein expression, such as codon adaptability, mRNA secondary structure, and various negative *cis-acting* elements in the transcription and translation processes. To do this, the complete DNA sequence and the DNA fragment coding the protein of interest, cloned into a plasmid vector, are sent for optimization to a commercial company with the necessary software and gene synthesis facilities. In addition, codon optimization can be tailored to a particular host (e.g., insect cell Sf9) for maximum expressions. Companies offering gene design system and DNA engineering include Gene-Script's OptimumGene™ (Piscataway, NJ, USA) and GENEART AG, Regensburg (Germany). In our experience codon optimization is a worthwhile practice for the large-scale protein production that is required for testing in both *in vitro* and *in vivo* models of diseases.

9.5
Choice of Host for Expression

Having constructed and optimized the codon sequence, the next step is to decide on the choice of a host for the production of the fusion protein. Options for this purpose include prokaryotes such as *E .coli* and mammalian cells such as CHO, COS-7, yeast or insect cells. Below we will highlight some of the advantages and disadvantages of each.

9.5.1
Expression in Bacteria

The gram-negative bacterium *E. coli* is a good expression host, which, in shake-flask cultures, can produce several milligram of fusion protein per liter of culture medium. Simmons has produced large amounts of IgG in *E. coli* [94]. However, expression in *E. coli* has its drawbacks, especially when the proteins are large (>30 kDa). One disadvantage of *E. coli* as host is that proteins <5 kDa and >80 kDa, including multimeric proteins which require post-translational modification for activity are more difficult to express efficiently. In addition, the disulfide bonds, formed between two cysteine residues in an expressed protein, are produced inefficiently in the reducing environment of the *E. coli* cytoplasm. Furthermore, using the *E. coli* system for the expression of a protein of interest can lead to the formation of inclusion bodies (insoluble protein aggregates) which can complicate the purification step and reduce the overall yield [95].

To reduce protein degradation, strains carrying mutations which eliminate the production of proteases can sometimes enhance accumulation by reducing proteolytic degradation. One example is a strain of *E. coli* (BL21), which is deficient in two proteases encoded by the lon (cytoplasmic) and ompT (periplasmic) genes. Other prokaryotic expression systems are also available and include several gram-positive strains such as *Bacillus subtilis* [96] and *Streptomyces lividans* [97].

9.5.2
Expression in Mammalian Cells

Mammalian cell expression can be undertaken in cells either in monolayers or in suspension cultures. The latter is more efficient in terms of protein production per ml of culture medium. This is particularly evident with use of the commercially available FreeStyle™ 293 (Invitrogen) system. Using this system, one can grow and transfect 293 cells in suspension, eliminating the need to carry multiple flasks of adherent cells when large numbers of cells are required. The FreeStyle 293 System comes with a transfection reagent (293fectin™) for the transfection, in suspension, of the 293 cells and a serum-free 293-expression medium (Gibco™ FreeStyle™). For large-scale protein expression within a short period of time, the FreeStyle 293 system is a good choice. Since large-scale protein expression requires large-scale fermentation vessels and CO_2 shaking incubators, it is best to pool resources with other laboratories with such facilities and expertise. This avoids unnecessary trial and error experiments and optimizes protein production.

9.5.3
Expression in Insect Cells

Expression systems based on insect cells for example, the Bac-to-Bac® or BaculoDirect™ Baculovirus Expression systems, from Invitrogen, are good and efficient options. Insect cells (Sf9 or Sf21) yield relatively high levels of protein because they can be grown in large scale in suspension cultures. In addition, protein expression in insect cells is accompanied by post-translational modification, although not as well as in mammalian expression systems. Protein expression in insect cells may be either intra- or extracellular. If extracellular expression is desired, the addition of a signal peptide sequence in the expression vector will result in the secretion of the protein into the growth medium. The Baculovirus expression system however requires several cloning and subcloning steps into specially designed plasmid vectors, which are time consuming. In addition, plaque assays need to be performed to confirm protein expression, a process which requires some expertise to achieve good results. To avoid these steps, one can use a mammalian expression system in which an Fc-based expression vector can be directly transfected into mammalian cells grown in suspension cultures such as FreeStyle 293 as outlined in Section 9.5.2 above.

9.5.4
Expression in Yeast

Another attractive option, especially for large-scale production of proteins, is to use lower eukaryotes such as yeasts. Protein expression using this single cell eukaryotic organism is a well-established technology, which is already used for large-scale production of several recombinant proteins. Furthermore, yeasts are genetically manipulable and regarded as safe. For more information concerning the expression of antibody fusion proteins by yeasts and fungi, refer to the review article by Joosten *et al.* [98].

9.6
Purification of Fc Fusion Proteins

Fusion proteins containing the Fc fragment from the human $\gamma 1$, $\gamma 2$, or $\gamma 4$ heavy chains, can easily be purified with affinity chromatography using either staphylo-coccal protein A, produced by *Staphylococcus aureus* or streptococcal protein G affinity matrix. Each isotype binds either protein A or G with different degrees of efficiency [99]. If protein is expressed intracellularly, the cells are centrifuged, lysed and the supernatant is then filtered (0.2 µm) to remove contaminants and passed through the appropriate chromatography column. Generally speaking the basic principle is to bind at neutral pH and to elute at acidic pH. The drawback with protein A is that that this approach results in binding of all IgG containing proteins indiscriminately. To avoid or minimize this effect, the media used in the expression protocol can be carefully formulated to decrease the presence of any IgG type molecules (e.g., serum free media). Low pH (<3.0) is normally used for elution of the bound Fc-fused proteins from the affinity matrix, potentially damaging the protein of interest. However, specially formulated buffers are now available commercially which can be used to protect proteins in a low pH environment.

Alternatively elution at neutral pH with mild chaotropic salt may be used. A number of studies have recently been performed to examine the underlying mechanism whereby various antibodies and Fc fusion proteins bind to protein A. As a result many differences in binding capacities have been observed indicating that not all human IgG molecules bind to protein A with the same affinity. It should be noted that murine IgG interacts poorly with protein A and also with protein G. To address this issue, selected amino acid residues in the mouse IgG Fc region (T252M, T254S, and T252M-T254S) were replaced with the corresponding human IgG amino acid residues [100]. This interchange resulted in a significant improvement in the affinity of the modified murine IgG for protein A. Protein modifications have also been very useful in the production of chicken antibodies with a higher affinity for protein A. In their native configuration chicken antibodies have a very low affinity for protein A. However, once fused with the IgG-Fc fragment, they can be purified to a high degree.

9.7
Demonstration of Biological Activity in Fc Constructs *In Vitro* and *In Vivo*

Once the production and purification steps are completed it is essential to show that the Fc fusion proteins are folded correctly and that they are biologically active. This is achieved in several ways. A SDS-PAGE should first be run to confirm protein purity, followed then by a Western blot analysis, which confirms protein identity using an antibody against the protein of interest or an anti-IgG antibody. The Fc region serves as a convenient molecular tag sequence for the identification and purification of fusion proteins with varying specificities. Reducing and non-reducing gels can also be used to check the proteins in their monomeric and dimeric configurations. The successful completion of the purification step using protein A is an indication that the Fc region has folded correctly, because it is recognized by antibodies and by protein A. Additionally, the protein would be expected to retain important effector functions characteristic of antibodies, such as binding to Fc receptors.

Biological activities can be determined in several ways depending on the nature of the protein. If the protein has enzymatic activity, it can be tested for this activity using an appropriate substrate or an ELISA assay. Some biological activity tests are undertaken using cell-based proliferation assays using specialized cell lines.

9.8
Pharmacokinetics

In vivo pharmacokinetic (PK) experiments are usually carried out to determine the serum stability and half-life of the fusion proteins. This is normally done in the mouse by injecting the fusion protein subcutaneously, taking blood samples at specified time intervals (1–48 h) and assaying the samples with an ELISA kit for the presence of the protein of interest.

It is worth noting that many fusion proteins are unable to fold correctly during overexpression and instead form insoluble aggregates in the form of tetramers or hexamers rather than the expected homodimers. This affects their PKs. One important consequence is a decrease in the serum half-life of the fusion protein, which is exactly the opposite result to that desired, since one of the main reasons for constructing fusion proteins is to increase the half-life of the protein of interest [80] [R. Lumry and H. Eyring, *J. Phys. Chem.* 58 (1954), p. 110. **Full Text** via Cross-Ref | View Record in Scopus | Cited By in Scopus (323) [80]]. Possible strategies to reduce aggregate formation include the culture of cells at reduced temperatures or to increase the activity of molecular chaperones; that is, proteins adapted to facilitate protein folding thereby stabilizing the correct folding of fusion proteins. Such options, although not entirely satisfactory, can in some cases, increase the yield of correctly folded overexpressed proteins and in turn achieve an increased serum half-life.

9.9
Applications of Fc Fusion Proteins

9.9.1
Therapeutic Proteins

Due to their increased serum-half-life, which in turn results in reduced frequency of administration, Fc constructs have become a very popular alternative to such techniques as PEGylation. Moreover, in practice, they have been found to be highly effective therapeutics (Table 9.3). Examples of such constructs are increasing every year. For example, antagonists of tumor necrosis factor-α (TNF-α) such as etanercept (TNFR-1 [p55]-Fc), have been shown to be highly effective for the treatment of rheumatoid arthritis, ankylosing spondylitis, and psoriatic arthritis [5] (see Section 9.3.1). Another example is a modified human IFN-α-2b linked to the Fc portion of IgG1. This construct has therapeutic potential for the treatment of hepatitis C virus infection (Table 9.4).

9.9.2
Protein/Cytokine Traps

Fc constructs have been used in the production of "cytokine traps". The methodology is based on the fact that some cytokines have two binding domains, each of which can bind to a separate receptor chain. As a consequence of this dual receptor chain interaction the cytokine binds much more tightly to the two-receptor chain complex than it does to a single protein receptor chain. Rilonacept (Arcalyst®) (IL-1 trap) for example is an engineered dimeric fusion protein consisting of the ligand-binding domains of the extracellular portions of the human IL-1 receptor (IL-1RI) and IL-1 receptor accessory protein (IL-1RAcP) linked in-line to the Fc portion of human IgG1 [101]. Rilonacept was approved as an orphan drug for treatment of cryopyrin-associated periodic syndrome (CAPS), one of the rare auto-inflammatory syndromes. Rilonacept has a serum half-life of eight days and can thus be administered once a week. This is considerably longer than the half-life for anakinra (4–6 h), which is a rhIL-1R antagonist.

Another example of an Fc-based cytokine trap is IL-6Rα-Fc: gp130-Fc, which is a heterodimeric trap created by fusing the constant region of IgG and the extracellular domains of two distinct cytokine receptor components (IL-6Rα and gp130) [101]. This construct has the potential to antagonize IL-6 and possibly other members of the IL-6 cytokine subfamily.

9.9.3
Gene Therapy

For expression of viral interleukin 10 (vIL-10) *in vivo*, Adachi *et al.* developed a plasmid vector in which vIL-10 was fused with the IgG1–Fc fragment creating vIL-10-Fc. The construct was delivered into mouse muscle via electroporation [102]. The fusion construct was expressed and its concentration in serum was shown to be 100-fold higher than non-fusion vIL-10. The Fc construct also was shown to suppress phytohemagglutinin-induced IFN-δ expression by human peripheral blood mononuclear cells.

Glucagon-like peptide (GLP) is a physiological incretin with many important roles. It has been proposed for the treatment of type II diabetes; however one of its drawbacks is that in its native form it has a very short half-life. To overcome this limitation, the molecule was fused to IgG1-Fc generating GLP-Fc. Upon intramuscular injection of the plasmid vector into db/db mice, it was shown that gene transfer of the GLP-1-Fc construct led to normalized glucose tolerance by enhancing insulin secretion and suppressing glucagon [35].

9.9.4
Drug Delivery

The neonatal Fc receptor (FcRn) is used to carry aerosolized therapeutic proteins conjugated to the Fc domain and has been shown to be able to carry proteins across epithelial cells of the lung, successfully delivering protein molecules to the bloodstream.

Pulmonary delivery of an erythropoietin–Fc fusion molecule (Epo–Fc) has also been demonstrated in non-human primates using the FcRn pathway [103]. This pathway also delivers antibodies across epithelial cell barriers that are present in the lungs and intestines. Examples of applications of such methods include Syn-Fusion™ and Transceptor™ drugs. SynFusion technology fuses the Fc region of an antibody to a drug, proteins, peptides, small molecules, and antisense, resulting in active receptor-dependent uptake of these drugs. This facilitates the recirculation of proteins through the FcRn pathway, and thus the extension of their serum half-life. Transceptor technology uses the FcRn transport pathway to enable the pulmonary delivery of Fc fusion drugs.

9.9.5
Research Tool

Fc fusion proteins have been used to (i) identify novel receptor ligand interactions, (ii) identify binding determinants and (iii) study protein function.

i)   *Identification of novel receptor ligand interactions.* Fc fusion constructs have been used to identify cells or tissues expressing molecules that bind the non-Fc portion of the fusion protein for example, cell adhesion molecules like selectins or CAMs and CD44, which are involved in cell–cell interactions and cell adhesion. CD44-Fc has been used to show that the cell–surface glycoprotein CD44 is a receptor for the glycosaminoglycan known as hyaluronan [104]. Also CD6-Fc was used to identify its ligand (ALCAM) [105].

ii) *Identification of binding determinants.* Fc fusion proteins with mutations have been used to locate residues involved in binding. CD22, CD40 and gp39 for example were used to identify key residues in receptor/ligand interactions [106].

iii) ***Study of protein function*** [37].
In an attempt to block initial HIV virus propagation, Binley *et al.* [6] fused wild-type gp120 (a glycoprotein with spikes that project outwards from the HIV virus particle and which bind to CD4 in human cells) and several variable loop-deleted gp120s, to the IgG1-Fc domain generating gp120-Fc. They then used these constructs to probe the interaction of the HIV envelope (Env) with different cellular receptors believed to control virus attachment (CD4, CCR5, DC-SIGN and syndecans). The inhibitory properties of various anti-Env mAbs were investigated through inhibition of Env binding to cell lines expressing these receptors. With these binding assays it was possible to show that a broadly neutralizing mAb (2G12), directed to a unique epitope of gp120, could block the binding of Env to CCR5 and partially block Env-DC-SIGN binding, however there was no effect on ENV-syndecan.

9.9.6
Tumor Targeting

Fusing proteins with whole antibodies and cytokines (immunocytokines) such as interleukin 2 (IL-2) has been shown to be effective in several mouse tumor models. Recently Liu *et al.* [15], fused the extracellular domains of a T-cell co-stimulatory ligand from the B7 family (B7.1) with the IgG1-Fc. The B7.1-Fc was shown to have potency in stimulating immune responses and to be an effective antitumor agent, particularly in combination with regulatory T-cell depletion. The antitumor properties of B7.1 have also been exploited in a similar construct, notably an Ab fusion protein in which B7.1 was fused with an anti-HER/neu IgG3 Ab generating B7.her2.IgG3. This fusion construct targets breast cancer cells in which her2/neu is overexpressed [88].

Another example is aflibercept (VEGF Trap), which is a fully humanized recombinant fusion protein. Aflibercept is composed of the second Ig domain of vascular endothelial growth factor receptor 1 (VEGFR1) and the third Ig domain of vascular endothelial growth factor receptor 2 (VEGFR2), both of which are fused to the Fc region of human IgG1. Aflibercept binds to all VEGF-A isoforms as well as placental growth factor (PlGF), a homodimeric glycoprotein and an angiogenic factor of 46–50 kDa in size, belonging to the vascular endothelial growth factor (VEGF) sub-family, thereby preventing these factors from stimulating angiogenesis.

9.10
Immunogenicity

Immunogenicity of Fc fusion proteins is largely dependent on their structural similarities to native proteins. It is possible that new epitopes in the Fc fusion proteins can be recognized as foreign by the immune system. An immunogenicity survey on current Fc-based cytokines in the market, such as etanercept (TNF-a75-Fc) for example, which is a fully human molecule, has shown limited immunogenicity. In a European study in which 500 patients with rheumatoid arthritis were treated with etanercept, only 16% tested positive for anti-etanercept antibodies at any point in time during the 5-year follow-up, with only 5% being consistently positive at all visits [107]. On the other hand, little is known about antibodies to abatacept (CTLA4: Fc), however, to date, relatively few infusion reactions have been reported. Since Fc-based constructs represent a relatively new class of therapeutics, longer-term observations are still required to determine the clinical significance of their immunogenic potential.

Increased rates of infections with micro-organisms including reactivation of latent tuberculosis and diverse opportunistic infections have been observed with TNFα-receptor-Fc constructs [108]. In addition, cutaneous malignancies, but so far not visceral malignancy, possibly increased rates of hematologic malignancy, and rare allergic and even anaphylactic reactions have also been observed with these agents [109]. Demyelination syndromes have been reported in approximately 5–10 per 100 000 patients receiving TNF blockers [109].

Three deaths due to progressive multifocal encephalopathy (PML) have been reported in multiple sclerosis (MS) patients treated with Natalizumab (an α4, β1, integrin-specific mAb) when this drug was used in combination with Interferon β-1a [110]. Thus as with all therapeutics that offer the prospect of significant clinical improvement, the risk/benefit ratio must be weighed and patients should be carefully evaluated for increased risk, screened appropriately and informed in advance of the potential risks to which they will be exposed.

9.11
Conclusion

Fc-based fusion proteins have an increased serum half-life compared with the native or recombinant protein of interest, which in turn results in a reduction in the frequency of drug administration and thus simplifies therapeutic administration. Accordingly the development of this technology represents a major advance in drug delivery. However, like any technology, this approach has its limitations, including the restriction of targets to those in the circulation/extracellular fluid phase and those on the surface of cells. The latter limitation may be overcome by the production of fusion proteins, which are expressed within cells, so called *intrabodies* (intracellular antibodies). Other limitations include high manufacturing costs, poor stability, incorrect folding due to conformational changes in the structure of the fusion protein, immunogenicity and unforeseeable adverse reactions. For the most part, Fc-based fusion proteins and other biologic therapies have been shown to confer significant clinical benefits with generally low risk for serious adverse reactions. As the tools of protein and antibody engineering become increasingly widespread and more sophisticated, it is likely that there will be increased opportunities to develop new, potentially safer and yet more effective derivatives of this technology for an expanding range of clinical applications.

References

1 Jazayeri, J.A., and Carroll, G.J. (2008) Fc-based cytokines: prospects for engineering superior therapeutics. *BioDrugs*, **22**, 11–26.

2 Greenwald, R.B. (2001) PEG drugs: an overview. *J. Control. Release*, **74**, 159–171.

3 Huang, C. (2009) Receptor-Fc fusion therapeutics, traps, and MIMETIBODY™ technology. *Curr. Opin. Biotechnol.*, **20**, 692–699.

4 Dwyer, M.A., Huang, A.J., Pan, C.Q., and Lazarus, R.A. (1999) Expression and characterization of a DNase I-Fc fusion enzyme. *J. Biol. Chem.*, **274**, 9738–9743.

5 Feldmann, M. (2002) Development of anti-TNF therapy for rheumatoid arthritis. *Nat. Rev. Immunol.*, **2**, 364–371.

6 Binley, J.M., Ngo-Abdalla, S., Moore, P., Bobardt, M., *et al.* (2006) Inhibition of HIV Env binding to cellular receptors by monoclonal antibody 2G12 as probed by Fc-tagged gp120. *Retrovirology*, **3**, 39.

7 Meier, W., Gill, A., Rogge, M., Dabora, R., *et al.* (1995) Immunomodulation by LFA3TIP, an LFA-3/IgG1 fusion protein: cell line dependent glycosylation effects on pharmacokinetics and pharmacodynamic markers. *Ther. Immunol.*, **2**, 159–171.

8 Davis, P.M., Abraham, R., Xu, L., Nadler, S.G., and Suchard, S.J. (2007) Abatacept binds to the Fc receptor CD64 but does not mediate complement-dependent cytotoxicity or antibody-dependent cellular cytotoxicity. *J. Rheumatol.*, **34**, 2204–2210.

9 Hendeles, L., Asmus, M., and Chesrown, S. (2004) Evaluation of cytokine modulators for asthma. *Paediatr. Respir. Rev.*, **5** (Suppl. A), S107–S112.

10 Forrest, K., Logan, B., Strange, J., Roszman, T.L., and Goebel, J. (2005) Daclizumab therapy in kidney transplantation – different mechanisms of action in-vivo versus ex-vivo? *Transpl. Immunol.*, **14**, 43–47.

11 Zhang, M., Zhang, J., Yan, M., Li, H., Yang, C., and Yu, D. (2008) Recombinant anti-vascular endothelial growth factor fusion protein efficiently suppresses choroidal neovascularization in monkeys. *Mol. Vis.*, **14**, 37–49.

12 Sugiura, T. (2003) Baculoviral expression of correctly processed ADAMTS proteins fused with the human IgG-Fc region. *J. Biotechnol.*, **100**, 193–201.

13 Gao, L.H., Hu, X.W., Chen, W., Xu, J.J., Zhao, J., and Chen, H.P. (2005) Expression of ATR-Fc fusion protein in CHO cells. *Sheng Wu Gong Cheng Xue Bao*, **21**, 826–831.

14 Mitchell, D., Pobre, E.G., Mulivor, A.W., Grinberg, A.V., Castonguay, R., Monnell, T.E., Solban, N., Ucran, J.A., Pearsall, R.S., Underwood, K.W., Seehra, J., and Kumar, R. (2010) ALK1-Fc inhibits multiple mediators of angiogenesis and suppresses tumor growth. *Mol. Cancer Ther.*, **9**, 379–388.

15 Liu, A., Hu, P., Khawli, L.A., and Epstein, A.L. (2005) Combination B7-Fc fusion protein treatment and Treg cell depletion therapy. *Clin. Cancer Res.*, **11**, 8492–8502.

16 Kong, Q.L., Guan, Y.Z., Jing, X.F., Li, C., Guo, X.H., Lu, Z., and An, Y.Q. (2006) BPI700-Fc gamma1(700) chimeric gene expression and its protective effect in a mice model of the lethal E. coli infection. *Chin. Med. J.*, **119**, 474–481.

17 Vugmeyster, Y., Seshasayee, D., Chang, W., Storn, A., *et al.* (2006) A soluble BAFF antagonist, BR3-Fc, decreases peripheral blood B cells and lymphoid tissue marginal zone and follicular B cells in cynomolgus monkeys. *Am. J. Pathol.*, **168**, 476–489.

18 Gazit, R., Rechnitzer, H., Achdout, H., Katzenell, A., *et al.* (2004) Recognition of

mycoplasma hyorhinis by CD99-Fc molecule. *Eur. J. Immunol.*, **34**, 2032–2040.

19 Nitschke, L., Floyd, H., Ferguson, D.J., and Crocker, P.R. (1999) Identification of CD22 ligands on bone marrow sinusoidal endothelium implicated in CD22-dependent homing of recirculating B cells. *J. Exp. Med.*, **189**, 1513–1518.

20 Daub, K., Siegel-Axel, D., Schonberger, T., Leder, C., *et al.* (2010) Inhibition of foam cell formation using a soluble CD68-Fc fusion protein. *J. Mol. Med.*, **88**, 909–920.

21 Doggrell, S.A. (2008) CD47-Fc fusion proteins as putative immunotherapeutic agents for the treatment of immunological and inflammatory diseases. *Expert Opin. Ther. Pat.*, **18**, 555–561.

22 Martin, P.L., Sachs, C., Hoberman, A., Jiao, Q., and Bugelski, P.J. (2010) Effects of CNTO 530, an erythropoietin mimetic-IgG4 fusion protein, on embryofetal development in rats and rabbits. *Birth Defects Res. B Dev. Reprod. Toxicol.*, **89**, 87–96.

23 Ma, Q., DeMarte, L., Wang, Y., Stanners, C.P., and Junghans, R.P. (2004) Carcinoembryonic antigen-immunoglobulin Fc fusion protein (CEA-Fc) for identification and activation of anti-CEA immunoglobulin-T-cell receptor-modified T cells, representative of a new class of Ig fusion proteins. *Cancer Gene Ther.*, **11**, 297–306.

24 Xu, Y., Zhang, C., Jia, L., *et al.* (2009) A novel approach to inhibit HIV-1 infection and enhance lysis of HIV by a targeted activator of complement. *Virol. J.*, **6**, 123.

25 Manjunath, N., Shankar, P., Wan, J., Weninger, W., *et al.* (2001) Effector differentiation is not prerequisite for generation of memory cytotoxic T lymphocytes. *J. Clin. Invest.*, **108**, 871–878.

26 Way, J.C., Lauder, S., Brunkhorst, B., Kong, S.M., *et al.* (2005) Improvement of Fc-erythropoietin structure and pharmacokinetics by modification at a disulfide bond. *Protein Eng. Des. Sel.*, **18**, 111–118.

27 Ogiwara, K., Nagaoka, M., Cho, C.S., and Akaike, T. (2005) Construction of a novel extracellular matrix using a new genetically engineered epidermal growth factor fused to IgG-Fc. *Biotechnol. Lett.*, 27, 1633–1637.

28 Vu, J.R., Fouts, T., Bobb, K., Burns, J., McDermott, B., Israel, D.I., Godfrey, K., and DeVico, A. (2006) An immunoglobulin fusion protein based on the gp120-CD4 receptor complex potently inhibits human immunodeficiency virus type 1 *in vitro*. *AIDS Res. Hum. Retroviruses*, **22**, 477–490.

29 Gurbaxani, B.M., and Morrison, S.L. (2006) Development of new models for the analysis of Fc-FcRn interactions. *Mol. Immunol.*, **43**, 1379–1389.

30 Low, S.C., Nunes, S.L., Bitonti, A.J., and Dumont, J.A. (2005) Oral and pulmonary delivery of FSH-Fc fusion proteins via neonatal Fc receptor-mediated transcytosis. *Hum. Reprod.*, **20**, 1805–1813.

31 Zhiqiang, A.N. (2009) *Therapeutic Monoclonal Antibodies: From Bench to Clinic*, John Wiley.

32 Peters, R.T., Low, S.C., Kamphaus, G.D., Dumont, J.A., Amari, J.V., Lu, Q., Zarbis-Papastoitsis, G., Reidy, T.J., Merricks, E.P., Nichols, T.C., and Bitonti, A.J. (2010) Prolonged activity of factor IX as a monomeric Fc fusion protein. *Blood*, **115**, 2057–2064.

33 Carr, M.E. Jr (2010) Future directions in hemostasis: normalizing the lives of patients with hemophilia. *Thromb. Res.*, 125, S78–S81.

34 Massberg, S., Konrad, I., Bultmann, A., Schulz, C., *et al.* (2004) Soluble glycoprotein VI dimer inhibits platelet adhesion and aggregation to the injured vessel wall *in vivo*. *FASEB J.*, **18**, 397–399.

35 Kumar, M., Hunag, Y., Glinka, Y., Prud'homme, G.J., and Wang, Q. (2007) Gene therapy of diabetes using a novel GLP-1/IgG1-Fc fusion construct normalizes glucose levels in db/db mice. *Gene Ther.*, **14**, 162–172.

36 Lee, C.H., Woo, J.H., Cho, K.K., Kang, S.H., Kang, S.K., and Choi, Y.J. (2007) Expression and

characterization of human growth hormone-Fc fusion proteins for transcytosis induction. *Biotechnol. Appl. Biochem.*, **46**, 211–217.

37 Zheng, X.X., Steele, A.W., Hancock, W.W., Kawamoto, K., *et al.* (1999) IL-2 receptor-targeted cytolytic IL-2/Fc fusion protein treatment blocks diabetogenic autoimmunity in nonobese diabetic mice. *J. Immunol.*, **163**, 4041–4048.

38 Kim, Y., Maslinski, W., Zheng, X., and Stevens, A.C. (1998) Targeting the IL-15 receptor with an antagonist IL-15 mutant/Fc gamma2a protein blocks delayed-type hypersensitivity. *J. Immunol.*, **15**, 5742–5748.

39 Nickerson, P., Zheng, X.X., Steiger, J., Steele, A.W., *et al.* (1997) Prolonged islet allograft acceptance in the absence of expression of interleukin-4. *Adv. Nephrol. Necker Hosp.*, **26**, 171–180.

40 Zheng, X.X., Steele, A.W., Nickerson, P.W., Steurer, W., Steiger, J., and Strom, T.B. (1995) Administration of noncytolytic IL-10/Fc in murine models of lipopolysaccharide-induced septic shock and allogeneic islet transplantation. *J. Immunol.*, **154**, 5590–5600.

41 Sivakumar, P.V., Westrich, G.M., Kanaly, S., Garka, K., Born, T.L., Derry, J.M., and Viney, J.L. (2002) Interleukin 18 is a primary mediator of the inflammation associated with dextran sulphate sodium induced colitis: blocking interleukin 18 attenuates intestinal damage. *Gut*, **50**, 812–820.

42 Dumont, J.A., Low, S.C., Peters, R.T., and Bitonti, A.J. (2006) Monomeric Fc fusions: impact on pharmacokinetic and biological activity of protein therapeutics. *BioDrugs*, **20**, 151–160.

43 Booth, V., Keizer, D.W., Kamphuis, M.B., Clark-Lewis, I., and Sykes, B.D. (2002) The CXCR3 binding chemokine IP-10/CXCL10: structure and receptor interactions. *Biochemistry*, **41**, 10418–10425.

44 Bush, K.A., Farmer, K.M., Walker, J.S., and Kirkham, B.W. (2002) Reduction of joint inflammation and bone erosion in rat adjuvant arthritis by treatment with interleukin-17 receptor IgG1 Fc fusion protein. *Arthritis Rheum.*, **46**, 802–805.

45 Kim-Schulze, S., Scotto, L., Vlad, G., Piazza, F., Lin, H., Liu, Z., Cortesini, R., and Suciu-Foca, N. (2006) Recombinant Ig-like transcript 3-Fc modulates T cell responses via induction of Th anergy and differentiation of CD8+ T suppressor cells. *J. Immunol.*, **176**, 2790–2798.

46 Jazayeri, J.A., De Weerd, N., Raye, W., Velkov, T., Santos, L., Taylor, D., and Carroll, G.J. (2007) Generation of mutant leukaemia inhibitory factor (LIF)-IgG heavy chain fusion proteins as bivalent antagonists of LIF. *J. Immunol. Methods*, **323**, 1–10.

47 Zocher, M., Baeuerle, P.A., Dreier, T., and Iglesias, A. (2003) Specific depletion of autoreactive B lymphocytes by a recombinant fusion protein *in vitro* and *in vivo*. *Int. Immunol.*, **15**, 789–796.

48 Zubairi, S., Sanos, S.L., Hill, S., and Kaye, P.M. (2004) Immunotherapy with OX40L-Fc or anti-CTLA-4 enhances local tissue responses and killing of Leishmania donovani. *Eur. J. Immunol.*, **34**, 1433–1440.

49 Mazanet, M.M., and Hughes, C.C. (2002) B7-H1 is expressed by human endothelial cells and suppresses T cell cytokine synthesis. *J. Immunol.*, **169**, 3581–3588.

50 Qin, W., Feng, J., Li, Y., Lin, Z., and Shen, B. (2006) Fusion protein of CDR mimetic peptide with Fc inhibit TNF-alpha induced cytotoxicity. *Mol. Immunol.*, **43**, 660–666.

51 Ono, K., Kamihira, M., Kuga, Y., Matsumoto, H., *et al.* (2003) Production of anti-prion scFv-Fc fusion proteins by recombinant animal cells. *J. Biosci. Bioeng.*, **95**, 231–238.

52 Bobardt, M.D., Chatterji, U., Schaffer, L., de Witte, L., and Gallay, P.A. (2010) Syndecan-Fc hybrid molecule as a potent *in vitro* microbicidal anti-HIV-1 agent. *Antimicrob. Agents Chemother.*, **54**, 2753–2766.

53 Croll, S.D., Chesnutt, C.R., Rudge, J.S., Acheson, A., *et al.* (1998) Co-infusion with a TrkB-Fc receptor body carrier enhances BDNF distribution in the adult rat brain. *Exp. Neurol.*, **152**, 20–33.

54 Sandalon, Z., Bruckheimer, E.M., Lustig, K.H., Rogers, L.C., Peluso, R.W.,

and Burstein, H. (2004) Secretion of a TNFR:Fc fusion protein following pulmonary administration of pseudotyped adeno-associated virus vectors. *J. Virol.*, **78**, 12355–12365.

55 Sanchez-Fueyo, A., Tian, J., Picarella, D., Domenig, C., Zheng, X.X., Sabatos, C.A., Manlongat, N., Bender, O., Kamradt, T., Kuchroo, V.K., Gutierrez-Ramos, J.C., Coyle, A.J., and Strom, T.B. (2003) Tim-3 inhibits T helper type 1-mediated auto- and alloimmune responses and promotes immunological tolerance. *Nat. Immunol.*, **4**, 1093–1101.

56 Moon, E.Y., and Ryu, S.K. (2007) TACI:Fc scavenging B cell activating factor (BAFF) alleviates ovalbumin-induced bronchial asthma in mice. *Exp. Mol. Med.*, **39**, 343–352.

57 Rosengard, A.M., Alonso, L.C., Korb, L.C., Baldwin, W.M., 3rd, Sanfilippo, F., Turka, L.A., and Ahearn, J.M. (1999) Functional characterization of soluble and membrane-bound forms of vaccinia virus complement control protein (VCP). *Mol. Immunol.*, **36**, 685–697.

58 Michaelsen, T.E., Frangione, B., and Franklin, E.C. (1977) The amino acid sequence of a human immunoglobulin G3m(g) pFc' fragment. *J. Immunol.*, **119**, 558–563.

59 Plaut, M., Pierce, J.H., Watson, C.J., Hanley-Hyde, J., Nordan, R.P., and Paul, W.E. (1989) Mast cell lines produce lymphokines in response to cross-linkage of Fc epsilon RI or to calcium ionophores. *Nature*, **339**, 64–67.

60 Ioan-Facsinay, A., de Kimpe, S.J., Hellwig, S.M., van Lent, P.L., Hofhuis, F.M., *et al.* (2002) FcgammaRI (CD64) contributes substantially to severity of arthritis, hypersensitivity responses, and protection from bacterial infection. *Immunity*, **16**, 391–402.

61 Shibuya, A., and Honda, S.-I. (2006) Molecular and functional characteristics of the Fcα/μR, a novel Fc receptor for IgM and IgA. *Springer Semin. Immunopathol.*, **28**, 377–382.

62 Wines, B.D., Gavin, A., Powell, M.S., Steinitz, M., Buchanan, R.R., and Mark Hogarth, P. (2003) Soluble FcgammaRIIa inhibits rheumatoid

factor binding to immune complexes. *Immunology*, **109**, 246–254.

63 Hippen, K.L., Buhl, A.M., D'Ambrosio, D., Nakamura, K., Persin, C., and Cambier, J.C. (1997) Fc gammaRIIB1 inhibition of BCR-mediated phosphoinositide hydrolysis and Ca2+ mobilization is integrated by CD19 dephosphorylation. *Immunity*, **7**, 49–58.

64 Chen, L., Pielak, G.J., and Thompson, N.L. (1999) The cytoplasmic region of mouse Fc gamma RIIb1, but not Fc gamma RIIb2, binds phospholipid membranes. *Biochemistry*, **38**, 2102–2109.

65 Li, P., Jiang, N., Nagarajan, S., *et al.* (2007) Molecular basis of cell and developmental biology. *J. Biol. Chem.*, **282**, 6210–6221.

66 Daeron, M. (1997) Fc receptor biology. *Annu. Rev. Immunol.*, **15**, 203–234.

67 Lobo, E.D., Hansen, R.J., and Balthasar, J.P. (2004) Antibody pharmacokinetics and pharmacodynamics. *J. Pharm. Sci.*, **93**, 2645–2668.

68 Spiegelberg, H.L., and Grey, H.M. (1968) Catabolism of human {gamma}G immunoglobulins of different heavy chain subclasses: II. Catabolism of {gamma}G myeloma proteins in heterologous species. *J. Immunol.*, **101**, 711–716.

69 Smith, M.R. (2003) Rituximab (monoclonal anti-CD20 antibody): mechanisms of action and resistance. *Oncogene*, **22**, 7359–7368.

70 Jefferis, R., Lund, J., and Pound, J.D. (1998) IgG-Fc-mediated effector functions: molecular definition of interaction sites for effector ligands and the role of glycosylation. *Immunol. Rev.*, **163**, 59–76.

71 Isakov, N. (1998) Role of immunoreceptor tyrosine-based activation motif in signal transduction from antigen and Fc receptors. *Adv. Immunol.*, **69**, 183–247.

72 Shields, R.L. (2002) Lack of fucose on human IgG1 N-linked oligosaccharide improves binding to human Fcgamma RIII and antibody-dependent cellular toxicity. *J. Biol. Chem.*, **277**, 26733–26740.

73 Kontermann, R., Dübel, S., Kenanova, V.E., Olafsen, T., Andersen, J.T., Sandlie, I., and Wu, A.M. (2010) *Engineering of the Fc Region for Improved PK (FcRn Interaction). Antibody Engineering*, Springer Berlin Heidelberg, pp. 411–430.

74 Dall'Acqua, W.F., Woods, R.M., Ward, E.S., and Palaszynski, S.R. (2002) Increasing the affinity of a human IgG1 for the neonatal Fc receptor: biological consequences. *J. Immunol.*, **169**, 5171–5180.

75 Hinton, P.R., Xiong, J.M., Johlfs, M.G., Tang, M.T., Keller, S., and Tsurushita, N. (2006) An engineered human IgG1 antibody with longer serum half-life. *J. Immunol.*, **176**, 346–356.

76 Fridman, W.H., and Sautes, C. (1990) Fc gamma receptors (Fc gamma R) and IgG-binding factors (IgG-BF): their relationships. *Res. Immunol.*, **141**, 60–64. discussion 105–108.

77 Gillies, S.D., Lan, Y., Lo, K.M., Super, M., and Wesolowski, J. (1999) Improving the efficacy of antibody-interleukin 2 fusion proteins by reducing their interaction with Fc receptors. *Cancer Res.*, **59**, 2159–2166.

78 Vaccaro, C., Zhou, J., Ober, R.J., and Ward, E.S. (2005) Engineering the Fc region of immunoglobulin G to modulate *in vivo* antibody levels. *Nat. Biotechnol.*, **23**, 1283–1288.

79 Kenanova, V., Olafsen, T., Crow, D.M., Sundaresan, G., Subbarayan, M., Carter, N.H., Ikle, D.N., Yazaki, P.J., Chatziioannou, A.F., Gambhir, S.S., Williams, L.E., Shively, J.E., Colcher, D., Raubitschek, A.A., and Wu, A.M. (2005) Tailoring the pharmacokinetics and positron emission tomography imaging properties of anti-carcinoembryonic antigen single-chain Fv-Fc antibody fragments. *Cancer Res.*, **65**, 622–631.

80 Peppel, K., Crawford, D., and Beutler, B. (1991) A tumor necrosis factor (TNF) receptor-IgG heavy chain chimeric protein as a bivalent antagonist of TNF activity. *J. Exp. Med.*, **174**, 1483–1489.

81 Garrison, L., and McDonnell, N.D. (1999) Etanercept: therapeutic use in patients with rheumatoid arthritis. *Ann. Rheum. Dis.*, **58**, 165–169.

82 Lo, K.M., Sudo, Y., Chen, J., Li, Y., Lan, Y., Kong, S.M., Chen, L., An, Q., and Gillies, S.D. (1998) High level expression and secretion of Fc-X fusion proteins in mammalian cells. *Protein Eng.*, **11**, 495–500.

83 Liu, J., Wei, D., Qian, F., Zhou, Y., Wang, J., Ma, Y., and Han, Z. (2003) pPIC9-Fc: a vector system for the production of single-chain Fv-Fc fusions in Pichia pastoris as detection reagents *in vitro*. *J. Biochem. (Tokyo)*, **134**, 911–917.

84 Jones, T.D., Hanlon, M., Smith, B.J., Heise, C.T., *et al.* (2004) The development of a modified human IFN-alpha2b linked to the Fc portion of human IgG1 as a novel potential therapeutic for the treatment of hepatitis C virus infection. *J. Interferon Cytokine Res.*, **24**, 560–572.

85 Bitonti, A.J., Dumont, J.A., Low, S.C., *et al.* (2004) Pulmonary delivery of an erythropoietin Fc fusion protein in non-human primates through an immunoglobulin transport pathway. *Proc. Natl. Acad. Sci. U.S.A.*, **101**, 9763–9768.

86 Tiu, R.V., and Sekeres, M.A. (2008) The role of AMG-531 in the treatment of thrombocytopenia in idiopathic thrombocytopenic purpura and myelodysplastic syndromes. *Expert Opin. Biol. Ther.*, **8**, 1021–1030.

87 Kuter, D.J., Bussel, J.B., Lyons, R.M., Pullarkat, V., *et al.* (2008) Efficacy of romiplostim in patients with chronic immune thrombocytopenic purpura: a double-blind randomised controlled trial. *Lancet*, **371**, 395–403.

88 Challita-Eid, P.M., Penichet, M.L., Shin, S.-U., Poles, T., Mosammaparast, N., Mahmood, K., Slamon, D.J., Morrison, S.L., and Rosenblatt, J.D. (1998) A B7.1-antibody fusion protein retains antibody specificity and ability to activate via the T cell costimulatory pathway. *J. Immunol.*, **160**, 3419–3426.

89 Holzer, W., Petersen, F., Strittmatter, W., Matzku, S., and von Hoegen, I. (1996) A fusion protein of IL-8 and a Fab antibody fragment binds to IL-8 receptors and induces neutrophil activation. *Cytokine*, **8**, 214–221.

90 Holliger, P., Prospero, T., and Winter, G. (1993) "Diabodies": small bivalent and bispecific antibody fragments. *Proc. Natl. Acad. Sci. U.S.A.*, **90**, 6444–6448.

91 Bernhard, H., Salazar, L., Schiffman, K., Smorlesi, A., et al. (2002) Vaccination against the HER-2/neu oncogenic protein. *Endocr. Relat. Cancer*, **9**, 33–44.

92 Czerniecki, B.J., Koski, G.K., Koldovsky, U., Xu, S., Cohen, P.A., et al. (2007) Targeting HER-2/neu in early breast cancer development using dendritic cells with staged interleukin-12 burst secretion. *Cancer Res.*, **67**, 1842–1852.

93 Chiquito, J., Crasto, C., and Feng, J.-A. (1999) LINKER: a program to generate linker sequences for fusion proteins. *Protein Eng. Des. Sel.*, **13**, 309–312.

94 Simmons, L.C. (2002) Expression of full-length immunoglobulins in Escherichia coli: rapid and efficient production of aglycosylated antibodies. *J. Immunol. Methods*, **263**, 133–147.

95 Bessette, P.H., Aslund, F., Beckwith, J., and Georgiou, G. (1999) Efficient folding of proteins with multiple disulfide bonds in the *Escherichia coli* cytoplasm. *Proc. Natl. Acad. Sci. U.S.A.*, **96**, 13703–13708.

96 Wu, S.C., and Wong, S.L. (2002) Engineering of a *Bacillus subtilis* strain with adjustable levels of intracellular biotin for secretory production of functional streptavidin. *Appl. Environ. Microbiol.*, **68**, 1102–1108.

97 Binnie, C., Jenish, D., Cossar, D., Szabo, A., et al. (1997) Expression and characterization of soluble human erythropoietin receptor made in *Streptomyces lividans* 66. *Protein Expr. Purif.*, **11**, 271–278.

98 Joosten, V., Lokman, C., Van Den Hondel, C.A., and Punt, P.J. (2003) The production of antibody fragments and antibody fusion proteins by yeasts and filamentous fungi. *Microb. Cell Fact.*, **2**, 1.

99 Ghose, S., Hubbard, B., and Cramer, S.M. (2006) Binding capacity differences for antibodies and Fc-fusion proteins on protein A chromatographic materials. *Biotechnol. Bioeng.*, **96**, 768–779.

100 Nagaoka, M., and Akaike, T. (2003) Single amino acid substitution in the mouse IgG1 Fc region induces drastic enhancement of the affinity to protein A. *Protein Eng.*, **16**, 243–245.

101 Economides , A.N., Carpenter, L.R., Rudge, J.S., Wong, V., et al. (2003) Cytokine traps: multi-component, high-affinity blockers of cytokine action.*Nat. Med.*, **9**, :47–52..

102 Adachi, O., Nakano, A., Sato, O., Kawamoto, S., et al. (2002) Gene transfer of Fc-fusion cytokine by *in vivo* electroporation: application to gene therapy for viral myocarditis. *Gene Ther.*, **9**, 577–583.

103 Bitonti, A.J., and Dumont, J.A. (2006) Pulmonary administration of therapeutic proteins using an immunoglobulin transport pathway. *Adv. Drug Deliv. Rev.*, **58**, 1106–1118.

104 Aruffo, A., Stamenkovic, I., Melnick, M., Underhill, C.B., and Seed, B. (1990) CD44 is the principal cell surface receptor for hyaluronate. *Cell*, **61**, 1303–1313.

105 Bowen, M.A., Patel, D.D., Li, X., et al. (1995) Cloning, mapping, and characterization of activated leukocyte-cell adhesion molecule (ALCAM), a CD6 ligand. *J. Exp. Med.*, **181**, 2213–2220.

106 Vinson, M., van der Merwe, P.A., Kelm, S., May, A., Jones, E.Y., and Crocker, P.R. (1996) Characterization of the sialic acid-binding site in sialoadhesin by site-directed mutagenesis. *J. Biol. Chem.*, **271**, 9267–9272.

107 Haraoui, B., Pelletier, J.P., and Martel-Pelletier, J. (2007) Immunogenicity of biologic agents: a new concern for the practicing rheumatologist? *Curr. Rheumatol. Rep.*, **9**, 265–267.

108 Keane, J., Gershon, S., Wise, R.P., Mirabile-Levens, E., et al. (2001)

Tuberculosis associated with infliximab, a tumor necrosis factor alpha-neutralizing agent. *N. Engl. J. Med.*, **345**, 1098–1104.

109 Castro-Rueda, H., and Kavanaugh, A. (2008) Biologic therapy for early rheumatoid arthritis: the latest evidence. *Curr. Opin. Rheumatol.*, **20**, 314–319.

110 Rudick, R.A., Stuart, W.H., Calabresi, P.A., *et al.* (2006) Natalizumab plus interferon beta-1a for relapsing multiple sclerosis. *N. Engl. J. Med.*, **354**, 911–923.

10
Monomeric Fc Fusion Technology: An Approach to Create Long-Lasting Clotting Factors

Jennifer A. Dumont, Xiaomei Jin, Robert T. Peters, Alvin Luk, Glenn F. Pierce, and Alan J. Bitonti

10.1
Introduction

Therapeutic proteins are one of the most important and rapidly growing segments of the pharmaceutical market. Since the approval of insulin in 1982, numerous recombinant protein drugs have been approved and become available as valuable therapeutic options. However, failed protein drugs have far outnumbered the successful ones. Among many of the hurdles for protein therapy such as parenteral delivery, solubility, immunogenicity, and large-scale production, protein drug stability *in vivo* is one of the great challenges facing protein therapeutics. The instability of a protein drug *in vivo*, caused by proteolytic degradation and/or other clearance mechanisms, often necessitates relatively frequent invasive dosing, resulting in unwanted side-effects and/or limited therapeutic benefit and suboptimal patient compliance. Frequent patient dosing also means an extra burden on manufacturing capacity and treatment costs. Therefore, great effort has been devoted to explore different technologies for prolonging the half-lives of protein drugs by circumventing or interfering with their usual clearance pathway(s). Protein drug modifications including PEGylation, glycoPEGylation, mutation of protease recognition sites, and fusion to other proteins such as albumin and the Fc domain of IgG, have been approaches to extend the half-life of many therapeutic protein drugs [1–5]. This review will focus on the use of Fc monomer fusion technology as a strategy to prolong the half-life of therapeutic proteins with a focus on clotting factors.

10.2
Neonatal Fc Receptor and Interaction with Immunoglobulin G

Serum proteins that are too large for renal filtration (>50–70 kDa) have a half-life of approximately three days. However, immunoglobulin G (IgG) has a uniquely

Therapeutic Proteins: Strategies to Modulate Their Plasma Half-Lives, First Edition. Edited by Roland Kontermann.
© 2012 Wiley-VCH Verlag GmbH & Co. KGaA. Published 2012 by Wiley-VCH Verlag GmbH & Co. KGaA.

long half-life of approximately 21 days in humans [6, 7]. The mechanism for the prolonged half-life of this plasma protein has been attributed to its interaction with an intracellular receptor called the neonatal Fc receptor (FcRn).

Nearly 50 years ago, Brambell proposed a link between the transmission of passive immunity in the form of transfer of maternal antibodies to the fetus, and the protection of IgG, from catabolism [8]. He suggested that a single receptor might mediate both processes and that the Fc domain of IgG was involved in the transport of IgG, as well as its protection from degradation [9]. This receptor was isolated years later from the intestinal epithelium of neonatal rodents, and so was named the neonatal Fc receptor = [10] . Numerous studies have now proven the hypothesis that FcRn regulates IgG homeostasis. These include animal models in which FcRn has been genetically deleted, resulting in a reduced plasma IgG half-life and low plasma IgG concentrations [11–13]. The only report in humans of a defect in FcRn is that of two siblings diagnosed with familial hyper-catabolic hypoproteinemia in the early 1960s [14]. The clinical manifestation was that of low plasma IgG levels and the genetic defect was a single nucleotide trans-version from alanine to proline at the midpoint of the signal sequence of β_2-microglobulin (β_2m), which is a subunit of FcRn and is common to all MHC class I molecules [15]. The result was a half-life of approximately three days in these siblings. IgG synthesis was shown to be normal, but they each showed hyperca-tabolism of IgG. Hence, the contribution of FcRn to IgG homeostasis in humans has been demonstrated in what is essentially a naturally occurring knockout for FcRn.

FcRn comprises two subunits, one of which is a membrane-bound heavy chain that is structurally related to the major histocompatibility complex (MHC) class I molecules, and the other subunit is a soluble light chain β_2m [16]. FcRn mediates the transport of IgG from mother to fetus across the small intestine and yolk sac of rodents, and across the placenta in humans thereby conferring humoral immu-nity. In rodents, this is accomplished through expression of the receptor at high levels during the first few weeks of life in the epithelial cells lining the small intestine, as well as in the yolk sac during gestation [10, 17–19]. Thus, the impor-tance of FcRn in rodents is clear during the neonatal period when it facilitates the absorption and transfer of IgG from ingested milk across the intestine. At the time of weaning, FcRn expression in the intestine is considerably diminished (approximately 1000-fold) in rodents [20]. In contrast, passive immunity in humans is accomplished through FcRn that is expressed in the syncytiotrophoblast cells of the placenta, where it mediates the transport of maternal IgG across the placenta to the fetus [19]. After the initial isolation of FcRn, it soon became evident that FcRn expression was not limited to gestational or neonatal periods in development, but rather was found in many cell and tissue types throughout adult life.

The major site for the maintenance of IgG homeostasis is thought to be the vascular endothelium, resulting in circulating levels of approximately $12\,\text{mg}\cdot\text{mL}^{-1}$ IgG and an exceptionally long half-life in humans [21, 22]. FcRn has been found to be expressed in the vascular endothelium of the large vascular beds in muscle and skin of mice where the contact area with blood is great [22–24]. Here, IgG can be readily internalized along with other plasma proteins by fluid-phase pinocytosis as described below. FcRn is also expressed in various organs in humans, including the kidney, where the receptor is thought to play a role in the salvage of IgG [23, 25], and the liver, where the receptor may further contribute to maintaining levels of circulating IgG [26, 27]. Furthermore, FcRn is found in the mucosal epithelial cells, including the lung in humans, where expression has been shown to be relatively constant throughout life, and functional transport of IgG has been demonstrated [28–30].

FcRn is located in intracellular compartments of most cell types, reportedly the sorting endosome [31, 32]. The fate of plasma proteins taken up by cells is typically proteolytic degradation in lysosomes. However, circulating IgG taken up by cells binds specifically to FcRn in slightly acidic endosomes in a pH-dependent manner, and is recycled with FcRn back to the plasma membrane, where physiologic pH triggers release of IgG back into the circulation. As a result, IgG is able to avoid degradation in lysosomal compartments and is thereby protected from degradation. Directional recycling occurs as a result of the pH-dependent binding of IgG to FcRn. It is well documented that IgG binds with high affinity to FcRn at low pH (<6.5) and not at physiological, neutral pH (7.4) [33–35].

The regions of IgG that bind to FcRn and are responsible for the control of IgG catabolism are localized to the Fc region, in the CH2 and CH3 domains [36]. Site-directed mutagenesis identified three amino acids (Ile 253, His 310, and His 435) that are critical in the binding to FcRn and play an important role in regulation of the plasma half-life of IgG [37]. The interaction of IgG with FcRn is strictly pH-dependent, as binding occurs with high affinity at acidic pH (<6.5) but not at neutral pH (7 to 7.4). This is due to the presence of the conserved histidine residues. The imidazole group of histidine is protonated at acidic pH and nonprotonated at neutral or physiological pH. Hydrophobic Ile 253 is also involved [34, 35, 38]. The more precise interactions of IgG with FcRn have been revealed by crystal structures of FcRn with Fc for both rat and human [39, 40] .

10.3
Traditional Fc Fusion Proteins

Many biologically active proteins or peptides have relatively short plasma half-lives because of rapid proteolytic degradation and/or renal clearance, which limit their exposure to the target tissue and consequently their pharmacologic effects. Since Fc, the constant region of IgG, is responsible for the relatively long plasma half-life of IgG, it has been coupled to various proteins and/or peptides to be used as research tools and also as therapeutic agents. These include Fc fusions that are approved biotherapeutics (e.g., etanercept, alefacept, abatacept, rilonocept, romiplostim) [41–45], as well as several others in various phases of clinical development [46]. The functions of these Fc fusion drugs fall into various categories, including cytokines, hormones, enzymes, receptors, ligands, peptides, single chain antibodies, and blood coagulation factors.

10.4
Monomeric Fc Fusion Proteins Show Improved Biologic Properties

Fc fusion proteins were originally generated as homodimers containing two effector molecules fused to a dimer of Fc (Figure 10.1a). Depending on the therapeutic protein fused to the Fc dimer, these molecules may be very large and because of this, steric hindrance between the two effector molecules or between the effector molecules and the Fc dimer may occur [30, 47, 48]. In addition, therapeutic protein homodimers may be heavily charged as a result of glycosylation on the therapeutic protein portion of the molecule. All of these properties present possible challenges for proper dimer formation and could adversely affect the binding to both effector protein receptors and FcRn. Thus, a second generation Fc fusion, the monomeric Fc fusion protein, consisting of one effector molecule fused to a dimer of Fc (Figure 10.1b) was developed. This monomeric Fc fusion is reduced in size and charge, both changes advantageous for reducing steric hindrance introduced by two effector molecules. While the traditional dimeric Fc fusion molecule has an increased half-life compared with the unconjugated effector molecule, the monomer configuration has been shown to result in an even greater extension of the circulating half-life [5, 30, 47]. This monomeric Fc fusion configuration has been shown to have improved properties, including pharmacokinetic parameters and biologic activity, and has the potential to achieve less frequent dose administration regimens for a broad range of protein therapeutics.

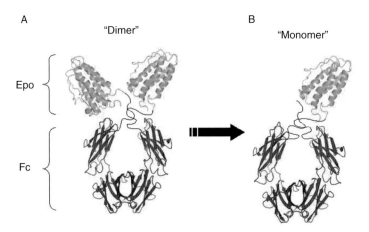

Figure 10.1 Schematic diagram for erythropoietin Fc (EPOFc) dimer and monomer. The dimer is composed of two molecules of EPO joined to dimeric Fc and the monomer has a single molecule of EPO joined to dimeric Fc.

10.4.1
EPOFc as a Prototype Construct

An Fc fusion molecule comprising erythropoietin (EPO) linked to the Fc domain of IgG1 was chosen as the prototype molecule to test the concept of delivery of a large therapeutic molecule by noninvasive means through the lung by leveraging the FcRn pathway. EPO is a potent cytokine (30 kDa) that stimulates red blood cell production and is used in the treatment of anemia [49]. The EPOFc monomer and dimer (Figure 10.1a,b) were created by recombinantly coupling human EPO to the Fc domain (hinge-CH2-CH3) of IgG1. They were produced in Chinese hamster ovary (CHO) cells by transfection with a plasmid expressing EPOFc to create the dimer and by cotransfection of plasmids expressing EPOFc and Fc to create the monomer [30].

EPOFc monomer and dimer pulmonary delivery was targeted to the upper and central airways of nonhuman primates, since FcRn expression was demonstrated to be highest in these regions of the human and nonhuman primate lung, with very little expression in the alveolar regions. This was accomplished by administering aerosolized protein with a particle size of 4 to 6 μm to intubated, anesthetized cynomolgus monkeys that were allowed to breathe spontaneously. Uptake of EPOFc monomer was markedly better than that of the EPOFc dimer (Figure 10.2, Table 10.1). In comparison to EPOFc dimer, EPOFc monomer demonstrated a four- to five fold increase in the C_{max}, and a six- to tenfold increase in the area under the plasma concentration-time curve (AUC). The circulating half-life of the monomer was 25 h compared with 16 h for the dimer. As a result, the bioavailability of EPOFc monomer was >30%, compared with intravenous (i.v.) administration, which is a significant improvement over the 5 to 10% bioavailability observed

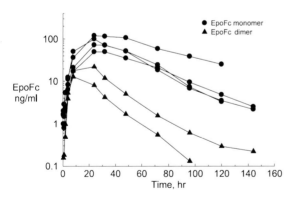

Figure 10.2 Pulmonary delivery of EPOFc monomer and dimer in nonhuman primates. Equivalent doses (20 mg·kg⁻¹) of EPOFc dimer or EPOFc monomer were aerosolized with an Aeroneb Pro® nebulizer and delivered directly into the lungs of intubated cynomol- gus monkeys using shallow breathing maneuvers. Serum concentrations of the dimer and monomer were measured with an EPO-specific ELISA. The number of animals per group were $n = 2$ for EPOFc dimer and $n = 4$ for EPOFc monomer.

Table 10.1 Summary of pharmacokinetics for dimeric and monomeric EPOFc and rFIXFc.

	Species and route	Construct	Dose (ug·kg^{-1})	C$_{max}$ (µg·mL^{-1})	AUC (µg h mL^{-1})	T$_{1/2}$ (h)
EPOFc	Monkey/ pulmonary	Dimer	20	0.017	0.557	16
		Monomer	20	0.086	5.279	25
rFIXFc	Neonatal rat/ oral	Dimer	1370	0.590	9570	14
		Monomer	1000	5.83	121 250	19
	Adult rat/IV	Dimer	5000	7.50	109	22
		Monomer	5000	33.0	509	35
	Adult FIX-deficient mouse/IV	Dimer	5000	10.1	167	53
		Monomer	5000	33.4	761	46

for EPOFc dimer. Therefore, reducing the size and charge of the molecule resulted in a striking increase in transport of the molecule across the lung epithelium. Moreover, this uptake was achieved almost completely via the upper and central airways since forced delivery of the EPOFc dimer to the deep lung resulted in less transport into the bloodstream. Similarly, the transport was proven to be FcRn-mediated since mutation of three amino acids in the Fc domain required for FcRn binding (I253A, H310A, H435A) resulted in poor absorption when administered either to the upper airways or deep lung [30].

The Fc monomer technology has since been applied to several other biotherapeutics, including interferon β (IFNβ) and interferon α (IFNα), as well as clotting factor IX (FIX) [5, 47, 48]. IFNβ showed an improvement in both the bioactivity and PK as an Fc-monomer compared with the dimer, while IFNα had remarkable improvement in PK as a dimer, and showed an increase in bioactivity as an Fc monomer [47, 48]. This article focuses on the application of the Fc monomer technology to clotting factors IX and VIII in an effort to create long-lasting clotting factors.

10.4.2
Clotting Factor Fc Fusions for the Treatment of Hemophilia

Hemophilia A and hemophilia B are bleeding disorders linked to the X chromosome, resulting from deficiency in factor VIII (FVIII) or factor IX (FIX, respectively. Patients with severe hemophilia A or hemophilia B (<1 IU·dL^{-1} of endogenous factor) in particular have repeated frequent bleeding episodes into joints and muscles resulting in hemophilic arthropathy. Internal bleeding episodes, including CNS bleeds, if untreated can result in death. Treatment for hemophilia A and hemophilia B has been on-demand, that is, patients are administered plasma-derived or recombinant DNA-produced preparations of the deficient factor when hemorrhage occurs or before a surgical procedure. Replacement therapy can be given at regular intervals as prophylaxis in order to maintain FVIII or FIX levels greater than $1\,IU \cdot dL^{-1}$ (1% of normal FVIII or FIX plasma level), thereby converting a severe phenotype into a milder one, and as a result, potentially preventing life-threatening hemorrhages and musculoskeletal bleeding events. The aim of FVIII or FIX replacement therapy is to maintain proper coagulation factor levels for sufficient duration whether the treatment is prophylactic, on-demand to manage spontaneous or traumatic bleeds, or in a surgical setting. The prophylactic regimens have been shown to prevent crippling arthropathy [50, 51]. However, to maintain hemostasis, multiple injections are usually required because of the relatively short half-lives of currently available FVIII (approximately 12 h) [52, 53] and FIX (approximately 18 h) [54] products. In most cases, coagulation factors are administered at home by i.v. injection and repeat venous access is necessary, which is particularly challenging in a pediatric population, and may necessitate the use of central venous access devices (ports) such as Port-A-Cath. New generations of recombinant FVIII and FIX with longer half-lives are expected to permit fewer injections, thus reducing the need for repeated peripheral venous access, and potentially improving the acceptance of and compliance with prophylactic regimens. Another potential benefit of reduced i.v. injections are decreased need for central venous ports, reduced risk of infections due to port placement, and decreased repeated dosing in the treatment of break through bleeds or for surgeries. Fc monomer fusion technology has been applied to create recombinant FVIII (rFVIII) and FIX (rFIX) monomeric Fc fusion proteins for the treatment of hemophilia A and hemophilia B.

10.4.2.1 Recombinant Factor IX-Fc Fusion Protein (rFIXFc)

FIX is a 55 kDa serine protease and is an essential component of the coagulation cascade [55–57]. Hemophilia B or Christmas disease is X-linked and affects 1 in 30 000 male births resulting in a deficiency of FIX [58]. Current treatment includes replacement therapy with plasma-derived or recombinant FIX given on-demand for the treatment of bleeds or prophylactically two to three times weekly [59]. Less frequent dosing and prolonged protection from bleeding remain significant unmet needs.

FIX has been recombinantly coupled with Fc as an approach to extend the half-life of this clotting factor and thereby reduce the frequency of injections for hemophilia B patients. The rFIXFc dimer was created by expression of human FIX (hFIX) linked to the Fc domain of IgG1 in HEK293 cells without an intervening linker sequence. This construct is homodimeric since it contains two rFIXFc molecules. The rFIXFc monomer was produced by cotransfection of hFIX linked to Fc along with Fc alone, resulting in a single (monomeric) molecule of hFIX attached to dimeric Fc [5].

Pharmacokinetics (PK) of rFIXFc Dimer and Monomer The PK of the rFIXFc dimer and rFIXFc monomer were compared in neonatal rats after oral administration.

FcRn expression and IgG transport is relatively high during the first few weeks of life in rodents, while secreted acid in the stomach and the level of digestive enzymes are low. As a result, this model is useful in studying the transport and PK of Fc fusion proteins from the epithelial cells in the intestine into the circulation [11, 60]. The uptake of the rFIXFc monomer was much greater compared with the rFIXFc dimer after a similar molar dose, with a 10-fold greater C_{max} and greater than 12-fold increase in AUC (Figure 10.3a, Table 10.1). Furthermore, there was an increase in half-life for the monomer compared with the dimer (19h vs. 14h, respectively).

The PK of rFIXFc monomer and dimer were also studied in adult rats and adult FIX-deficient mice after a single i.v. dose. The FIX-deficient mice are of a knockout strain that was produced using selective gene targeting [61] and as a result, these mice have no endogenous clotting activity, making them a good model for severe human hemophilia B. In both the rat and mouse studies, the exposure, measured by C_{max} and AUC, for the monomer were considerably greater (three- to fourfold and fivefold, respectively) compared with that of the dimer (Tables 10.1 and 10.2). In adult rats, the terminal half-life of the monomer was greater compared with the dimer (Figure 10.3b,c). However, the half-life for the two constructs was similar in the FIX-deficient mice (Table 10.1). It is interesting to note that in all three models, the recovery of the monomer appeared to be greater compared with the dimer. Moreover, the half-life for both Fc fusion molecules were markedly longer compared with rFIX (22 and 35h vs 5.8h). Overall, the rFIXFc monomer construct has distinctly improved PK after oral or i.v. administration in rodents in comparison to the dimer construct.

Pharmacokinetic Comparison of rFIXFc Monomer with rFIX The PK of rFIXFc monomer clotting activity was compared with recombinant FIX (BeneFIX®, rFIX) in FIX-deficient mice. Activity was measured by two different methods: (i) a one-stage assay known as activated partial thromboplastin time, (aPTT) and (ii) whole blood clotting time (WBCT).

Hemophilia B mice were administered a single i.v. dose of $200 \, \text{IU} \cdot \text{kg}^{-1}$ rFIXFc monomer or rFIX and the PK was measured using a modified aPTT assay. The plasma FIX activity declined in a bi-exponential fashion following injection of both rFIXFc and rFIX (Figure 10.4a). The systemic exposure was markedly greater for rFIXFc compared with rFIX, that is, the C_{max} was $135 \, \text{IU} \cdot \text{dL}^{-1}$ and $106 \, \text{IU} \cdot \text{dL}^{-1}$,

---

Figure 10.3 PK of rFIXFc dimer compared with rFIXFc monomer in neonatal rats, adult rats and adult FIX-deficient mice. (a) The PK of rFIXFc dimer and monomer were compared in 10-day-old rats administered a single equimolar oral dose (~10 nmol·kg^{-1}). The concentration of rFIXFc dimer or monomer in plasma was measured using a FIXFc-specific ELISA ($n = 4$ animals per time point, curves represent the mean ± SD). (b) PK of rFIXFc dimer and monomer were compared in adult rats after a single i.v. dose (5 mg·kg^{-1}). The concentration of FIXFc was determined using a specific ELISA ($n = 4$ animals per time point, curves represent the mean ± SD). (c) PK of rFIXFc dimer and monomer compared in adult FIX-deficient mice. Mice were administered a single i.v. dose (5 mg·kg^{-1}) and the concentration of FIXFc in plasma was determined using a specific ELISA ($n = 4$ animals per time point, curves represent mean ± SD).

a) Oral PK in neonatal rats

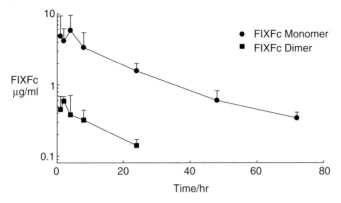

b) IV PK in adult rats

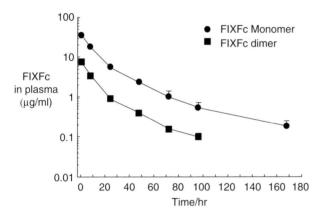

c) IV PK in adult FIX-deficient mice

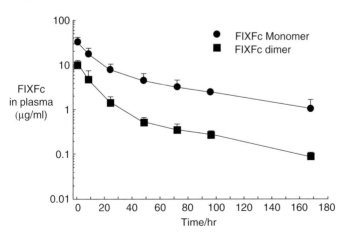

respectively, and the AUC was 1792 and 520 h IU · mL^{-1}, respectively. This represents a 27% increase in C_{max} and 3.4-fold increase in AUC. Thus, the PK parameters measured by coagulation activity are greatly improved for the rFIXFc compared with rFIX. Of note, the increased C_{max} indicates improved recovery for rFIXFc compared with rFIX, which is recognized to have low recovery compared with plasma-derived FIX.

WBCT in Hemophilia B Mice The functional activity of rFIXFc was evaluated *ex vivo* by measuring the WBCT. rFIXFc monomer and rFIX were administered as a single i.v. dose to hemophilia B mice and WBCT was measured as described by Peters *et al.* [5]. Fifteen minutes after dosing, normal clotting was observed in all mice treated with either rFIXFc or rFIX. WBCT returned to baseline by 72 h in all mice treated with rFIX, whereas blood from rFIXFc mice still clotted normally. It wasn't until 144 h that all mice treated with rFIXFc returned to baseline. The percentage of animals able to clot blood at each time point is shown in Figure 10.4b.

PK of rFIXFc in FIX-Deficient Dogs and Monkeys The PK of rFIXFc and rFIX have also been studied in two other species in addition to mice and rats. In FIX-deficient dogs and cynomolgus monkeys, the half-life for rFIXFc was approximately three- to fourfold greater compared with rFIX (Table 10.2). The dog model was one of severe hemophilia B, for which the mutation is similar to that found in humans.

Table 10.2 Summary of terminal half-lives of rFIXFc monomer and rFIX after a single i.v. dose.

Species	rFIX, h	rFIXFc, h
Normal mice	12.3	47.2 ± 4.8
FIX-deficient mice	13.2	46.2 ± 10.1 (47)
Rats	5.8	34.8 ± 5.3
FIX-deficient dogs	(17–18)[a]	47.5 (38.3)
Monkey	12.7[b]	47.3 ± 9.1
FcRn/β$_2$m knockout (KO) mice	16.5 ± 3.0	16.9 ± 2.1
hFcRn/hβ$_2$m transgenic mice	14.2 ± 2.9	53.0 ± 6.6

a) Published values of half-life for recombinant FIX, based on activity data (Brinkhous 1996; McCarthy 2002)
b) Published values of $t_{1/2}$ for recombinant FIX, based on ELISA data (McCarthy 2002)
Animals were administered a single dose of ~200 IU · kg^{-1} rFIXFc in normal mice, $n = 8$), FIX-deficient mice ($n = 11$), FcRn KO mice ($n = 4$), hFcRn/hβ$_2$m transgenic mice ($n = 4$), and rats ($n = 9$); 140 IU · kg^{-1} rFIXFc for FIX-deficient dogs ($n = 2$); and at 25, 100, or 500 IU · kg^{-1} rFIXFc for monkeys ($n = 2, 3,$ or 3, respectively). For rFIX animals were administered a single dose of 100 IU · kg^{-1} in normal mice ($n = 5$) and FIX-deficient mice ($n = 5$), and 200 IU · kg^{-1} in FcRn KO mice ($n = 4$), FcRn Tg32B mice ($n = 4$), and rats ($n = 5$). Concentration of rFIXFc or FIX was measured by a specific ELISA. FIX activity data was also obtained from FIX-deficient mice ($n = 4$/time point) and FIX-deficient dogs ($n = 2$) and ($t_{1/2}$) based on activity was calculated (listed in parentheses).

a)

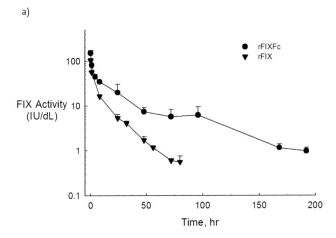

b) WBCT of rFIXFc versus rFIX in FIX-deficient mice

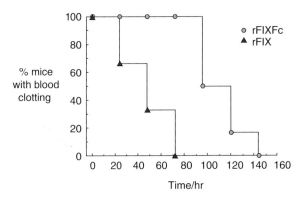

Figure 10.4 Functional activity of rFIXFc and rFIX in FIX-deficient mice. (a) Clotting activity measured by modified aPTT assay in FIX-deficient mice. FIX-deficient mice were administered a single i.v. dose of rFIXFc (200 IU·kg^{-1}, $n = 4$ mice per time point). Blood samples were collected at predetermined time points and analyzed for FIX activity (mean ± SD). (b) FIX-deficient mice (6 per group) were dosed intravenously with 50 IU·kg^{-1} rFIXFc or rFIX. Blood samples were collected before dosing and at various times after dosing. Blood samples were incubated at 37 °C and were visually inspected for the presence of a blood clot once per minute. The time needed for a clot to form was recorded and once the clotting activity returned to baseline (i.e., no clot formation), no additional samples were obtained (samples were collected 15 min to 144 h for rFIXFc or 15 min to 72 h for rFIX).

As a result, the concentration of rFIXFc in the circulation could be readily measured by both activity and antigen levels, since there was no endogenous FIX activity in these animals. The half-life for rFIXFc (47.5 h) was markedly longer compared with that reported for rFIX (17 to 18 h) in FIX-deficient dogs [62, 63]. Similarly, in normal cynomolgus monkeys, the half-life for rFIXFc was 47 h compared with 12.7 h for rFIX observed in this study.

PK in FcRn Knockout Mice and hFcRn Transgenic Mice To confirm the mechanism of action for prolongation of half-life, the PK of rFIXFc was assessed in FcRn knockout mice. As expected, the elimination half-life of rFIXFc was similar to rFIX in the FcRn knockout mice (16.9 h for rFIXFc and 16.5 h for rFIX) (Table 10.2). However, in mice that lacked murine FcRn (mFcRn -/-, mβ_2m -/-) but were transgenic for huFcRn (hFcRn +/+, hβ_2m +/+), the half-life of FIXFc was 53.0 h while rFIX remained unchanged (14.2 h) compared with the knockout mice. These data confirm that FcRn mediates the prolongation of rFIXFc half-life [5].

Clinical Study of rFIXFc rFIXFc was evaluated in an open-label Phase 1/2a dose escalation study in previously-treated patients ($n = 14$) with severe hemophilia B at i.v. doses ranging from 1 to 100 IU · kg^{-1} [64] (ClinicalTrials.gov identifier NCT00716716). The lowest two dose levels (1 and 5 IU · kg^{-1}, 1 subject in each group) were assessed for safety only. In addition to safety, the PK of rFIXFc was assessed in the remaining dose levels (12.5, 25, 50 and 100 IU · kg^{-1}). rFIXFc was well-tolerated in this single dose escalation study. PK analysis showed approximately a threefold increase in half-life, reduced clearance, and improved incremental recovery compared with historical values reported for rFIX [65, 66]. The results from the Phase 1/2a rFIXFc study supported development of the Phase 3 study to evaluate the safety, pharmacokinetics, and efficacy of rFIXFc in previously-treated patients with severe hemophilia B (ClinicalTrials.gov identifier NCT01027364).

10.4.2.2 Recombinant Factor VIII-Fc Fusion Protein (rFVIIIFc)

Hemophilia A or FVIII deficiency is also an X-linked disease and affects 1 in 5000 males [58]. FVIII has a molecular weight of approximately 280 kDa and is comprised of a heavy (200 kDa) and light chain (80 kDa) with several distinct domains (A1-A2-A3-B-C1-C2). It was discovered that the B-domain which contains 19 of 25 potential glycosylation sites has no impact on the function of when FVIII is activated [67] . In fact, B-domain deleted FVIII is a normal intermediate during the process of coagulation *in vivo*. Current treatment for hemophilia A patients includes injection of FVIII product given on-demand in response to a bleed or as a prophylactic regimen in a majority of cases every other day or three times weekly. Prophylactic treatment in hemophilia A patients has clearly demonstrated reduction in hemarthroses and alleviation of subsequent arthropathy [50, 53]. However, there is an unmet medical need for long-lasting rFVIII products to address less frequent dosing.

A recombinant B-domain-deleted (BDD) factor VIII-Fc fusion protein (rFVIIIFc) was created as an approach to extend the plasma half-life of FVIII. rFVIIIFc is a heterodimeric protein comprised of BDD FVIII fused recombinantly to the Fc domain of human IgG1 as described above (single FVIII molecule attached to dimeric Fc). For this Fc fusion therapeutic protein, the Fc monomer technology was absolutely enabling, as we were unable to produce the dimeric Fc fusion protein, presumably because the large size-limited secretion from the cells (~380 kDa). The PK and clotting activity of the rFVIIIFc monomer were studied in mouse and dog models of hemophilia A. Due to the difficulty in making the dimer of rFVIIIFc, the PK comparisons were made only with unconjugated rFVIII. The results of these studies are described below.

Clotting Activity of rFVIIIFc and rFVIII in Hemophilia A Mice   The ability of rFVIIIFc to clot blood was evaluated in hemophilia A mice by measuring WBCT after a single dose of rFVIIIFc or BDD-rFVIII (ReFacto®) at $50 \, IU \cdot kg^{-1}$. At selected times after dosing, blood was collected and visually inspected once per minute for the formation of a clot. Baseline clotting in these mice was >60 min, but after administration of rFVIIIFc or ReFacto, WBCT was immediately restored to normal. Blood from mice treated with ReFacto lost the ability to clot after 42 h. In contrast, all of the mice treated with rFVIIIFc ($50 \, IU \cdot kg^{-1}$) maintained effective clotting through 96 h, but lost the ability to clot at 120 h after dosing (Figure 10.5). These data suggest that there is an approximate threefold improvement in WBCT for rFVIIIFc compared with ReFacto.

rFVIIIFc and rFVIII PK and PD in Hemophilia A Dogs   The PK and pharmacodynamics of rFVIIIFc were also evaluated in the in-bred Chapel Hill colony of hemophilia

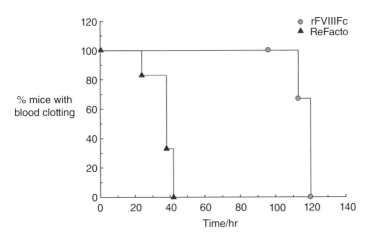

Figure 10.5   WBCT of rFVIIIFc and rFVIII (ReFacto) in FVIII-deficient mice after i.v. administration. FVIII-deficient mice (6 per group) were injected intravenously with a single dose of rFVIIIFc or rFVIII at $50 \, IU \cdot kg^{-1}$. Blood was collected from each mouse before dosing and post-dosing at various time points. The blood samples were incubated at 37 °C and visually inspected once per minute for the presence of a clot. Time of clot formation was recorded. Once clotting returned to baseline, no additional samples were collected.

A dogs [68, 69]. These dogs have a severe hemophilic phenotype that is comparable to the severe form of human disease with FVIII:C <1%. A single i.v. dose (125 IU · kg^{-1} rFVIIIFc) was administered to four dogs, which immediately corrected the blood clotting to normal as measured by WBCT and aPTT [48]. The WBCT remained below 20 min, the time consistent with FVIII:C >1%, through approximately 96 h. The range of WBCT in our normal dogs is 8 to 12 min. The concentration of rFVIIIFc in the plasma was measured by both ELISA and chromogenic activity assay, which showed that the concentration vs. time curves were similar using both methods. Furthermore, the terminal half-life was similar 15.7 ± 1.7 h (ELISA) and 15.4 ± 0.3 h (activity).

Two of the dogs also received a single dose of ReFacto, 114 IU · kg^{-1} for one dog and 120 IU · kg^{-1} for the other, prior to receiving rFVIIIFc (125 IU · kg^{-1}) 72 h later in a cross over design. A washout period of three days was chosen to allow the clotting to return to baseline. Clotting was corrected to normal immediately after dosing with both ReFacto and rFVIIIFc (determined by WBCT and clotting activity measured using an aPTT assay). However, the WBCT normalization after rFVIIIFc lasted for approximately twice as long compared with ReFacto. Moreover, the half-life determined by ELISA or activity was approximately half that determined for ReFacto (7.0 h and 6.7 h). No adverse clinical signs were detected with any of the infusions. Therefore, construction of a monomeric Fc fusion of FVIII produces a molecule with a defined mechanism of action that has an increased half-life and the potential to provide prolonged protection from bleeding.

10.5
Summary

Protein therapeutics have emerged as a major source of innovative drug development in modern medicine. Nonetheless, they still face many issues and one of the most challenging ones is caused by poor half-life *in vivo*. Great efforts have been made to discover and develop approaches to prolong the half-life of protein or peptide drugs. Fc fusion technology, by which the Fc fragment of IgG is coupled to a protein or peptide drug to extend its half-life *in vivo* by means of the FcRn recycling mechanism, is one of the successful examples. Well-known and validated traditional Fc fusion drugs, such as etanercept for the treatment of rheumatoid arthritis, consist of two copies of protein or peptide molecules linked to the Fc region to improve PK, solubility, and production efficiency. A new class of Fc fusion proteins that links a single copy of protein drug to the Fc region has demonstrated improved biologic properties, including extended half-life and increased bioavailability. Recombinant blood coagulation factor-Fc fusions (rFVIIIFc and rFIXFc) were constructed with this technology to create long-lasting clotting factors for the treatment of hemophilia A and hemophilia B. Both rFVIIIFc and rFIXFc have been shown to have increased systemic half-lives in preclinical models and are currently being evaluated in global pivotal clinical trials.

Acknowledgments

The authors are associated with Biogen Idec Hemophilia, a company focused on the development of long-lasting clotting factors for therapeutic use. The work with rFIXFc and rFVIIIFc were performed in partnership with Swedish Orphan Biovitrum. No other sources of funding were used to assist in the preparation of this article.

References

1 Mei, B., Pan, C., Jiang, H., Tjandra, H., Strauss, J., Chen, Y., Liu, T., Zhang, X., Severs, J., Newgren, J., Chen, J., Gu, J., Subramanyam, B., Fournel, M.A., Pierce, G.F., and Murphy, J.E. (2010) Rational design of a fully active, long-acting PEGylated factor VIII for haemophilia A treatment. *Blood*, **116** (2), 270–279.

2 DeFrees, S., Wang, Z.G., Xing, R., Scott, A.E., Wang, J., Zopf, D., Gouty, D.L., Sjoberg, E.R., Panneerselvam, K., Brinkman-Van der Linden, E.C., Bayer, R.J., Tarp, M.A., and Clausen, H. (2006) GlycoPEGylation of recombinant therapeutic proteins produced in *Escherichia coli*. *Glycobiology*, **16** (9), 833–843.

3 Chuang, V.T., Kragh-Hansen, U., and Otagiri, M. (2002) Pharmaceutical strategies utilizing recombinant human serum albumin. *Pharm. Res.*, **19** (5), 569–577.

4 Ashkenazi, A., Capon, D.J., and Ward, R.H. (1993) Immunoadhesins. *Int. Rev. Immunol.*, **10** (2–3), 219–227.

5 Peters, R.T., Low, S.C., Kamphaus, G.D., Dumont, J.A., Amari, J.V., Lu, Q., Zarbis-Papastoitsis, G., Reidy, T.J., Merricks, E.P., Nichols, T.C., and Bitonti, A.J. (2010) Prolonged activity of factor IX as a monomeric Fc fusion protein. *Blood*, **115** (10), 2057–2064.

6 Waldmann, T.A., and Strober, W. (1969) Metabolism of immunoglobulins. *Prog. Allergy*, **13**, 1–110.

7 Morell, A., Terry, W.D., and Waldmann, T.A. (1970) Metabolic properties of IgG subclasses in man. *J. Clin. Invest.*, **49** (4), 673–680.

8 Brambell, F.W., Hemmings, W.A., and Morris, I.G. (1964) A theoretical model of gamma-globulin catabolism. *Nature*, **203**, 1352–1354.

9 Brambell, F.W. (1966) The transmission of immunity from mother to young and the catabolism of immunoglobulins. *Lancet*, **2** (7473), 1087–1093.

10 Simister, N.E., and Rees, A.R. (1985) Isolation and characterization of an Fc receptor from neonatal rat small intestine. *Eur. J. Immunol.*, **15** (7), 733–738.

11 Ghetie, V., Hubbard, J.G., Kim, J.K., Tsen, M.F., Lee, Y., and Ward, E.S. (1996) Abnormally short serum half-lives of IgG in beta 2-microglobulin-deficient mice. *Eur. J. Immunol.*, **26** (3), 690–696.

12 Israel, E.J., Wilsker, D.F., Hayes, K.C., Schoenfeld, D., and Simister, N.E. (1996) Increased clearance of IgG in mice that lack beta 2-microglobulin: possible protective role of FcRn. *Immunology*, **89** (4), 573–578.

13 Junghans, R.P., and Anderson, C.L. (1996) The protection receptor for IgG catabolism is the beta 2-microglobulin-containing neonatal intestinal transport receptor. *Proc. Natl. Acad. Sci. U.S.A.*, **93**, 5512–5516.

14 Waldmann, T.A., and Terry, W.D. (1990) Familial hypercatabolic hypoproteinemia. A disorder of endogenous catabolism of albumin and immunoglobulin. *J. Clin. Invest.*, **86** (16), 2093–2098.

15 Wani, M.A., Haynes, L.D., Kim, J., Bronson, C.L., Chaudhury, C., Mohanty, S., Waldmann, T.A., Robinson, J.M., and Anderson, C.L. (2006) Familial hypercatabolic hypoproteinemia caused by deficiency of the neonatal Fc receptor, FcRn, due to a mutant beta2-microglobulin

gene. *Proc. Natl. Acad. Sci. U.S.A.*, **103** (13), 5084–5089.

16 Simister, N.E., and Mostov, K.E. (1989) An Fc receptor structurally related to MHC class I antigens. *Nature*, **337** (6203), 184–187.

17 Rodewald, R., and Kraehenbuhl, J.P. (1984) Receptor-mediated transport of IgG. *J. Cell Biol.*, **99** (1)Pt 2):159s–164s.

18 Roberts, D.M., Guenthert, M., and Rodewald, R. (1990) Isolation and characterization of the Fc receptor from the fetal yolk sac of the rat. *J. Cell Biol.*, **111** (5)Pt 1):1867–1876.

19 Simister, N.E., Story, C.M., Chen, H.L., and Hunt, J.S. (1996) An IgG-transporting Fc receptor expressed in the syncytiotrophoblast of human placenta. *Eur. J. Immunol.*, **26** (7), 1527–1531.

20 Martin, M.G., Wu, S.V., and Walsh, J.H. (1997) Ontogenetic development and distribution of antibody transport and Fc receptor mRNA expression in rat intestine. *Dig. Dis Sci*, **42** (5), 1062–1069.

21 Ward, E.S., Zhou, J., Ghetie, V., and Ober, R.J. (2003) Evidence to support the cellular mechanism involved in serum IgG homeostasis in humans. *Int. Immunol.*, **15** (2), 187–195.

22 Borvak, J., Richardson, J., Medesan, C., Antohe, F., Radu, C., Simionescu, M., Ghetie, V., and Ward, E.S. (1998) Functional expression of the MHC class I-related receptor, FcRn, in endothelial cells of mice. *Int. Immunol.*, **10** (9), 1289–1298.

23 Akilesh, S., Huber, T.B., Wu, H., Wang, G., Roopenian, B., Kopp, J.B., Miner, J.H., Roopenian, D.C., Unanue, E.R., and Shaw, A.S. (2008) Podocytes use FcRn to clear IgG from the glomerular basement membrane. *Proc. Natl. Acad. Sci. U.S.A.*, **105** (3), 967–972.

24 Christianson, G.J., Brooks, W., Vekasi, S., Manolfi, E.A., Niles, J., Roopenian, S.L., Roths, J.B., Rothlein, R., and Roopenian, D.C. (1997) Beta2-microglobulin-deficient mice are protected from hypergammaglobulinemia and have defective antibody responses because of increased IgG catabolism. *J. Immunol.*, **159** (10), 4781–4792.

25 Haymann, J.P., Levraud, J.P., Bouet, S., Kappes, V., Hagege, J., Nguyen, G., Xu, Y., Rondeau, E., and Sraer, J.D. (2000)

Characterization and localization of the neonatal Fc receptor in adult human kidney. *J. Am. Soc. Nephrol.*, **11** (4), 632–639.

26 Blumberg, R.S., Koss, T., Story, C.M., Barisani, D., Polischuk, J., Lipin, A., Pablo, L., Green, R., and Simister, N.E. (1995) A major histocompatibility complex class I-related Fc receptor for IgG on rat hepatocytes. *J. Clin. Invest.*, **95** (5), 2397–2402.

27 Telleman, P., and Junghans, R.P. (2000) The role of the Brambell receptor (FcRB) in liver: protection of endocytosed immunoglobulin G (IgG) from catabolism in hepatocytes rather than transport of IgG to bile. *Immunology*, **100** (2), 245–251.

28 Israel, E.J., Taylor, S., Wu, Z., Mizoguchi, E., Blumberg, R.S., Bhan, A., and Simister, N.E. (1997) Expression of the neonatal Fc receptor, FcRn, on human intestinal epithelial cells. *Immunology*, **92** (1), 69–74.

29 Spiekermann, G.M., Finn, P.W., Ward, E.S., Dumont, J., Dickinson, B.L., Blumberg, R.S., and Lencer, W.I. (2002) Receptor-mediated immunoglobulin G transport across mucosal barriers in adult life: functional expression of FcRn in the mammalian lung. *J. Exp. Med.*, **196** (3), 303–310.

30 Bitonti, A.J., Dumont, J.A., Low, S.C., Peters, R.T., Kropp, K.E., Palombella, V.J., Stattel, J.M., Lu, Y., Tan, C.A., Song, J.J., Garcia, A.M., Simister, N.E., Spiekermann, G.M., Lencer, W.I., and Blumberg, R.S. (2004) Pulmonary delivery of an erythropoietin Fc fusion protein in non-human primates through an immunoglobulin transport pathway. *Proc. Natl. Acad. Sci. U.S.A.*, **101** (26), 9763–9768.

31 Ober, R.J., Martinez, C., Vaccaro, C., Zhou, J., and Ward, E.S. (2004) Visualizing the site and dynamics of IgG salvage by the MHC class I-related receptor, FcRn. *J. Immunol.*, **172** (4), 2021–2029.

32 Ward, E.S., and Ober, R.J. (2009) Chapter 4: Multitasking by exploitation of intracellular transport functions the many faces of FcRn. *Adv. Immunol.*, **103**, 77–115.

33 Jones, E.A., and Waldmann, T.A. (1972) The mechanism of intestinal uptake and transcellular transport of IgG in the neonatal rat. *J. Clin. Invest.*, **51** (11), 2916–2927.

34 Rodewald, R. (1976) pH-dependent binding of immunoglobulins to intestinal cells of the neonatal rat. *J. Cell Biol.*, **71** (2), 666–669.

35 Raghavan, M., Gastinel, L.N., and Bjorkman, P.J. (1993) The class I major histocompatibility complex related Fc receptor shows pH-dependent stability differences correlating with immunoglobulin binding and release. *Biochemistry*, **32** (33), 8654–8660.

36 Kim, J.K., Tsen, M.F., Ghetie, V., and Ward, E.S. (1994) Localization of the site of the murine IgG1 molecule that is involved in binding to the murine intestinal Fc receptor. *Eur. J. Immunol.*, **24** (10), 2429–2434.

37 Kim, J.K., Firan, M., Radu, C.G., Kim, C.H., Ghetie, V., and Ward, E.S. (1999) Mapping the site on human IgG for binding of the MHC class I-related receptor, FcRn. *Eur. J. Immunol.*, **29** (9), 2819–2825.

38 Vaughn, D.E., and Bjorkman, P.J. (1998) Structural basis of pH-dependent antibody binding by the neonatal Fc receptor. *Structure*, **6** (1), 63–73.

39 Burmeister, W.P., Gastinel, L.N., Simister, N.E., Blum, M.L., and Bjorkman, P.J. (1994) Crystal structure at 2.2 A° resolution of the MHC-related neonatal Fc receptor. *Nature*, **372** (6504), 336–343.

40 West, A.P., Jr., and Bjorkman, P.J. (2000) Crystal structure and immunoglobulin G binding properties of the human major histocompatibility complex-related Fc receptor. *Biochemistry*, **39** (32), 9698–9708.

41 Garrison, L., and McDonnell, N.D. (1999) Etanercept–therapeutic use in patients with rheumatoid arthritis. *Ann. Rheum. Dis.*, **58** (Suppl. I), I65–I69.

42 Frampton, J.E., and Wagstaff, A.I. (2003) Alefacept. *Ann. J. Dermatol.*, **4** (4), 227–286.

43 Kremer, J.M., Westhovens, R., Leon, M., Di Giorgio, E., Alten, R., Steinfeld, S., Russell, A., Dougados, M., Emery, P., Nuamah, I.F., Williams, G.R., Becker, J.C., Hagerty, D.T., and Moreland, L.W. (2003) Treatment of rheumatoid arthritis by selective inhibition of T-cell activation with fusion protein CTLA4Ig. *N. Engl. J. Med.*, **349** (20), 1907–1915.

44 Goldbach-Mansky, R., Shroff, S., Wilson, M., Snyder, C., Plehn, S., Barham, B., Pham, T.H., Kastner, D.L., Pucino, F., Wesley, R.A., Papadopoulos, J., Weinstein, S., and Mellis, S. (1976) A pilot study to evaluate the safety and efficacy of the long-acting IL-1 inhibitor rilonacept (IL-1 Trap) in patients with familial cold autoinflammatory syndrome. *Arthritis Rheum.*, **58** (8), 2432–2442.

45 Kuter, D.J., Rummel, M., Boccia, R., Macik, B.G., Pabinger, I., Selleslag, D., Rodeghiero, F., Chong, B.H., Wang, X., and Berger, D.P. (2010) Romiplostim or standard of care in patients with immune thrombocytopenia. *N. Engl. J. Med.*, **363** (20), 1889–1899.

46 Roopenian, D.C., and Akilesh, S. (2007) FcRn: the neonatal Fc receptor comes of age. *Nat. Rev. Immunol.*, **7** (9), 715–725.

47 Dumont, J.A., Low, S.C., Peters, R.T., and Bitonti, A.J. (2006) Monomeric Fc fusions: impact on pharmacokinetic and biological activity of protein therapeutics. *BioDrugs*, **20** (3), 151–160.

48 Dumont, J.A., Kamphaus, G.D., Fraley, C., Ashworth, T., Franck, H., Merricks, E.P., Nichols, T.C., and Bitonti, A.J. (2009) Factor VIII-Fc fusion protein shows extended half-life and hemostatic activity in hemophilia A dogs. *Blood*, **114**, #545.

49 Graber, S.E., and Krantz, S.B. (1978) Erythropoietin and the control of red cell production. *Annu. Rev. Med.*, **29**, 51–66.

50 Aladort, L.M. (1992) Prophylaxis: the next haemophilia treatment. *J. Intern. Med.*, **232** (1), 1–2.

51 Manco-Johnson, M. (2007) Comparing prophylaxis and episodic treatment in hemophilia A-implications for clinical practice. *Haemophilia*, **13** (Suppl. 2), 4–9.

52 Björkman, S., and Berntorp, E. (2001) Pharmacokinetics of coagulation factors: clinical relevance for patients with haemophilia. *Clin. Pharmacokinet.*, **40** (11), 815–832.

53 Björkman, S., Folkesson, A., and Jönsson, S. (2009) Pharmacokinetics and dose requirements of factor VIII over the age range 3–74 years: a population analysis based on 50 patients with long-term prophylactic treatment for haemophilia A. *Eur. J. Clin. Pharmacol.*, **65** (10), 989–998.

54 White, G., Shapiro, A., Ragni, M., Garzone, P., Goodfellow, J., Tubridy, K., and Courter, S. (1998) Clinical evaluation of recombinant factor IX. *Semin. Hematol.*, **35** (2 Suppl 2), 33–38.

55 Limentani, S.A., Roth, D.A., Furie, B.C., and Furie, B. (1993) Recombinant blood clotting proteins for hemophilia therapy. *Semin. Thromb. Hemost.*, **19** (1), 62–72.

56 Green, P.M., Naylor, J.A., and Giannelli, F. (1995) The hemophilias. *Adv. Genet.*, **32**, 99–139.

57 Bolton-Maggs, P.H., and Pasi, K.J. (2003) Haemophilias A and B. *Lancet*, **361**, (9371) 1801–1809.

58 Mannucci, P.M., and Tuddenham, E.G. (2001) The haemophilias – from royal genes to gene therapy. *N. Engl. J. Med.*, **344**, 1773–1779.

59 Shapiro, A.D., Korth-Bradley, J., and Poon, M.C. (2005) Use of pharmacokinetics in the coagulation factor treatment of patients with haemophilia. *Haemophilia*, **11** (6), 571–582.

60 Morris, B., and Morris, R. (1974) The absorption of ^{125}I-labelled immunoglobulin G by different regions of the gut in young rats. *J. Physiol.*, **241** (3), 761–770.

61 Lin, H.F., Maeda, N., Smithies, O., Straight, D.L., and Stafford, D.W. (1997) A coagulation factor IX-deficient mouse model for human hemophilia B. *Blood*, **90** (10), 3962–3966.

62 Brinkhous, K.M., Sigman, J.L., Read, M.S., Stewart, P.F., McCarthy, K.P., Timony, G.A., Leppanen, S.D., Rup, B.J., Keith, J.C., Jr., Garzone, P.D., and Schaub, R.G. (1996) Recombinant human factor IX: replacement therapy, prophylaxis, and pharmacokinetics in canine hemophilia B. *Blood*, **88** (7), 2603–2610.

63 McCarthy, K., Stewart, P., Sigman, J., Read, M., Keith, J.C., Jr., Brinkhous, K.M., Nichols, T.C., and Schaub, R.G. (2002) Pharmacokinetics of recombinant factor IX after intravenous and subcutaneous administration in dogs and cynomolgus monkeys. *Thromb. Haemost.*, **87** (5), 824–830.

64 Shapiro, A.D., Ragni, M., Valentino, L., Key, N., Josephson, N., Powell, J., Cheng, G., Tubridy, K.L., Peters, R., Dumont, J., Luk, A., Hallen, B., Gozzi, P., Bitonti, A., and Pierce, G.F. (2010) Safety and prolonged biological activity following a single administration of a recombinant molecular fusion of native human coagulation factor IX and the Fc region of immunoglobulin G (IgG) (rFIXFc) to subjects with hemophilia B. *Haemophilia*, **16** (Suppl. 4), 1–158. #07FP07.

65 Roth, D.A., Kessler, C.M., Pasi, K.J., Rup, B., Courter, S.G., Tubridy, K.L., and Recombinant Factor IX Study Group. (2001) Human recombinant factor IX: safety and efficacy studies in hemophilia B patients previously treated with plasma-derived factor IX concentrates. *Blood*, **98**, 3600–3606.

66 BeneFIX® Summary of Product Characteristics for BeneFIX. Electronic Medicines Compendium 2010.

67 Sandberg, H., Almstedt, A., Brandt, J., Gray, E., Holmquist, L., Oswaldsson, U., Sebring, S., and Mikaelsson, M. (2001) Structural and functional characteristics of the B-domain-deleted recombinant factor VIII protein, r-VIII SQ. *Thromb. Haemost.*, **85** (1), 93–100.

68 Graham, J.B., Buckwalter, J.A., Hartley, L.J., and Brinkhous, K.M. (1949) Canine hemophilia; observations on the course, the clotting anomaly, and the effect of blood transfusions. *J. Exp Med.*, **90** (2), 97–111.

69 Lozier, J.N., Dutra, A., Pak, E., Zhou, N., Zheng, Z., Nichols, T.C., Bellinger, D.A., Read, M., and Morgan, R.A. (2002) The Chapel Hill hemophilia A dog colony exhibits a factor VIII gene inversion. *Proc. Natl. Acad. Sci. U.S.A.*, **99** (20), 12991–12996.

11

The Diverse Roles of FcRn: Implications for Antibody Engineering

E. Sally Ward and Raimund J. Ober

11.1
Introduction

Over the last 15 years it has become evident that the MHC Class I-related receptor, FcRn, serves multiple functions through its ability to transport immunoglobulin G (IgG) within and across cells (reviewed in [1–3]). FcRn is expressed throughout life in a diverse array of tissues and cell types (e.g., [4–11]). As an IgG transporter, this receptor not only maintains constant IgG concentrations in the body [4, 12, 13], but also delivers antibodies of this class across cellular barriers [5, 14–21]. In this chapter we first give a brief overview of the early studies of FcRn followed by the molecular details of FcRn–IgG interactions and a description of FcRn behavior at the level of intracellular trafficking. Finally, we discuss how FcRn–IgG interactions can be engineered to alter both the pharmacokinetic properties of an IgG and the concentrations of endogenous IgGs. In addition, although more recent reports have demonstrated that FcRn binds and maintains the homeostasis of albumin [22, 23], this topic will not be discussed in detail here. Instead, the reader is referred to Chapter 10.

11.2
FcRn: Early Characterization and Diverse Expression Patterns

The role of FcRn as its namesake, the neonatal Fc receptor, in transporting maternal IgG from mother to young across the neonatal intestine was characterized in the 1980s [24–27]. These studies included the seminal work of Simister and Mostov describing the cloning of the rat FcRn α-chain gene and the identification of this polypeptide as a member of the MHC Class I receptor family [27]. Subsequent analyses have resulted in the isolation of FcRn from human placenta [28, 29] and the demonstration that it is an essential player in the transport of maternal IgG from mother to young during the third trimester of pregnancy [15].

Therapeutic Proteins: Strategies to Modulate Their Plasma Half-Lives, First Edition. Edited by
Roland Kontermann.
© 2012 Wiley-VCH Verlag GmbH & Co. KGaA. Published 2012 by Wiley-VCH Verlag GmbH & Co. KGaA.

FcRn α–chain genes have been isolated from many different species [27, 29–36], indicating that the protein is broadly expressed in mammals. Although FcRn genes show some conservation across species, there are differences in both putative interaction residues with IgG and regions involved in trafficking. The functional impact of these differences for IgG interactions have been investigated for mouse/rat and human FcRn (discussed below), but in many cases, the way these variations impact on activity is unknown.

11.3
The Molecular Details of FcRn–IgG Interactions

A combination of mutagenesis and structural studies has been used to map the interaction site of FcRn on IgG in rodents and humans [37–41]. IgG residues Ile253, His310, His435 are centrally involved in the interactions [37–41]. These residues are conserved across most human and mouse/rat IgG isotypes and are located at the CH2–CH3 domain interface [42] (Figure 11.1). His436 (present in most mouse IgG isotypes) or Tyr436 (present in most human IgG isotypes) play a lesser but significant role in binding to FcRn [38, 43]. Multiple studies have shown that the FcRn binding site on IgG does not overlap with the interaction sites for the classical FcγRs and complement C1q [37, 43–45]. The distinct interaction sites have implications for the engineering of antibodies that are discussed further below.

Mutagenesis and structural studies have also been used to identify the residues of rat FcRn that are involved in IgG binding [41, 46]. These residues include Ile1 of β2-microglobulin and Glu117, Glu118, Glu132, Trp133, Glu135 and Asp137 of the FcRn α-chain. The interaction of acidic residues of FcRn with histidines on IgG represents the primary contributor to the pH dependence of the interaction,

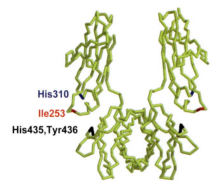

His310

Ile253

His435,Tyr436

Figure 11.1 Structure (α-carbon trace) of the Fc region of human IgG1 [42] with the location of the key residues that are involved in binding to mouse or human FcRn indicated [15, 39, 43] (drawn using Rasmol, courtesy of Roger Sayle). The same residues of mouse IgG1 are also involved in FcRn binding [37, 38] except that Tyr436 is replaced by histidines in most mouse IgG isotypes.

with binding at acidic pH that for most IgG isotypes becomes undetectable at near neutral pH [25, 40, 47, 48].

Importantly, human and mouse FcRn exhibit marked differences in binding specificity [49]. Although mouse FcRn binds to IgGs from a wide array of species, human FcRn is much more selective [49]. For example, mouse IgGs bind very poorly, if at all, to human FcRn. Conversely, mouse FcRn binds to human IgG1 with higher affinity relative to the corresponding human FcRn-human IgG1 interaction. This cross-species difference can confound the interpretation of data when (engineered) human IgGs are tested preclinically in mice [50] (discussed further below). Site-directed mutagenesis of human FcRn has been used to determine the molecular basis for the difference in binding between human and mouse FcRn [51, 52]. These studies have shown that an amino acid difference at residue 137 (Leu in human FcRn, Glu or Asp in mouse or rat FcRn, respectively) is the major contributor, with other regions playing more minor roles.

FcRn has two possible binding sites on IgG, and these sites are not equivalent [53–55]. This has led to the concept that binding of FcRn to one site reduces the affinity of the second site [53, 56] through either steric effects or longer range conformational changes. This raises the question as to whether two functional sites per IgG molecule are necessary for activity in FcRn-mediated functions. This has been addressed by generating hybrid Fc fragments comprising one wild-type Ig heavy chain associated with a mutated heavy chain that does not bind to FcRn [37, 57, 58]. Such hybrid molecules are poorly delivered across cellular barriers [37, 58] and have reduced *in vivo* half-lives [57]. Whether this lower activity is due to effects on FcRn trafficking (through lack of dimerization of the receptor) or reduced avidity for interactions with membrane-bound FcRn, or a combination of both, is currently unknown.

11.4
FcRn Is Expressed Ubiquitously throughout the Body Where It Serves Multiple Functions

In addition to the overexpression of FcRn in the neonatal gut and the human placenta, this receptor is expressed in endothelial, epithelial and many, but not all, hematopoietic cells throughout life (e.g., [4–11, 18]). Although FcRn is expressed in all "professional" antigen presenting cells (dendritic cells, macrophages and B cells), it is not detectable in T cells [7, 59]. FcRn is also present in more specialized cells/sites such as corneal epithelium/endothelium and podocytes in the kidney [10, 11, 60]. This raises the question as to which functions it might serve at these different cellular/body sites.

To identify which sites are important for the control of IgG levels, we have generated mice containing a "floxed" FcRn allele that can be conditionally deleted in different cell subsets by crossing with appropriate Cre recombinase expressing strains [59]. To date, we have characterized mice that lack FcRn in all endothelial and hematopoietic cells by using a Cre recombinase strain in which expression is driven by the Tie2 promoter [61]. These studies, combined with others using bone marrow transfers [8, 9], have shown that hematopoietic and endothelial cells constitute the primary sites for the regulation of IgG persistence *in vivo* [59].

In addition to an important contribution of hematopoietic cells for the control of IgG concentrations and pharmacokinetics [8, 9], expression of FcRn in antigen presenting cells has been shown to play a role in presentation to cognate T cells [9, 62]. Specifically, presentation of antigen derived from immune complexes is more efficient in FcRn-sufficient dendritic cells relative to their knockout counterparts [9]. This has been demonstrated to be due to enhanced, FcRn-dependent trafficking of such complexes into lysosomes in dendritic cells. In addition, invariant chain drives FcRn into lysosomes in antigen presenting cells [63]. This invariant chain enhancement of lysosomal expression could provide an explanation for the differences in fate of immune complexes across cell types: in invariant chain expressing antigen presenting cells, immune complexes enter lysosomes [9], whereas in epithelial cells these complexes are transcytosed [21].

Recent studies have also explored the role of FcRn expression in the kidney [10, 60]. FcRn is present in proximal tubular epithelial cells, podocytes and renal vascular endothelia [10, 60]. Studies in FcRn α-chain deficient mice have demonstrated that FcRn in podocytes is involved in removing IgG from the glomerular basement membrane, thereby enhancing removal of IgG from the kidneys and reducing glomerular damage [10]. Interestingly, using the approach of transplanting kidneys from wild-type mice into FcRn-/- recipients, renal FcRn has been shown to be a key player in the maintenance of serum albumin levels while assisting in the elimination of IgG [60]. This indicates that by analogy with *in vitro* studies using rat FcRn [58], rodent FcRn handles albumin and IgG differently. How this is achieved is poorly understood but is an area that deserves attention.

The function of FcRn expression at the blood–brain barrier (BBB) and other immune privileged sites, such as the eye, has attracted significant interest [11, 64–67]. In some studies, FcRn has been reported to enhance the egress of IgG from the brain [65, 66], whereas others have shown that FcRn plays no role in the disposition of IgG at this site [67]. Defining whether FcRn is involved, and can even be modulated, in IgG transport at the BBB has obvious implications for drug delivery and the treatment of neurological diseases.

11.5
The Cell Biology of FcRn and Its Intracellular Transport of IgG

Over the past decade, studies have been directed towards defining how IgG cargo behaves at the level of subcellular trafficking in both endothelial and epithelial cells (e.g., [14, 20, 58, 68–72]). A model for FcRn-mediated trafficking of IgGs is shown in Figure 11.2. Recent studies have given insight into the intracellular sorting, recycling and exocytic processes that result in FcRn-mediated recycling and in some cases, transcytosis, of IgG. For example, live cell imaging of endothelial cells has demonstrated that wild-type IgGs are recycled from sorting endo-

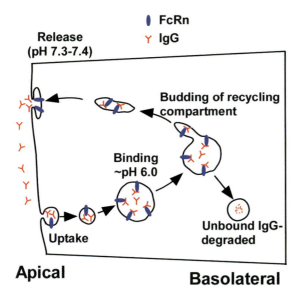

Figure 11.2 Model for FcRn-mediated trafficking of IgG in polarized endothelial cells. IgG is taken up into the cells by fluid phase pinocytosis and enters early, acidic endosomes in which it can bind to FcRn. Bound IgG is recycled (or transcytosed; not shown) and released at the cell surface due to the change in pH. Unbound IgG enters lysosomes and is degraded.

somes (i.e., large endosomal structures of around 1–2 μm) into the recycling pathway [68]. By contrast, IgGs that are mutated so that they no longer bind to FcRn are not sorted. By default, these IgGs remain in the vacuole of the sorting endosome as it matures to form a late endosome. The contents of these late endosomes are ultimately delivered to lysosomes where they are degraded.

Recycling of IgG by FcRn is followed by exocytic release that can involve complete fusion of the FcRn positive compartment with the plasma membrane and release of IgG [20, 69]. Alternatively, other types of exocytic processes can occur, ranging from partial fusion to prolonged release in which IgG is released in fusion events that resemble kiss-and-run or kiss-and-linger interactions similar to those described for neurotransmitter release at synaptic junctions [20, 73–75]. The visualization of distinct types of fusion events at the plasma membrane using total internal reflection fluorescence microscopy (TIRFM; [76]), without understanding which intracellular pathways preceded them, led to the development of a multifocal imaging modality in our laboratory [69, 77]. This multifocal plane microscopy (MUM) involves the simultaneous collection of fluorescent signal from multiple focal planes within a cell in addition to the use of TIRFM. This allows rapidly moving tubulovesicular transport containers (TCs) to be tracked as they move, for example, from sorting endosomes to exocytic sites at the plasma membrane. Similarly, the analysis of endocytic processes followed by intracellular tracking of vesicular TCs on the early endosomal pathway is possible [70]. The tubulovesicular

TCs observed in these analyses are analogous to those described by Bjorkman and colleagues using electron tomography to study rat jejunal sections [78].

The use of MUM has resulted in the identification of different pathways for both recycling/exocytosis and endocytosis [69, 70]. In the most direct pathway of endosomal recycling and exocytosis, tubules can extend from sorting endosomes and fuse with the plasma membrane while still attached to the originating compartment [69]. Alternatively, recycling compartments can take more indirect itineraries within the cell before they undergo exocytosis (Figure 11.3). By analogy, FcRn is endocytosed from the plasma membrane and enters sorting endosomes via direct and indirect pathways [70]. This raises the question as to what determines the pathways that are taken? Understanding these processes in molecular terms could open up avenues for modulating antibody behavior *in vivo*.

11.6
The Molecular Determinants of FcRn Trafficking

Studies in several laboratories have analyzed the determinants of FcRn that regulate intracellular trafficking [80–83]. Both dileucine and tyrosine based motifs are involved in controlling endocytosis, transcytosis and basolateral targeting [80, 82]. In addition, biochemical studies have demonstrated that the cytosolic tail of FcRn

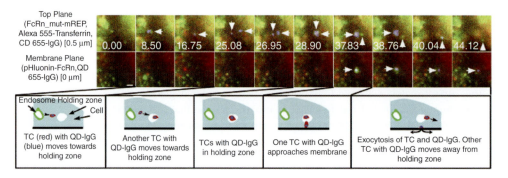

Figure 11.3 An example of the use of multifocal plane microscopy (MUM) to analyze the trafficking pathways taken by IgG and FcRn following sorting in endosomes. Endothelial (HMEC-1) cells were cotransfected with pHluorin-FcRn and FcRn-mRFP (a mutated, high affinity variant of FcRn [52] was used) and incubated in medium (pH 7.3) with quantum dot (QD-655)-human IgG1 mutant (MST-HN [79]) complexes and Alexa 555-labeled transferrin before and during imaging. FcRn and transferrin are indicated by green and QD-IgG complexes by red. The events of interest are highlighted in blue. An FcRn/transferrin-positive TC with IgG (leftward arrow) leaves a sorting endosome at 8.50 s. Later, another TC containing QD-IgG (downward arrows) enters the same region of the cell that we have designated a "holding zone" (28.90 s). One of the TCs exocytoses (38.76 s), releasing IgG (rightward arrows) on the membrane plane. The second TC (upward arrows) moves away from the holding zone in the top plane. Further details are described in Prabhat *et al* [69]. Bar = 1 μm.

associates with the μ chain of clathrin [81], indicating that endocytic uptake occurs through clathrin-coated pits. This is also consistent with more recent electron tomography studies demonstrating the presence of clathrin at both endocytic and exocytic sites [78]. The molecular basis for the dominance of apical to basolateral transcytosis of IgG by rat FcRn, which is reversed for human FcRn, has also been investigated [83]. Interestingly, this difference can be accounted for by variations in the numbers of N-linked glycans on the ectodomain (i.e., exposed to endosomal lumen) of FcRn: in humans there is one glycosylation site, whereas rat/mouse FcRn has four sites [27, 29, 30]. Generation of a mutated human FcRn with four glycans results in a redistribution of the rodentized variant to the apical surface and reversal of the predominant direction of transport to that observed for rodent FcRn, that is, apical to basolateral delivery [83].

Towards the goal of defining which molecular effectors determine the trafficking pathways taken by FcRn and its bound ligand, several studies have investigated possible associations, and even direct involvement, of different Rab GTPases in intracellular trafficking [71, 72, 84]. Rab GTPases represent effector proteins that play a major role in regulating the trafficking behavior of proteins on the pathways that include secretion and endocytosis. For example, Rab4, Rab5, Rab11 and Rab25 are involved in regulating endocytic recycling and transcytosis in different cell types (e.g., [85–91]). In endothelial cells, we have shown that although both Rab4a and Rab11a are present on "leaving" FcRn+ TCs as they segregate from sorting endosomes, only Rab11a approaches the plasma membrane during exocytic events [71]. This suggests that Rab11a is an important player in exocytosis, whereas Rab4a functions at an earlier point in the recycling pathway. In polarized epithelial cells, the role of Rab11a and Rab25 in recycling and transcytosis has also been investigated [72]. These studies have shown that Rab25 and the actin motor myosin Vb play an important role in transcytosis. By contrast, Rab11a does not affect transcytosis but is involved in the regulation of recycling at the basolateral membrane. Unraveling the role of different Rab GTPases in the regulation of FcRn and IgG trafficking could have important implications for the modulation of intracellular trafficking pathways which in turn could affect IgG distribution and transport.

11.7
Engineering IgG–FcRn Interactions

The identification of FcRn as the receptor that regulates IgG levels and transport suggests ways of modulating the *in vivo* half-life and delivery of IgG across cellular barriers. Increasing the persistence becomes particularly relevant for therapeutic antibodies [92–96], where production and delivery costs represent a major financial burden and clinical overhead. In 1997 we showed that the engineering of a mouse Fc fragment for increased affinity for FcRn at pH 6.0 generated an antibody fragment with increased *in vivo* half-life relative to its parent [92]. Multiple subsequent reports have described similar engineering for human IgGs with consequent increased persistence in non-human primates or mice that transgenically express human FcRn [93–95, 97, 98]. Through these studies the importance of the pH dependence of FcRn–IgG interactions has become apparent: antibodies that gain significant binding at near neutral pH (by contrast with their parent wild-type antibodies that do not bind detectably) have reduced *in vivo* half-lives [50, 98, 99]. Antibodies of this class accumulate to high levels in cells due to FcRn-mediated uptake and inefficient release during exocytosis [79]. Microscopy studies demonstrate that these antibodies follow the constitutive degradation pathway of FcRn into lysosomes [100].

The effects of alterations in the pH dependence of FcRn–IgG interactions has led to controversy concerning possible correlations between changes in affinity and *in vivo* persistence [101–103]. Some of these controversies can be explained if the gain of binding at near neutral pH is taken into consideration. For example, the comparison of the half-lives of two human IgG1 molecules that have the same affinity for murine FcRn at pH 6.0, but 10-fold differences at near neutral pH, exemplifies how increased affinity at near neutral pH reduces persistence [1]. Similarly, a mutated antibody (N434A) that has 4-fold increased affinity for human FcRn at pH 6.0 while retaining low affinity at pH 7.4 has a longer half life in non-human primates (cynomolgus monkeys) than a mutant (N434W) with 80-fold enhancement at pH 6.0 but significant binding at pH 7.4 [98]. Since increases in affinity at pH 6.0 are generally accompanied by parallel increases at near neutral pH [98, 99], the intrinsic nature of FcRn–IgG interactions will most likely impose an upper limit to the half-life extension that can be achieved.

The cross-species differences for FcRn binding properties between man and mouse have major implications for the use of mice as preclinical models [50]. This becomes particularly relevant when engineering antibodies for increased *in vivo* persistence. Specifically, the higher affinity of mouse FcRn for (human) IgGs relative to human FcRn can result in significant binding of an "affinity enhanced" antibody for mouse FcRn at near neutral pH. By contrast, this antibody could retain very low affinity for human FcRn at this pH as a consequence of cross-species differences. Thus, although such an engineered antibody is predicted to have increased longevity in humans (or non-human primates), it will have a shorter half-life than its parent wild-type molecule in mice. An example of this effect has been described for a mutated human IgG1 that transports better than its wild-type parent across the human *ex vivo* placenta (indicating improved transport and longevity), but has a half life of around 63 hours ($t_{1/2}$ = ~250 hours for wild-type) in mice [50]. The difference in binding specificity across species has motivated the development of engineered "humanized" mice that transgenically express human FcRn and are deleted for endogenous FcRn [97, 104]. These mice provide valuable preclinical tools for the analysis of human antibodies, and have recently been used to demonstrate the increased efficacy of an antitumor antibody with extended half-life [96].

Interestingly, recent studies have also shown that the isoelectric point (pI) of an antibody can impact its *in vivo* persistence [105]. Specifically, IgG molecules with high pI are eliminated more rapidly than their counterparts with lower pI. It has been suggested that the effect of pI is due to the repulsion of more negatively charged proteins (with lower pI) from the cell surface of pinocytic/endocytic cells that are involved in IgG degradation. This mechanism therefore represents an FcRn-independent pathway through which the persistence of an antibody can be modulated. However, antibodies with lower pI are expected to have fundamentally different properties *in vivo* relative to those engineered for half-life extension through modification of FcRn binding: antibodies with lower pI would be predicted to be transported across cellular barriers less efficiently relative to their analogs with higher pI. By contrast, engineering for increased FcRn binding with maintenance of low affinity at near neutral pH results in improved transcellular delivery [50].

11.8
Inhibitors of FcRn Function

Antibodies that have gained significant binding for FcRn at near neutral pH are highly efficient inhibitors of FcRn [79]. Specifically, they are taken into FcRn-expressing cells by receptor-mediated endocytosis and compete (due to their higher affinity at acidic pH) with wild-type IgGs for binding to receptors in endosomes. We showed that these antibodies can be used to lower the levels of endogenous IgGs *in vivo* [79], and more recently, that they can be used to treat arthritis in a serum transfer model [106]. Other inhibitors of FcRn, including synthetic dimeric peptides, have also been described [107]. In addition, antibodies that bind through their variable regions to FcRn can block immune thrombocytopenia and myasthenia gravis in rodent models of disease [108, 109]. As such, these FcRn inhibitors have significant potential for the treatment of IgG-mediated autoimmunity. It is also possible that they will have uses in other indications such as the induction of rapid clearance of toxin-antibody complexes and blockade of transport of pathogenic antibodies across the placenta during pregnancy.

Intravenous gammaglobulin (IVIG) is used in high doses for a wide array of indications, including the therapy of antibody-mediated autoimmunity (e.g., [110, 111]). Despite its widespread use, the mechanism of action of this reagent remains a matter of debate. In some studies, evidence in support of FcRn blockade has been presented [112–115], whereas others have argued that effects on FcRn do not contribute to anti-inflammatory sequelae that result from the induction of FcγRIIb upregulation by sialylated IgG molecules [116–118]. These mechanisms are not mutually exclusive, and it is clear that at the high doses of IVIG used (approaching the whole body load of endogenous IgG), competition for FcRn binding will occur. The combination of recombinant, sialylated Abdegs that have both properties of FcγRIIb upregulation and potent inhibition could therefore represent an attractive approach for the treatment of inflammatory, IgG-mediated disease. However, several studies in mouse models of autoimmunity have indicated that FcRn-inhibition alone is sufficient to ameliorate disease [106, 108, 109, 119], suggesting that the combined effects of sialylation and FcRn blockade might not be necessary in all therapeutic situations.

11.9
Engineering Mice with Altered FcRn Function

In addition to the generation of mice that are deleted for FcRn, the realization that transgenic overexpression of FcRn might be useful for the production of high levels of antibodies motivated the development of mice overexpressing bovine FcRn [120]. Consistent with the increased activity of FcRn, these mice have abnormally high levels of serum IgG in the steady state. Following immunization, they show substantial increases in immunogen-specific antibody levels, in addition to greater numbers of antigen specific B cells and plasma cells. Perhaps surprisingly, these mice do not show indications of antibody-mediated autoimmunity or renal damage. Consequently such mice provide powerful tools for the efficient production of antibodies.

11.10
Concluding Remarks

The multiple activities of FcRn, emanating from its ability to transport IgG within and across cells of many different types, have become apparent during the past decade. This reveals new opportunities for targeting FcRn activity that can be exploited to generate "new generation" therapeutics. In the future, we anticipate that an improved understanding of the molecular determinants of the intracellular trafficking of FcRn and IgG will lead to an ability to modulate FcRn-mediated transport of IgGs for therapeutic benefit.

Acknowledgments

We are indebted to our many coworkers and collaborators who have contributed to our studies related to FcRn. Our research in this area has been supported in part by grants from the NIH to E.S.W (**RO1 AI039167, RO1 AI055556, RO1 AR056478**) and R.J.O. (**RO1 GM071048 and RO1 GM085575**) and National Multiple Sclerosis Society to E.S.W. (**RG 2411 and RG 4308**).

References

1 Ward, E.S., and Ober, R.J. (2009) Multitasking by exploitation of intracellular transport functions: the many faces of FcRn. *Adv. Immunol.*, **103**, 77–115.

2 Kuo, T.T., *et al.* (2010) Neonatal Fc receptor: from immunity to therapeutics. *J. Clin. Immunol.*, **30**, 777–789.

3 Roopenian, D.C., and Sun, V.Z. (2010) Clinical ramifications of the MHC family Fc receptor FcRn. *J. Clin. Immunol.*, **30**, 790–797.

4 Ghetie, V., *et al.* (1996) Abnormally short serum half-lives of IgG in beta 2-microglobulin-deficient mice. *Eur. J. Immunol.*, **26**, 690–696.

5 Dickinson, B.L., *et al.* (1999) Bidirectional FcRn-dependent IgG transport in a polarized human intestinal epithelial cell line. *J. Clin. Invest.*, **104**, 903–911.

6 Kobayashi, N., *et al.* (2002) FcRn-mediated transcytosis of immunoglobulin G in human renal proximal tubular epithelial cells. *Am. J. Physiol. Renal Physiol.*, **282**, F358–F365.

7 Zhu, X., *et al.* (2001) MHC class I-related neonatal Fc receptor for IgG is functionally expressed in monocytes, intestinal macrophages, and dendritic cells. *J. Immunol.*, **166**, 3266–3276.

8 Akilesh, S., *et al.* (2007) Neonatal FcR expression in bone marrow-derived cells functions to protect serum IgG from catabolism. *J. Immunol.*, **179**, 4580–4588.

9 Qiao, S.W., *et al.* (2008) Dependence of antibody-mediated presentation of antigen on FcRn. *Proc. Natl. Acad. Sci. U.S.A.*, **105**, 9337–9342.

10 Akilesh, S., *et al.* (2008) Podocytes use FcRn to clear IgG from the glomerular basement membrane. *Proc. Natl. Acad. Sci. U.S.A.*, **105**, 967–972.

11 Kim, H., *et al.* (2008) Mapping of the neonatal Fc receptor in the rodent eye. *Invest. Ophthalmol. Vis. Sci.*, **49**, 2025–2029.

12 Junghans, R.P., and Anderson, C.L. (1996) The protection receptor for IgG catabolism is the beta2-microglobulin-containing neonatal intestinal transport receptor. *Proc. Natl. Acad. Sci. U.S.A.*, **93**, 5512–5516.

13 Israel, E.J., *et al.* (1996) Increased clearance of IgG in mice that lack beta 2-microglobulin: possible protective role of FcRn. *Immunology*, **89**, 573–578.

14 McCarthy, K.M., *et al.* (2000) Bidirectional transcytosis of IgG by the rat neonatal Fc receptor expressed in a rat kidney cell line: a system to study protein transport across epithelia. *J. Cell. Sci.*, **113**, 1277–1285.

15 Firan, M., *et al.* (2001) The MHC class I related receptor, FcRn, plays an essential role in the maternofetal transfer of gammaglobulin in humans. *Int. Immunol.*, **13**, 993–1002.

16 Antohe, F., *et al.* (2001) Expression of functionally active FcRn and the differentiated bidirectional transport of IgG in human placental endothelial cells. *Hum. Immunol.*, **62**, 93–105.

17 Haymann, J.P., *et al.* (2000) Characterization and localization of the neonatal Fc receptor in adult human kidney. *J. Am. Soc. Nephrol.*, **11**, 632–639.

18 Spiekermann, G.M., *et al.* (2002) Receptor-mediated immunoglobulin G transport across mucosal barriers in adult life: functional expression of FcRn in the mammalian lung. *J. Exp. Med.*, **196**, 303–310.

19 Claypool, S.M., *et al.* (2004) Bidirectional transepithelial IgG transport by a strongly polarized basolateral membrane Fc-g receptor. *Mol. Biol. Cell*, **15**, 1746–1759.

20 Ober, R.J., *et al.* (2004) Exocytosis of IgG as mediated by the receptor, FcRn: an analysis at the single-molecule level. *Proc. Natl. Acad. Sci. U.S.A.*, **101**, 11076–11081.

21 Yoshida, M., *et al.* (2004) Human neonatal Fc receptor mediates transport of IgG into luminal secretions for delivery of antigens to mucosal dendritic cells. *Immunity*, **20**, 769–783.

22 Chaudhury, C., *et al.* (2003) The major histocompatibility complex-related Fc receptor for IgG (FcRn) binds albumin and prolongs its lifespan. *J. Exp. Med.*, **197**, 315–322.

23 Andersen, J.T., *et al.* (2006) The conserved histidine 166 residue of the human neonatal Fc receptor heavy chain is critical for the pH-dependent binding to albumin. *Eur. J. Immunol.*, **36**, 3044–3051.

24 Rodewald, R., and Abrahamson, D.R. (1982) Receptor-mediated transport of IgG across the intestinal epithelium of the neonatal rat. *Ciba Found. Symp.*, **92**, 209–232.

25 Wallace, K.H., and Rees, A.R. (1980) Studies on the immunoglobulin-G Fc-fragment receptor from neonatal rat small intestine. *Biochem. J.*, **188**, 9–16.

26 Simister, N.E., and Rees, A.R. (1985) Isolation and characterization of an Fc receptor from neonatal rat small intestine. *Eur. J. Immunol.*, **15**, 733–738.

27 Simister, N.E., and Mostov, K.E. (1989) An Fc receptor structurally related to MHC class I antigens. *Nature,* **337,** 184–187.

28 Simister, N.E., *et al.* (1996) An IgG-transporting Fc receptor expressed in the syncytiotrophoblast of human placenta. *Eur. J. Immunol.,* **26,** 1527–1531.

29 Story, C.M., *et al.* (1994) A major histocompatibility complex class I-like Fc receptor cloned from human placenta: possible role in transfer of immunoglobulin G from mother to fetus. *J. Exp. Med.,* **180,** 2377–2381.

30 Ahouse, J.J., *et al.* (1993) Mouse MHC class I-like Fc receptor encoded outside the MHC. *J. Immunol.,* **151,** 6076–6088.

31 Kandil, E., *et al.* (1995) Structural and phylogenetic analysis of the MHC class I-like Fc receptor gene. *J. Immunol.,* **154,** 5907–5918.

32 Adamski, F.M., *et al.* (2000) Expression of the Fc receptor in the mammary gland during lactation in the marsupial *Trichosurus vulpecula* (brushtail possum). *Mol. Immunol.,* **37,** 435–444.

33 Schnulle, P.M., and Hurley, W.L. (2003) Sequence and expression of the FcRn in the porcine mammary gland. *Vet. Immunol. Immunopathol.,* **91,** 227–231.

34 Kacskovics, I., *et al.* (2000) Cloning and characterization of the bovine MHC class I-like Fc receptor. *J. Immunol.,* **164,** 1889–1897.

35 Mayer, B., *et al.* (2002) Redistribution of the sheep neonatal Fc receptor in the mammary gland around the time of parturition in ewes and its localization in the small intestine of neonatal lambs. *Immunology,* **107,** 288–296.

36 Kacskovics, I., *et al.* (2006) Cloning and characterization of the dromedary (*Camelus dromedarius*) neonatal Fc receptor (drFcRn). *Dev. Comp. Immunol.,* **30,** 1203–1215.

37 Kim, J.K., *et al.* (1994) Localization of the site of the murine IgG1 molecule that is involved in binding to the murine intestinal Fc receptor. *Eur. J. Immunol.,* **24,** 2429–2434.

38 Medesan, C., *et al.* (1997) Delineation of the amino acid residues involved in transcytosis and catabolism of mouse IgG1. *J. Immunol.,* **158,** 2211–2217.

39 Kim, J.K., *et al.* (1999) Mapping the site on human IgG for binding of the MHC class I-related receptor, FcRn. *Eur. J. Immunol.,* **29,** 2819–2825.

40 Raghavan, M., *et al.* (1995) Analysis of the pH dependence of the neonatal Fc receptor/immunoglobulin G interaction using antibody and receptor variants. *Biochemistry,* **34,** 14649–14657.

41 Martin, W.L., *et al.* (2001) Crystal structure at 2.8 Å of an FcRn/ heterodimeric Fc complex: mechanism of pH dependent binding. *Mol. Cell.,* **7,** 867–877.

42 Deisenhofer, J. (1981) Crystallographic refinement and atomic models of a human Fc fragment and its complex with fragment B of protein A from *Staphylococcus aureus* at 2.9- and 2.8-Å resolution. *Biochemistry,* **20,** 2361–2370.

43 Shields, R.L., *et al.* (2001) High resolution mapping of the binding site on human IgG1 for FcγRI, FcγRII, FcγRIII, and FcRn and design of IgG1 variants with improved binding to the FcγR. *J. Biol. Chem.,* **276,** 6591–6604.

44 Duncan, A.R., *et al.* (1988) Localization of the binding site for the human high-affinity Fc receptor on IgG. *Nature,* **332,** 563–564.

45 Jefferis, R., *et al.* (1998) IgG-Fc-mediated effector functions: molecular definition of interaction sites for effector ligands and the role of glycosylation. *Immunol. Rev.,* **163,** 59–76.

46 Vaughn, D.E., *et al.* (1997) Identification of critical IgG binding epitopes on the neonatal Fc receptor. *J. Mol. Biol.,* **274,** 597–607.

47 Rodewald, R. (1976) pH-dependent binding of immunoglobulins to intestinal cells of the neonatal rat. *J. Cell Biol.,* **71,** 666–669.

48 Popov, S., *et al.* (1996) The stoichiometry and affinity of the interaction of murine Fc fragments with the MHC class I-related receptor, FcRn. *Mol. Immunol.,* **33,** 521–530.

49 Ober, R.J., *et al.* (2001) Differences in promiscuity for antibody-FcRn interactions across species: implications for therapeutic antibodies. *Int. Immunol.,* **13,** 1551–1559.

50 Vaccaro, C., *et al.* (2006) Divergent activities of an engineered antibody in

murine and human systems have implications for therapeutic antibodies. *Proc. Natl. Acad. Sci. U.S.A.*, **103**, 18709–18714.

51 Zhou, J., *et al.* (2003) Generation of mutated variants of the human form of the MHC class I-related receptor, FcRn, with increased affinity for mouse immunoglobulin G. *J. Mol. Biol.*, **332**, 901–913.

52 Zhou, J., *et al.* (2005) Conferring the binding properties of the mouse MHC class I-related receptor, FcRn, onto the human ortholog by sequential rounds of site-directed mutagenesis. *J. Mol. Biol.*, **345**, 1071–1081.

53 Weng, Z., *et al.* (1998) Computational determination of the structure of rat Fc bound to the neonatal Fc receptor. *J. Mol. Biol.*, **282**, 217–225.

54 Schuck, P., *et al.* (1999) Sedimentation equilibrium analysis of recombinant mouse FcRn with murine IgG1. *Mol. Immunol.*, **36**, 1117–1125.

55 Sanchez, L.M., *et al.* (1999) Stoichiometry of the interaction between the major histocompatibility complex-related Fc receptor and its Fc ligand. *Biochemistry*, **38**, 9471–9476.

56 Ghetie, V., and Ward, E.S. (1997) FcRn: the MHC class I-related receptor that is more than an IgG transporter. *Immunol. Today*, **18**, 592–598.

57 Kim, J.K., *et al.* (1994) Catabolism of the murine IgG1 molecule: evidence that both CH2-CH3 domain interfaces are required for persistence of IgG1 in the circulation of mice. *Scand. J. Immunol.*, **40**, 457–465.

58 Tesar, D.B., *et al.* (2006) Ligand valency affects transcytosis, recycling and intracellular trafficking mediated by the neonatal Fc receptor. *Traffic*, **7**, 1127–1142.

59 Perez-Montoyo, H., *et al.* (2009) Conditional deletion of the MHC Class I-related receptor, FcRn, reveals the sites of IgG homeostasis in mice. *Proc. Natl. Acad. Sci. U.S.A.*, **106**, 2788–2793.

60 Sarav, M., *et al.* (2009) Renal FcRn reclaims albumin but facilitates elimination of IgG. *J. Am. Soc. Nephrol.*, **20**, 1941–1952.

61 Kisanuki, Y.Y., *et al.* (2001) Tie2-Cre transgenic mice: a new model for endothelial cell-lineage analysis *in vivo*. *Dev. Biol.*, **230**, 230–242.

62 Mi, W., *et al.* (2008) Targeting the neonatal Fc receptor for antigen delivery using engineered Fc fragments. *J. Immunol.*, **181**, 7550–7561.

63 Ye, L., *et al.* (2008) The MHC class II-associated invariant chain interacts with the neonatal Fc gamma receptor and modulates its trafficking to endosomal/lysosomal compartments. *J. Immunol.*, **181**, 2572–2585.

64 Schlachetzki, F., *et al.* (2002) Expression of the neonatal Fc receptor (FcRn) at the blood-brain barrier. *J. Neurochem.*, **81**, 203–206.

65 Zhang, Y., and Pardridge, W.M. (2001) Mediated efflux of IgG molecules from brain to blood across the blood-brain barrier. *J. Neuroimmunol.*, **114**, 168–172.

66 Deane, R., *et al.* (2005) IgG-assisted age-dependent clearance of Alzheimer's amyloid beta peptide by the blood-brain barrier neonatal Fc receptor. *J. Neurosci.*, **25**, 11495–11503.

67 Wang, W., *et al.* (2008) Monoclonal antibody pharmacokinetics and pharmacodynamics. *Clin. Pharmacol. Ther.*, **84**, 548–558.

68 Ober, R.J., *et al.* (2004) Visualizing the site and dynamics of IgG salvage by the MHC class I-related receptor, FcRn. *J. Immunol.*, **172**, 2021–2029.

69 Prabhat, P., *et al.* (2007) Elucidation of intracellular recycling pathways leading to exocytosis of the Fc receptor, FcRn, by using multifocal plane microscopy. *Proc. Natl. Acad. Sci. U.S.A.*, **104**, 5889–5894.

70 Ram, S., *et al.* (2008) High accuracy 3D quantum dot tracking with multifocal plane microscopy for the study of fast intracellular dynamics in live cells. *Biophys. J.*, **95**, 6025–6043.

71 Ward, E.S., *et al.* (2005) From sorting endosomes to exocytosis: association of Rab4 and Rab11 GTPases with the Fc receptor, FcRn, during recycling. *Mol. Biol. Cell*, **16**, 2028–2038.

72 Tzaban, S., *et al.* (2009) The recycling and transcytotic pathways for IgG transport by FcRn are distinct and display an inherent polarity. *J. Cell Biol.*, **185**, 673–684.

73 Storrie, B., and Desjardins, M. (1996) The biogenesis of lysosomes: is it a kiss and run, continuous fusion and fission process? *Bioessays*, **18**, 895–903.

74 Gandhi, S.P., and Stevens, C.F. (2003) Three modes of synaptic vesicular recycling revealed by single-vesicle imaging. *Nature*, **423**, 607–613.

75 Ryan, T.A. (2003) Kiss-and-run, fuse-pinch-and-linger, fuse-and-collapse: the life and times of a neurosecretory granule. *Proc. Natl. Acad. Sci. U.S.A.*, **100**, 2171–2173.

76 Steyer, J.A., and Almers, W. (2001) A real-time view of life within 100 nm of the plasma membrane. *Nat. Rev. Mol. Cell Biol.*, **2**, 268–275.

77 Prabhat, P., et al. (2004) Simultaneous imaging of different focal planes in fluorescence microscopy for the study of cellular dynamics in three dimensions. *IEEE Trans. Nanobioscience*, **3**, 237–242.

78 He, W., et al. (2008) FcRn-mediated antibody transport across epithelial cells revealed by electron tomography. *Nature*, **455**, 542–546.

79 Vaccaro, C., et al. (2005) Engineering the Fc region of immunoglobulin G to modulate *in vivo* antibody levels. *Nat. Biotechnol.*, **23**, 1283–1288.

80 Wu, Z., and Simister, N.E. (2001) Tryptophan- and dileucine-based endocytosis signals in the neonatal Fc receptor. *J. Biol. Chem.*, **276**, 5240–5247.

81 Wernick, N.L., et al. (2005) Recognition of the tryptophan-based endocytosis signal in the neonatal Fc Receptor by the mu subunit of adaptor protein-2. *J. Biol. Chem.*, **280**, 7309–7316.

82 Newton, E.E., et al. (2005) Characterization of basolateral-targeting signals in the neonatal Fc receptor. *J. Cell. Sci.*, **118**, 2461–2469.

83 Kuo, T.T., et al. (2009) N-Glycan moieties in neonatal Fc receptor determine steady-state membrane distribution and directional transport of IgG. *J. Biol. Chem.*, **284**, 8292–8300.

84 Jerdeva, G.V., et al. (2010) Comparison of FcRn- and pIgR-mediated transport in MDCK cells by fluorescence confocal microscopy. *Traffic*, **11**, 1205–1220.

85 Van Der Sluijs, P., et al. (1992) The small GTP-binding protein rab4 controls an early sorting event on the endocytic pathway. *Cell*, **70**, 729–740.

86 Daro, E., et al. (1996) Rab4 and cellubrevin define different early endosome populations on the pathway of transferrin receptor recycling. *Proc. Natl. Acad. Sci. U.S.A.*, **93**, 9559–9564.

87 Ullrich, O., et al. (1996) Rab11 regulates recycling through the pericentriolar recycling endosome. *J. Cell Biol.*, **135**, 913–924.

88 Green, E.G., et al. (1997) Rab11 is associated with transferrin-containing recycling compartments in K562 cells. *Biochem. Biophys. Res. Commun.*, **239**, 612–616.

89 Sönnichsen, B., et al. (2000) Distinct membrane domains on endosomes in the recycling pathway visualized by multicolor imaging of Rab4, Rab5, and Rab11. *J. Cell Biol.*, **149**, 901–914.

90 Casanova, J.E., et al. (1999) Association of Rab25 and Rab11a with the apical recycling system of polarized Madin-Darby canine kidney cells. *Mol. Biol. Cell*, **10**, 47–61.

91 Wang, X., et al. (2000) Regulation of vesicle trafficking in Madin-Darby canine kidney cells by Rab11a and Rab25. *J. Biol. Chem.*, **275**, 29138–29146.

92 Ghetie, V., et al. (1997) Increasing the serum persistence of an IgG fragment by random mutagenesis. *Nat. Biotechnol.*, **15**, 637–640.

93 Hinton, P.R., et al. (2004) Engineered human IgG antibodies with longer serum half-lives in primates. *J. Biol. Chem.*, **279**, 6213–6216.

94 Hinton, P.R., et al. (2006) An engineered human IgG1 antibody with longer serum half-life. *J. Immunol.*, **176**, 346–356.

95 Dall'Acqua, W.F., et al. (2006) Properties of human IgG1s engineered for enhanced binding to the neonatal Fc receptor (FcRn). *J. Biol. Chem.*, **281**, 23514–23524.

96 Zalevsky, J., et al. (2010) Enhanced antibody half-life improves *in vivo* activity. *Nat. Biotechnol.*, **28**, 157–159.

97 Petkova, S.B., et al. (2006) Enhanced half-life of genetically engineered human IgG1 antibodies in a humanized

FcRn mouse model: potential application in humorally mediated autoimmune disease. *Int. Immunol.*, **18**, 1759–1769.

98 Yeung, Y.A., *et al.* (2009) Engineering human IgG1 affinity to human neonatal Fc receptor: impact of affinity improvement on pharmacokinetics in primates. *J. Immunol.*, **182**, 7663–7671.

99 Dall'Acqua, W., *et al.* (2002) Increasing the affinity of a human IgG1 to the neonatal Fc receptor: biological consequences. *J. Immunol.*, **169**, 5171–5180.

100 Gan, Z., *et al.* (2009) Analyses of the recycling receptor, FcRn, in live cells reveal novel pathways for lysosomal delivery. *Traffic*, **10**, 600–614.

101 Datta-Mannan, A., *et al.* (2007) Humanized IgG1 variants with differential binding properties to the neonatal Fc receptor: relationship to pharmacokinetics in mice and primates. *Drug Metab. Dispos.*, **35**, 86–94.

102 Gurbaxani, B., *et al.* (2006) Analysis of a family of antibodies with different half-lives in mice fails to find a correlation between affinity for FcRn and serum half-life. *Mol. Immunol.*, **43**, 1462–1473.

103 Gurbaxani, B.M., and Morrison, S.L. (2006) Development of new models for the analysis of Fc-FcRn interactions. *Mol. Immunol.*, **43**, 1379–1389.

104 Roopenian, D.C., *et al.* (2003) The MHC Class I-like IgG receptor controls perinatal IgG transport, IgG homeostasis, and fate of IgG-Fc-coupled drugs. *J. Immunol.*, **170**, 3528–3533.

105 Igawa, T., *et al.* (2010) Reduced elimination of IgG antibodies by engineering the variable region. *Protein Eng. Des. Sel.*, **23**, 385–392.

106 Patel, D.A., *et al.* (2011) Neonatal Fc receptor blockade by Fc engineering ameliorates arthritis in a murine model. *J.Immunol.*, **187**, 1015–1022.

107 Mezo, A.R., *et al.* (2008) Reduction of IgG in nonhuman primates by a peptide antagonist of the neonatal Fc receptor FcRn. *Proc. Natl. Acad. Sci. U.S.A.*, **105**, 2337–2342.

108 Liu, L., *et al.* (2007) Amelioration of experimental autoimmune myasthenia gravis in rats by neonatal FcR blockade. *J. Immunol.*, **178**, 5390–5398.

109 Getman, K.E., and Balthasar, J.P. (2005) Pharmacokinetic effects of 4C9, an anti-FcRn antibody, in rats: implications for the use of FcRn inhibitors for the treatment of humoral autoimmune and alloimmune conditions. *J. Pharm. Sci.*, **94**, 718–729.

110 Kazatchkine, M.D., and Kaveri, S.V. (2001) Immunomodulation of autoimmune and inflammatory diseases with intravenous immune globulin. *N. Engl. J. Med.*, **345**, 747–755.

111 Stangel, M., and Pul, R. (2006) Basic principles of intravenous immunoglobulin (IVIg) treatment. *J Neurol*, **253** (Suppl. 5), v18–v24.

112 Hansen, R.J., and Balthasar, J.P. (2002) Effects of intravenous immunoglobulin on platelet count and antiplatelet antibody disposition in a rat model of immune thrombocytopenia. *Blood*, **100**, 2087–2093.

113 Hansen, R.J., and Balthasar, J.P. (2002) Intravenous immunoglobulin mediates an increase in anti-platelet antibody clearance via the FcRn receptor. *Thromb. Haemost.*, **88**, 898–899.

114 Jin, F., and Balthasar, J.P. (2005) Mechanisms of intravenous immunoglobulin action in immune thrombocytopenic purpura. *Hum. Immunol.*, **66**, 403–410.

115 I-like Fc receptor promotes humorally mediated autoimmune disease. *J. Clin. Invest.*, **113**, 1328–1333.

116 Samuelsson, A., *et al.* (2001) Anti-inflammatory activity of IVIG mediated through the inhibitory Fc receptor. *Science*, **291**, 484–486.

117 Bruhns, P., *et al.* (2003) Colony-stimulating factor-1-dependent macrophages are responsible for IVIG protection in antibody-induced autoimmune disease. *Immunity*, **18**, 573–581.

118 Anthony, R.M., *et al.* (2008) Recapitulation of IVIG anti-

inflammatory activity with a recombinant IgG Fc. *Science*, **320**, 373–376.

119 Li, N., *et al.* (2005) Complete FcRn dependence for intravenous Ig therapy in autoimmune skin blistering diseases. *J. Clin. Invest.*, **115**, 3440–3450.

120 Cervenak, J., *et al.* (2011) Neonatal FcR overexpression boosts humoral immune response in transgenic mice. *J. Immunol.*, **186**, 959–968.

12

Half-Life Extension by Fusion to Recombinant Albumin

Hubert J. Metzner, Thomas Weimer, and Stefan Schulte

12.1
Introduction

Therapeutic recombinant proteins have been used in medicine for more than 25 years. Since the first recombinant protein, insulin, was approved for use in 1982, the number of available biopharmaceuticals has increased steadily and now includes cytokines, coagulation factors and monoclonal antibodies. However, the advent of recombinant DNA technology has pushed the field beyond the reproduction of naturally occurring proteins by enabling the design of novel polypeptides not found in nature [1]. Fusion proteins, created when two or more genes are fused to produce a novel polypeptide that retains the properties of the original gene products, are one such example. Fusion protein technology has multiple applications, such as targeted therapy by fusing proteins that target specific cells with toxins. Denileukin diftitox, an immunotoxin derived from human interleukin-2 (IL-2) and the catalytic and transmembrane domains of diphtheria toxin, was one of the first fusion proteins approved for medical use [2, 3]. Protein fusion technology also has the ability to prolong the half-life of a therapeutic protein by fusing it to a protein known to have a long half-life, such as the fragment crystallizable (Fc) domain of immunoglobulin G (IgG), albumin, or transferrin. The first fusion protein, approved for use in 1998, was etanercept, a drug that is derived from the Fc domain of IgG1 and a fragment of the receptor for tumor necrosis factor-alpha (TNFα) [4]. Etanercept has a prolonged half-life and binds reversibly to TNFα, producing anti-inflammatory and immunosuppressant effects that have led to a broad range of clinical applications.

The importance of optimizing the half-life of therapeutic proteins will increase as the number of biopharmaceuticals in development continues to increase. Many promising peptides and small proteins have been identified or generated in the laboratory, but have failed to produce the desired effect *in vivo* due to their short half-lives, often measured in minutes or hours [5]. The relatively small size of these peptides and molecules places them below the threshold for renal clearance. Moreover, clinical application of such proteins would require frequent injections, which could potentially reduce convenience and compliance while increasing

Therapeutic Proteins: Strategies to Modulate Their Plasma Half-Lives, First Edition. Edited by Roland Kontermann.
© 2012 Wiley-VCH Verlag GmbH & Co. KGaA. Published 2012 by Wiley-VCH Verlag GmbH & Co. KGaA.

treatment costs and the risk of adverse events [6]. Fusing these entities to well-characterized proteins with longer half-lives, such as recombinant albumin, may prolong their time in circulation while maintaining their physiologic activity. Theoretically, the prolonged half-life could permit less frequent administration, while potentially maintaining the same efficacy (Figure 12.1). Fusion may also lead to more consistent and sustained protein concentrations, thereby improving tolerance and safety by maintaining blood levels above the therapeutic level but below the toxic level. These features can make therapy more convenient for both clinicians and patients, leading to improvements in treatment compliance. In addition, the reduced dosing frequency and improved safety profile may enable prophylaxis in areas where this was previously not feasible.

Numerous strategies exist for prolonging the half-life of a therapeutic protein. One approach comprises the covalent attachment of polyethylene glycol (PEG) polymers to another molecule, a process that has been used extensively [7, 8]. However, PEGylation in general produces different monomeric isoforms with varying biological activity. It also introduces a synthetic polymer that is not entirely biodegradable, and can compromise biological activity

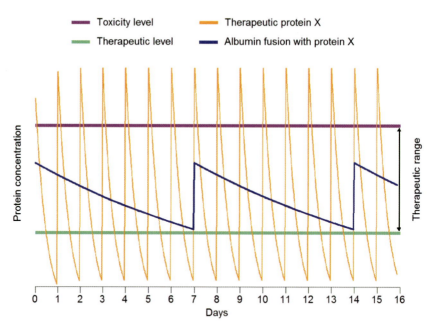

Figure 12.1 Hypothetical depiction of how albumin fusion can improve the pharmacokinetic properties of a therapeutic protein, permitting less frequent dosing while maintaining concentrations within the therapeutic range. The figure is based on an arbitrarily chosen set of data assuming a 25-fold difference in half-life between the therapeutic protein X and the corresponding albumin fusion protein (personal communication, Dr. Jochen Müller-Cohrs, CSL Behring GmbH, Marburg, Germany).

compared with the original protein due to structural changes caused by conjugation [7, 9].

Fusion protein technology, however, has several intrinsic advantages. Fusion proteins can be produced with single-step expression techniques, making production and purification relatively straightforward and economical [10]. This avoids the need for complicated chemical modifications during the manufacturing process [5, 10]. In most cases, fusion protein technology yields a homogenous product, and fusion to naturally occurring and abundant proteins such as albumin can improve the solubility and stability of a therapeutic protein while reducing renal clearance [11] and intracellular degradation. Lastly, with fusion protein technology, the genes can be fused in any order, permitting the construction of different variations of the same fusion protein. This design flexibility can help to optimize the biological activity of the product [11].

The choice of a fusion partner depends on the application. The Fc domain of IgG may be a particularly suitable fusion partner when bivalent interaction is desired, such as target interaction and Fc-mediated effector cell function (e.g., antibody-dependent cell-mediated cytotoxicity) [12]. In contrast, recombinant albumin fusion is particularly useful when the goal is monovalent interaction. Examples include fusion to cytokines [13], growth factors [14], and antibody molecules [15, 16]. Since transferrin crosses the blood–brain barrier, it may be particularly well suited for fusion to proteins that must be delivered to the brain [1].

This chapter will focus on recombinant albumin fusion proteins, which are produced as recombinant proteins composed of recombinant human albumin (rHA) genetically fused at the C- or N-terminus of a recombinant therapeutic protein. Several key aspects are addressed, including the rationale for using albumin as a fusion partner, its mode of action, practical applications, and the advantages and challenges of the recombinant albumin fusion platform. The therapeutic potential of recombinant albumin fusion technology is discussed with regard to several proteins and peptides that are in various stages of experimental, preclinical, and clinical development. The technology, characterization, and preclinical development of recombinant albumin fusion is discussed in the context of three proteins, the coagulation factor VIIa (FVIIa) and factor IX (FIX), and human growth hormone (hGH). Data on the clinical development of albumin fusion are provided for albinterferon alfa-2b (alb-IFN), the genetic fusion product of rHA and interferon alfa-2b (IFNα-2b). Current research provides a proof-of-principle for the concept of albumin fusion and indicates that the technology may have far-reaching clinical benefits.

12.2
Recombinant Albumin Fusion Proteins

Recombinant albumin fusion proteins are created by genetically fusing recombinant albumin to a therapeutic protein or peptide via the complementary DNA

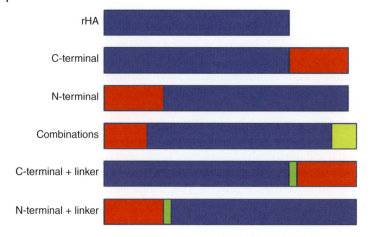

Figure 12.2 Possible configurations of fusion proteins using recombinant human albumin (rHA). Blue = rHA, red and yellow = therapeutic proteins/peptides, green = linker.

(cDNA) encoding each molecule. The fusion technology creates a hybrid polypeptide that makes use of the pharmacokinetic properties of albumin while retaining the biological activity of the therapeutic protein.

Orientation of the two genes to be fused can vary, providing flexibility in the design of the fusion protein (Figure 12.2). The therapeutic protein can be attached to the C-terminal or N-terminal end of the recombinant albumin gene or inserted within the albumin gene. The two genes can be contiguous or separated by a linker sequence, the length of which might optimize the interaction between the two protein moieties and other target molecules. Linkers can also contain cleavage sites that will physically separate the two protein moieties under certain conditions.

The cDNA construct is created in a prokaryotic system and is then transfected into eukaryotic cells, such as *Pichia pastoris* [11, 17], and, for complex proteins, mostly mammalian cell lines such as Chinese hamster ovary (CHO) cells [18, 19]. During fermentation, the fusion gene is transcribed and translated as a single polypeptide that is secreted into the fermentation broth. The broth can then be separated and clarified by centrifugation or filtration, and the protein can be purified using standard separation techniques [15, 17, 20–22].

rHa can be expressed in either prokaryotic or eukaryotic systems [6]. However, the eukaryotic expression systems are more efficient in producing the complex structure of human serum albumin (HSA), including 17 disulfide bridges [17]. The *P. pastoris* eukaryotic expression system is often used because, compared with other eukaryotic expression systems, this approach is economical and readily scaled-up for mass production [23].

12.2.1
Mode of Action

HSA is composed of 585 amino acids and has a molecular weight of approximately 67 kDa [24]. It is produced in the liver and is composed of three structurally homologous domains (DI, DII, and DIII), each of which contains two subdomains (A and B) connected by flexible loops [25].

HSA plays a role in regulating osmotic pressure and pH balance [12]. It also has a broad binding capacity and acts as a transporter for many insoluble and hydrophobic exogenous and endogenous ligands, including ions, fatty acids, amino acids, and many drugs [5]. However, HSA has no enzymatic activity and is not involved in immune functioning.

HSA is the most abundant plasma protein and, together with IgG, accounts for 80–90% of all plasma proteins, with a serum concentration of 35–50 mg mL^{-1} [5, 12]. It is highly stable and is found mainly in the vascular compartment.

The half-life of albumin in humans is almost three weeks, whereas most circulating plasma proteins have a half-life that is measured in days or hours [26]. This remarkably long half-life is due in part to pH-dependent recycling mediated by the neonatal Fc receptor (FcRn) [27]. FcRn resides in cells and on the surface of cells that have direct contact with circulating blood, such as vascular endothelial cells. Circulating albumin taken up by these cells binds to FcRn, as does IgG. Rather than undergoing lysosomal degradation, the complexes IgG–FcRn–albumin, FcRn–IgG, and FcRn–albumin return to the cell membrane where they are exposed to the physiologic pH of circulating blood, releasing the ligands IgG and albumin back into the bloodstream. As a result of this recycling process, the terminal half-life of HSA is 19 days [12].

For the fusion of albumin to large proteins, it is this recycling mechanism that likely contributes to the half-life extension of the fusion protein compared with the therapeutic protein alone. In the case of smaller peptides, both the recycling mechanism and the mere increase in the molecular weight above the clearance limit of the kidneys may be of relevance.

12.2.2
Practical Applications

In addition to prolonging the half-life of therapeutic proteins, recombinant albumin fusion technology may have other useful laboratory and clinical applications. Fusion proteins may be used to facilitate expression or purification of the target protein, minimize proteolysis of the target protein, increase therapeutic stability, or create reporter molecules for monitoring gene expression and protein localization [28]. Albumin fusion has been used to enhance *in vitro* expression of certain proteins that are typically difficult to express in mammalian cells, such as IL-15 and IL-17B [29]. In addition, sites of malignancy and inflammation have increased albumin accumulation due to leaky capillaries and poor lymphatic drainage, and therefore albumin may be beneficial in the targeted delivery of therapeutic drugs [5]. Covalent linkage of albumin to methotrexate, for example, improved drug delivery to sites of malignancy in patients with renal cell carcinoma [30] and in models of rheumatoid arthritis [31], and the same principle could be applied to recombinant albumin fusion proteins.

12.2.3
Advantages

Beyond its extraordinarily long half-life, albumin has several additional advantages as a fusion partner. Albumin is a natural carrier protein that is present at high concentrations in the plasma, has no effector function, and has a low risk of toxicity and immunogenicity [6, 32, 33]. The structure of albumin has been well characterized, which helps predict how it will interact with other protein moieties [33]. This flexibility has allowed recombinant albumin to be fused successfully to both small peptides and large, complex molecules such as cytokines [11] and coagulation factors [18, 19]. Albumin fusion proteins are technically efficient to produce [1]. Several recombinant albumin fusion proteins can be manufactured particularly efficiently in yeast expression systems on a commercial scale and at costs typically lower than for other methods of generating long-acting biopharmaceuticals [1]. This approach also avoids many of the secondary processing steps required for chemical modification or encapsulation [11]. In addition, in contrast to Fc fusion proteins, only monomeric fusions are formed, so purification and dosing are simplified [1].

12.2.4
Challenges

The primary challenges of recombinant albumin fusion are maintaining the biological activity of its fusion partner and minimizing the risk of toxicity and immunogenicity [1, 11]. While the use of linkers and the flexibility in orientation of the fused genes can help optimize biological activity, the presence of albumin can interfere with interactions between the therapeutic protein and its target(s). All novel polypeptides, particularly those containing nonhuman moieties, are potentially immunogenic, as are novel sequences in the area where the therapeutic protein and recombinant albumin are fused together, and this risk must be taken into account when developing recombinant albumin fusion proteins, despite the low immunogenicity of HSA. Steps can be taken to minimize the risk of immunogenicity during the design process (e.g., avoiding peptides that bind to T cell receptors or major histocompatibility complex class II molecules [34]) or production (e.g., choice of expression system, removal of potentially immunogenic aggregates during downstream processing, and avoidance of denaturation during formulation [35]). Lastly, future applications of albumin as a fusion partner for certain applications may be somewhat limited by the fact that it acts as a stabilizer and an enhancer of pharmacokinetic properties only and does not offer any secondary functions, such as cytotoxicity [1].

12.2.5
Therapeutic Potential

12.2.5.1 Fusion to Small Proteins and Peptides

Numerous small proteins and peptides have been successfully fused to albumin to prolong their half-lives (Table 12.1). The C-terminal fusion of rHA and glucagon-like peptide-1 (GLP-1), for example, is in clinical phase III testing as monotherapy or in combination with other medications for the treatment of type 2 diabetes mellitus [36]. This fusion protein, albiglutide (formerly known as albugon), was created by fusing the gene encoding a GLP-1 analog to the open reading frame of rHA [37]. Studies conducted *in vivo* indicate that albiglutide activates GLP-1-receptor-dependent central and peripheral pathways that regulate energy intake and glucose homeostasis [37]. Results from phase II clinical trials indicate that albiglutide improves fasting and postprandial plasma glucose levels in patients with diabetes, and that its extended half-life (approximately five days compared to around five minutes with native GLP-1), may permit once-weekly or less frequent dosing [57, 58].

A similar fusion protein was created by fusing recombinant albumin to the N-terminal end of GLP-1 [10]. A lysine residue was first added to the N-terminal end of GLP-1 to eliminate a cleavage site. The fusion protein, called KGLP-1/HSA, bound and activated GLP-1 receptors, albeit at a lower rate compared with KGLP-1 or native GLP-1. The lower binding affinity is likely due to spatial blockade caused by the albumin moiety. Nevertheless, KGLP-1/HSA retained much of the biological activity of GLP-1, and its duration of effect lasted longer, up to eightfold longer for its glucose-lowering effect and up to twofold longer for its insulin-releasing effect [10].

Albumin has also been genetically fused to the C-terminal end of the potent GLP-1 agonist exendin-4 [17]. As with albiglutide, the gene encoding exendin-4 was fused to the N-terminus of HSA to create a novel fusion protein (rEx-4/HSA). In addition, a peptide linker (GGGGS) was inserted between exendin-4 and HSA. Fusion to HSA retained the efficacy of exendin-4 but reduced its potency, possibly due to spatial blockade; however, a longer linker of 10 amino acids did not increase bioactivity further. In monkeys, the half-life of rEx-4/HSA was 70–80 hours – 33 times longer than that of exenatide (2.4 hours) [59], a synthetic version of exendin-4, and much longer than reported values for other GLP-1 modifications that employed fatty acid chain modification (11–15 hours) [60] or PEGylation (33.4 minutes) [61]. These results indicate that albumin fusion prolongs the half-life of exendin-4 while retaining its glucose-lowering effects, potentially permitting weekly dosing.

Other small peptides that have been successfully fused to albumin include insulin [38], the antithrombotic agents barbourin [39] and hirudin [40, 62]; the coagulation factor XIIa inhibitor infestin-4 [41]; B-type natriuretic peptide (BNP) [42, 43], used in the treatment of congestive heart failure; hGH, recombinant forms of which are used as replacement therapy for patients with hGH deficiency [14]; thioredoxin, a redox-active protein with anti-inflammatory effects [44];

Table 12.1 Application of recombinant albumin fusion technology to peptides and proteins in experimental, preclinical, or clinical development.

Therapeutic molecule	Therapeutic setting	Reference
GLP-1 (albiglutide; albugon)	Diabetes mellitus	[36–37]
Exendin-4	Diabetes mellitus	[17]
Insulin (Albulin)	Diabetes mellitus	[38]
Barbourin	Thrombosis	[39]
Hirudin	Thrombosis	[40, 62]
Infestin-4	Thrombosis	[41]
BNP (AlbuBNP; Cardeva)	Congestive heart failure	[42, 43]
hGH (Albutropin)	Growth hormone deficiency	[14]
Thioredoxin	Immunomodulation, anti-inflammatory, antioxidant, apoptosis	[44]
Thymosin-α1	Hepatitis B/C, anti-tumor	[21]
scFv	Anti-tumor	[15, 16]
taFv	Anti-tumor	[15]
scDb	Anti-tumor	[15]
IFNα (Albinterferon-α2b; Albuferon)	Hepatitis B/C, anti-tumor	[45–47]
TNFα	Immunomodulation, pro-inflammatory, anti-tumor	[48]
IL-2 (Albuleukin)	Anti-tumor	[49, 50]
G-CSF (Albugranin; Neugranin)	Neutropenia	[51]
Erythropoietin	Anemia	[52]
FVIIa	Hemophilia A with inhibitory antibodies	[18]
FIX	Hemophilia B	[19, 53]
BChE	Organophosphorus toxicity	[54]
Cocaine hydrolase	Cocaine toxicity/addiction	[55, 56]

BChE = butyrylcholinesterase; BNP = B-type natriuretic factor; FIX = factor IX; FVIIa = activated factor VII; G-CSF = granulocyte colony-stimulating factor; GLP-1 = glucagon-like peptide-1; hGH = human growth hormone; IFNα = interferon α; IL-2 = interleukin-2; scDb = single-chain diabodies; scFv = single-chain variable fragment; taFv = tandem scFv molecules; TNFα = tumor necrosis factor-alpha.

thymosin-α1, a naturally occurring thymic peptide with immunomodulatory functions [21]; and antibody fragments, such as the single-chain variable fragment (scFv), tandem scFv molecules (taFv), and single-chain diabodies (scDb) [15, 16]. This illustrates the therapeutic potential of albumin fusion proteins involving small molecules in a variety of therapeutic settings.

12.2.5.2 Fusion to Cytokines

Many cytokines have anti-inflammatory, immunomodulatory, or antiviral effects, but their clinical use is often limited by their short half-lives. The cytokine IFNα has a key role in regulating cell growth and differentiation by activating a cascade of intracellular pathways, resulting in antiviral, immunomodulatory, and antiproliferative effects [13]. Currently, IFNα is used to treat viral infections, such as hepatitis B and C, and certain malignancies, such as melanoma. The half-life of IFNα in humans is two to three hours, so treatment must be given daily or three times weekly [32]. Two PEGylated versions of IFNα are available (PEG-IFNα-2a and PEG-IFNα-2b) that permit weekly dosing and are the current standard of care for patients with chronic hepatitis C (CHC) [63–65]. Nevertheless, approximately 45% of CHC patients do not achieve a sustained virologic response (SVR) with these agents [66].

Albuferon (alb-IFN) combines the antiviral activity of IFNα with the prolonged plasma half-life of albumin. The fusion protein is expressed in genetically modified yeast strains (*Saccharomyces cerevisiae*), and the protein is purified from pooled supernatant using chromatography, yielding 95% intact, full-length protein [13, 67]. Preclinical studies indicate that alb-IFN has improved pharmacokinetics and activity that is similar to IFNα in terms of signaling pathway activation, transcriptional regulation, antiproliferative activity, and antiviral activity [11, 13].

In patients with CHC, the mean elimination half-life of alb-IFN was found to be around six days [66, 68]. This was approximately twofold greater than published results for PEG-IFNα-2a (77 hours) [69] and fourfold greater than results for PEG-IFNα-2b (40 hours) [65, 70]. Phase II trials conducted primarily in patients with genotype 1 disease who had failed prior IFNα [45] or were IFNα-naive [71] showed that alb-IFN had antiviral activity and an acceptable tolerability profile when given at doses of 900 or 1200 μg every two weeks or 1200 μg every four weeks. In a small phase II study in IFN-naive patients with genotype 2/3 CHC (*N* = 43), alb-IFN was well tolerated and had antiviral activity when given at a dose of 1500 μg every two or four weeks [72]. Results from two ongoing phase III trials evaluating alb-IFN (900 and 1200 μg) dosed every two or four weeks in IFN-naive patients with genotype 1 CHC (ACHIEVE 1) or genotype 2/3 CHC (ACHIEVE 2/3) have been reported [46, 47]. However, the higher two-week dosing regimen was recently withdrawn from the marketing authorization application by the sponsor due to regulatory feedback on the relative risk-to-benefit ratio [73]. Findings from the ACHIEVE 1 study showed that alb-IFN 1200 μg was associated with increased pulmonary adverse events, most probably due to the IFN moiety of the fusion molecule [46]. Consequently, all study patients receiving alb-IFN 1200 μg were switched to 900 µg given every two weeks. Final phase III clinical data have recently been reported [46]; see also Section 12.8).

Several variations and modifications have been made to alb-IFN to improve its homogeneity and stability. When expressed in *P. pastoris*, alb-IFN was found to migrate as doublets in sodium dodecyl sulfate polyacrylamide gel electrophoresis (SDS-PAGE) and was prone to form covalent aggregations in aqueous solution [74]. The observed heterogeneity and instability, caused by incomplete disulfide bridges between Cys1 and Cys98 of IFNα-2b, reduced the recovery rate (to 10%) and necessitated lyophilization. To restore disulfide pairing and improve the stability of alb-IFN, researchers in China created a unique protein by switching the orientation of the two proteins, placing IFNα-2b at the N-terminal end [74]. Compared with alb-IFN, the new protein (IFNα-2b-HSA) was relatively more homogenous in SDS-PAGE, easier to purify, and more stable (recovery was increased 2.5-fold). Antiviral activity of IFNα-2b-HSA *in vitro* was markedly increased compared with IFNα-2b and was similar, albeit slightly less, than that of alb-IFN. Other modifications, such as the use of non-ionic surfactants, site-directed mutagenesis to replace the free Cys34 residue with serine, and linker sequences of varying length and characteristics, have led to further improvements in the stability of this albumin fusion protein [75].

In addition to IFNα, recombinant albumin has been successfully fused to other cytokines, including TNFα, a multifunctional cytokine with a key role in apoptosis, cell survival, inflammation, and immunity [48]; IL-2, a leukocytotrophic hormone that is used to treat certain types of malignancy [49, 50]; granulocyte colony-stimulating factor, a colony-stimulating factor that increases white blood cell count and is therefore useful in patients with neutropenia [51]; and erythropoietin, a glycoprotein hormone that is used to treat anemia [52].

12.2.5.3 Fusion to Complex Proteins

Fusion to recombinant albumin has been used to successfully prolong the half-life of activated coagulation factor VII (FVIIa), which is used to treat patients with hemophilia A who have developed inhibitors against the replaced coagulation factor VIII, and factor IX (FIX), which is used to treat patients with hemophilia B. The management of hemophilia B, for example, includes treatment with recombinant FIX (rFIX; BeneFIX®). The half-life of rFIX is 18–34 hours [76–78] and requires intravenous administration every two to three days to prevent spontaneous bleeding [79]. The terminal half-life of the recombinant fusion protein linking coagulation factor IX with albumin (rIX-FP) was about three- to fivefold longer than that of rFIX in rat and rabbit animal models, which, if transferable to humans, could permit once-weekly dosing [19].

Albumin has also been successfully fused to other large, complex proteins, such as butyrylcholinesterase (BChE) and cocaine hydrolase. BChE is used to prevent nerve toxicity caused by exposure to organophosphorus (OP), a toxic compound that is used in various chemical weapons and pesticides [54]. Cocaine hydrolase is related to BChE and was developed as a more potent inhibitor of cocaine effects; however, its action is limited by its short half-life. An albumin–cocaine hydrolase fusion protein was developed [55] that, after a single injection in rats, had an elimination half-life of eight hours, suggesting ample stability for short-term applications and the potential for long-term use [56].

Albumin fusion technology is therefore a viable approach to improving the pharmacokinetics of important and complex protein therapeutics. Considerable interest in these complex proteins is growing, and will be addressed in more detail in the following section.

12.3
Albumin Fusion to Complex Proteins

12.3.1
Recombinant Fusion Protein Linking Coagulation Factor VIIa with Albumin (rVIIa-FP)

The advent of coagulation factor therapy has transformed the lives of patients with hemophilia A. Hemophilia A is a hereditary disorder caused by the lack of factor VIII (FVIII) [80]. Prior to the commercial production of FVIII, the life expectancy of patients with hemophilia A was less than 30 years, and many developed bleeding complications, such as arthrosis, and joint deformities and stiffness. Today, treatment with FVIII helps to reduce bleeding complications, improve quality of life, and even permit hemophilia patients to undergo major surgery. Nevertheless, 10–30% of patients develop inhibiting antibodies against exogenous FVIII [80]. For these patients, the use of recombinant coagulation factors with the ability to bypass FVIII in the coagulation cascade, such as activated recombinant FVII (rFVIIa, NovoSeven®), can effectively restore hemostasis [81].

rFVIIa became available in 1996 for the treatment of hemophilia A and B to reduce bleeding episodes in patients who had developed inhibitors to FVIII or FIX [81–82]. FVII is a member of the vitamin K-dependent coagulation factors with a molecular weight of approximately 50 kDa [83]. It is produced in the liver and secreted into the circulation as an inactive pro-enzyme (FVII) that is later converted to the active form (FVIIa).

The half-life of rFVIIa is approximately 2.5 hours, and therefore, in general, multiple injections are necessary to treat bleeding episodes [84]. For mild-to-moderate bleeding events, two to three doses must be given in two to three-hour intervals to achieve hemostasis; for serious bleeding events, doses may be required every two hours and continued for two to three weeks [33, 85]. For surgical procedures, injections are needed every two to three hours for at least two days, resulting in an average of 38 injections (range, 26–67) for minor surgical procedures and 81 injections (range, 71–128) for major surgery in one study of high-dose therapy [86]. The short half-life of rFVIIa also poses a challenge for long-term prophylaxis.

Preparations that prolong the half-life of rFVIIa while retaining its hemostatic function could reduce the number of injections required for each bleeding episode and make long-term prophylaxis feasible [87]. This prompted the development of the recombinant fusion protein linking coagulation factor VIIa with albumin known as rVIIa-FP [18, 87, 88].

12.3.2
Recombinant Fusion Protein Linking Coagulation Factor IX with Albumin (rIX-FP)

Hemophilia B is a severe bleeding disorder that arises from the deficiency of FIX. FIX has a molecular weight of approximately 57 kDa and is activated (FIXa) by either activated factor XI (FXIa) or by FVIIa in the presence of tissue factor (TF) [89]. FIXa forms an integral part of the tenase enzyme complex that activates factor X (FX). Treatment with plasma-derived or recombinant FIX requires intravenous administration every two to three days to prevent spontaneous bleeding [79]. The inconvenience of this approach can affect compliance. The need for a long-acting preparation of rFIX that would permit less frequent dosing prompted the development of the recombinant fusion protein linking coagulation factor IX with albumin (rIX-FP) [19, 53].

12.3.3
Butyrylcholinesterase (BChE)

Poisoning with OP agents is a severe problem facing military personnel who may encounter lethal doses of these compounds in chemical warfare situations [54]. The use of these agents in war and as pesticides has resulted in an increasing number of cases of both acute and delayed intoxication, resulting in damage to the central and peripheral nervous systems, myopathy, psychosis, general paralysis, and death. The primary mechanism of action of OP compounds is inhibition of acetylcholinesterase, an enzyme that hydrolyzes acetylcholine – a neurotransmitter found in the central and peripheral nervous systems, neuromuscular junctions, and red blood cells. The use of cholinesterases as a medical treatment for OP poisoning has many drawbacks, including limited supply and poor plasma stability. Efforts to improve the supply and properties of this enzyme led to the development of a recombinant BChE fused to HSA [54].

In addition to its ability to hydrolyze OP, BChE is the major detoxicating enzyme of cocaine [56]. Cocaine abuse and dependence are major health problems with devastating consequences. Currently, there is no reliable way to treat cocaine overdose or minimize the risk of relapse in recovering users. An ideal anti-cocaine therapy would be a drug that could accelerate cocaine hydrolysis by plasma BChE. The development of a recombinant BChE fused to HSA provides a promising approach to enhancing the pharmacokinetic properties of BChE [55, 56].

12.4
Recombinant Albumin Fusion Technology

12.4.1
Recombinant Fusion Protein Linking Coagulation Factor VIIa with Albumin (rVIIa-FP)

Given that initial attempts to prolong the half-life of coagulation factors such as FIX using albumin fusion technology had been only moderately successful [53], similar results were expected for rFVIIa, as the two proteins are structurally similar members of the same family of vitamin K-dependent coagulation factors [18]. Building upon this initial experience, however, we developed a novel recombinant fusion protein linking coagulation factor VIIa with albumin (rVIIa-FP) using a flexible glycine/serine linker fused between recombinant albumin and the C-terminal end of rFVIIa [18]. The length of the linker, with 31 amino acids, was experimentally optimized to minimize potential interactions between the two proteins and to optimize the activity of FVIIa [18, 33, 90]. The construct was transfected into mammalian cell lines [human embryonic kidney (HEK)-293 cells or CHO cells]. As a control, a human FVII cDNA was transfected [18]. A mammalian expression system was required to achieve the functionally important post-translational modifications of the fusion protein, including γ-carboxylation, *N*- and *O*-glycosylation, and β-hydroxylation [33]. The fusion protein had a molecular weight of approximately 120 kDa, whereas wild-type recombinant FVII (rFVII) had a molecular weight of 59 kDa when produced in the same cell line [33].

The FVII activity and molar specific activity of rVIIa-FP derived from HEK-293 or CHO cell lines were assessed and compared with those of wild-type rFVIIa [18]. FVII activity was determined using a commercially available chromogenic test kit with human plasma as reference. When expressed in HEK-293 cells, the specific FVII activity of rFVIIa was $2069\,IU\,mg^{-1}$ compared with $626\,IU\,mg^{-1}$ for rVIIa-FP. The molar specific activity, calculated from FVII antigen values as determined by FVII enzyme-linked immunosorbent assay (ELISA), was $104.8\,IU\,nmol^{-1}$ and $72.8\,IU\,nmol^{-1}$ for rFVIIa and rVIIa-FP, respectively. This difference in molar specific FVII activity between the two proteins was the same whether rVIIa-FP was derived from HEK-293 or CHO cells. Thus, the fusion protein containing a suitable linker retained much of the potency of the wild-type protein. In addition, using an *in vitro* model of human hemophilia A with anti-FVIII inhibitors using whole blood, rVIIa-FP produced a dose-dependent reduction in clot formation time in a manner similar to that of the wild-type protein.

12.4.2
Recombinant Fusion Protein Linking Coagulation Factor IX with Albumin (rIX-FP)

As previously mentioned, early attempts to prolong the half-life of FIX using recombinant albumin fusion technology only moderately improved the pharmacokinetics of FIX [53]. The incorporation of a glycine/serine linker, which was successfully used to produce rFVIIa-FP [18, 33], produced a protein with low FIX-related activity [19, 53, 91]. The low activity of this fusion protein was possibly caused by interference from the albumin portion of the protein with the interaction between FIX and other coagulation factors, such as FVIII and FX [91]. We therefore developed novel recombinant albumin FIX fusion proteins using linker sequences that contained the same cleavage site responsible for the proteolytic activation of FIX. In particular, a specifically designed linker was used based on a small peptide derived from the N-terminal region of the activation peptide of FIX [19, 92]. Activation of FIX by either FVIIa/TF or FXIa would also cleave the linker, separating the FIX and HSA moieties of the fusion protein. Thus, FIX would be liberated from albumin when activated, permitting enhanced biological activity compared to previous FIX fusion proteins [19]. Plasmids containing the fusion gene construct were transfected into HEK-293 and CHO cell lines [19]. The fusion proteins were purified using chromatography, and their FIX activity was compared with those of similar fusion proteins with noncleavable linkers, rFIX, and plasma-derived FIX (pdFIX) *in vitro* and *in vivo* [19].

FIX-albumin fusion proteins with cleavable linkers had 10- to 30-fold higher molar specific activity in a one-stage clotting assay than those with noncleavable linkers (Figure 12.3) [19]. Fusion proteins with cleavable linkers also showed higher activity in a thrombin generation assay than did those with noncleavable linkers. Despite the improvement in biological activity compared with fusion proteins having noncleavable linkers, the activity of the fusion proteins with cleavable linkers remained lower than that of rFIX or pdFIX *in vitro*. This suggests that the albumin moiety can interfere with the

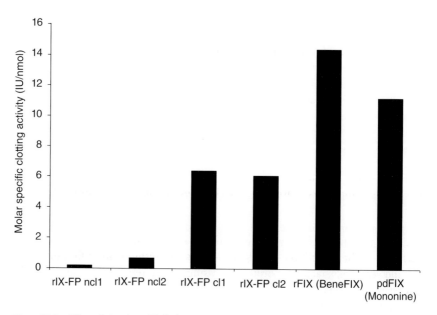

Figure 12.3 Effect of the cleavable linker on the molar specific factor IX (FIX) activity of recombinant FIX fusion protein (rIX-FP). Molar specific FIX activity was calculated based on the ratio of clotting activity and molar concentration, as determined by optical density. FIX clotting activity was based on an activated partial thromboplastin time assay. pd, plasma-derived; ncl1 and ncl2, two non-cleavable linkers; cl1 and cl2, two cleavable linkers [19].

activation process of rIX-FP to a certain extent, possibly due to steric hindrance [19].

12.4.3
Albutropin

hGH is responsible for various growth-promoting effects in the body. However, the hormone has a very short half-life of less than 20 minutes *in vivo* [93]. Thus, the treatment of hGH deficiency requires frequent dosing with hGH either three times a week intramuscularly or on a daily basis subcutaneously. Fusion of hGH to rHA is therefore a useful tool to increase the half-life of this hormone and to reduce its frequency of administration.

A yeast-based expression platform has been used in the production of a hGH albumin fusion protein called albutropin. It was constructed by genetically fusing rHA at the C-terminus to the N-terminus of recombinant hGH (rhGH) [14]. The recombinant fusion protein was expressed in a genetically modified form of the yeast *S. cerevisiae* laboratory strain AH22, with minimal post-translational modification. The fusion protein was produced using the modified yeast strain in a fermenter. Supernatants from multiple fermentations were pooled and then purified using conventional chromatography methods [94]. Albutropin was analyzed for purity and its degradation behavior compared with that of commercially available hGH and HSA.

Albutropin compared favorably with commercially available forms of hGH [94]. The rate of deamidation of the fusion protein was the same as that of rhGH. Fusion of HSA to rhGH increased solubility at pH 7.2, without increasing aggregation or inducing particle formation [94].

12.5
Technological Advantages and Challenges

The effectiveness of the recombinant albumin fusion technology has been demonstrated through its application to many different proteins, such as rFVIIa, rFIX, insulin, IFN, and GLP-1. Overall, the technology is a simple yet flexible platform for the production of proteins with an extended half-life. It permits high-quality production of a homogenous product while requiring fewer post-manufacturing modifications and purification steps, compared with other technologies, such as PEGylation. The use of well-established expression systems, including mammalian and yeast cells, facilitates production using robust, reproducible, and scalable processing methods.

One of the inherent challenges of albumin fusion technology is maintaining specific activity of the therapeutic protein, which is often reduced by the presence of the albumin moiety. As illustrated by the development of rVIIa-FP and rIX-FP, each molecule poses unique conditions and challenges. To retain as much specific activity as possible, it may be necessary to develop a specific concept for each fusion protein.

12.6
Pharmacokinetics of Recombinant Albumin Fusion Proteins

12.6.1
Recombinant Fusion Protein Linking Coagulation Factor VIIa with Albumin (rVIIa-FP)

The pharmacokinetics of rVIIa-FP was assessed in rats [18, 95], mice [95], and rabbits [88, 95]. In all animal species, the fusion protein demonstrated significant improvements in half-life, recovery, and area under the curve (AUC). In rats, the half-life of wild-type rFVIIa was increased from 39.5 minutes to 262.7 minutes by using the albumin fusion protein [18]. This corresponds to a nearly sevenfold increase in half-life. Moreover, the recovery of rVIIa-FP was significantly higher than that of rFVIIa and NovoSeven (Table 12.2) [18]. In mice and rabbits, the pharmacokinetic profile of the fusion protein was also significantly better than that of NovoSeven; thus, AUC, half-life, and recovery were all greater, while clearance was slower (Table 12.2) [88, 95].

12.6.2
Recombinant Fusion Protein Linking Coagulation Factor IX with Albumin (rIX-FP)

Similarly to rVIIa-FP, rIX-FP demonstrated improvements in pharmacokinetic parameters in rats [19, 96], rabbits [19, 96], and FIX$^{-/-}$ mice [19]. Specifically, the half-life, recovery, and AUC of rIX-FP were superior to those of rFIX (BeneFIX) (Table 12.3) [19, 96]. Of particular note, rIX-FP derived from CHO cells resulted in more pronounced improvements in recovery, half-life, and AUC compared with the same construct expressed in HEK-293 cells [19].

12.6.3
Albutropin

In rats and monkeys, the pharmacokinetic profile of albutropin was superior to that of hGH (Table 12.4) [14]. Following subcutaneous administration of the fusion protein in rats, plasma half-life and mean residence time were increased approxi-

Table 12.2 Pharmacokinetic parameters of the recombinant fusion protein linking coagulation factor VIIa with albumin (rVIIa-FP), compared with NovoSeven® in different animal species [18, 95]. Data are expressed as a ratio of rVIIa-FP to NovoSeven.

	Rats	Mice	Rabbits
AUC	14.5	11.4	14.2
Terminal half-life	5.8	4.1	8.1
In vivo recovery	2.4	2.0	2.0

AUC, area under the curve

Table 12.3 Pharmacokinetic parameters of the recombinant fusion protein linking coagulation factor IX with albumin (rIX-FP) expressed in CHO or HEK-293 cells compared with BeneFIX® in different animal species [19]. Data are expressed as a ratio of rIX-FP to BeneFIX.

	Rats		Rabbits	
	CHO	HEK-293	CHO	HEK-293
AUC	4.6	2.7	7.2	3.8
Terminal half-life	4.7	3.4	4.0	3.2
In vivo recovery	1.7	1.4	1.6	1.2

AUC, area under the curve; CHO, Chinese hamster ovary cells; HEK-293, human embryonic kidney 293 cells.

Table 12.4 Pharmacokinetic parameters of recombinant human growth hormone (rhGH) albumin fusion protein (rhGH-FP) compared with rhGH in different animal species [14]. Data are expressed as a ratio of rhGH-FP to rhGH.

	Rats	Monkeys
Mean residence time	8	5–10
Terminal half-life	8	6

mately eightfold and clearance was decreased twofold compared to hGH. In monkeys, subcutaneous administration of albutropin resulted in a sixfold longer terminal half-life, a five- to tenfold greater mean residence time, and an eightfold slower clearance than with hGH [14].

12.7
Preclinical Efficacy

12.7.1
Recombinant Fusion Protein Linking Coagulation Factor VIIa with Albumin (rVIIa-FP)

The activity of rVIIa-FP was evaluated *in vivo* using rats pre-treated with phenprocoumon, an anticoagulant that depletes FVII almost completely by 16 hours after application [18]. When administered at 15.75 hours after phenprocoumon administration, rVIIa-FP corrected whole blood clotting time determined by thrombelastography to a similar degree as did rFVIIa, confirming the hemostatic activity of the fusion protein. Moreover, when rVIIa-FP or rFVIIa were administered immediately after phenprocoumon, the fusion protein effectively corrected clotting time 16 hours later, but rFVIIa did not, due to its short half-life. Thus, the fusion protein was shown to be efficacious in an *in vivo* situation and its effects endured much longer than did those of rFVIIa [18]. The results suggest that a single injection of rVIIa-FP may be sufficient per bleeding episode [18].

12.7.2
Recombinant Fusion Protein Linking Coagulation Factor VIIa with Albumin (rVIIa-FP)

In a tail-tip bleeding model using FIX-deficient mice, cleavable rIX-FP reduced bleeding time in a dose-dependent manner that was comparable to that of rFIX [19]. The dose calculations were based on the FIX one-stage clotting assay. There were no significant differences between the groups. At a dose of $100\,IU\,kg^{-1}$ body weight, the bleeding time was reduced almost to that of normal mice, which had a mean bleeding time of around two minutes. Moreover, the cleavable rIX-FP minimizes the risk of generating excess FIXa fusion protein with extended half-life; thereby reducing the potential risk of entailing prothrombotic effects [19].

12.7.3
Albutropin

In GH-deficient hypophysectomized rats, albutropin administered daily, every other day, and every fourth day, significantly increased both cumulative body weight gain and tibial epiphyseal plate width [14]. These growth-promoting effects of albutropin were generally greater than those induced with equimolar doses of rhGH. However, unlike albutropin, rhGH failed to enhance these growth parameters when dosed every fourth day. In addition, when given as a single subcutaneous dose to monkeys, albutropin increased the plasma levels of insulin-like growth factor 1 (IGF-1) for up to seven days. This response was equivalent to that achieved by seven consecutive daily doses of rhGH. These findings suggest that albutropin could offer similar growth-promoting benefits to that of standard rhGH treatment regimens but with a reduced dosing frequency.

12.8
Clinical Efficacy

12.8.1
Albuferon

Final phase III data from the ACHIEVE 1 trial have recently been published [45]. A total of 1331 patients with chronic HCV genotype 1 were randomized to alb-IFN 900 or 1200 μg every two weeks or PEG-IFNα-2a 180 μg every week, all with oral weight-based ribavirin for a duration of 48 weeks. After a median of 28 weeks' treatment, all patients receiving alb-IFN 1200 μg had their dose reduced to 900 μg owing to the increased risk of pulmonary adverse events (see also Section 12.2.5.2).

Table 12.5 Selected safety summary of Albuferon in the ACHIEVE 1 trial [46].

	PEG-IFNα-2a %	alb-IFN 900 μg %	alb-IFN 1200 μg %
Completion rate	88.7	79.0	76.8
Discontinuation rate	6.8	14.9	17.0
Discontinuation due to AE	4.1	10.4	10.0
Discontinuation due to respiratory AE	0	0.9	1.6
Serious AE	10.9	11.1	13.6
Severe AE	19.5	20.6	24.1
Serious respiratory AE	0	0.7	2.0
Severe respiratory AE	0.5	2.3	2.5

AE = adverse event; alb-IFN = albinterferon alfa-2b; IFN = interferon; PEG-IFNα-2a = PEGylated interferon alfa-2a.

Based on an intent-to-treat analysis, alb-IFN met its primary efficacy endpoint of noninferiority to PEG-IFNα-2a for SVR: SVR rates were 51%, 48.2%, and 47.3% for PEG-IFNα-2a, and alb-IFN 900 μg and 1200 μg, respectively.

The safety profile of alb-IFN is shown in Table 12.5. Reports of adverse events were similar across all three arms, except for respiratory adverse events and rates of discontinuation, which were both higher in the two alb-IFN arms. No unexpected adverse events were observed.

12.9
Future Perspectives

Fusion protein technology, and recombinant albumin fusion proteins in particular, represent a promising area of research. The many recombinant albumin fusion proteins in various stages of preclinical and clinical development provide ample proof-of-principle that this approach is feasible. The fusion technology creates a hybrid polypeptide that makes use of the pharmacokinetic properties of albumin while retaining the biological activity of the therapeutic protein. The wide spectrum of current research on recombinant albumin fusion proteins indicates that these fusion proteins have the potential to positively impact treatment in a broad range of diseases. Future applications for fusion proteins will expand beyond the established areas of oncology, immunology, and inflammation.

12.10
Conclusion

Recombinant albumin fusion technology enables researchers to enhance the pharmacokinetic properties of candidate therapeutic proteins and therefore represents an important tool in the development of biopharmaceuticals. Albumin is an ideal fusion partner due to its remarkably long half-life, its abundance in human plasma, and its lack of enzymatic activity and immunogenic properties. Albumin fusion has been used to successfully extend the half-life of various therapeutic proteins, including small proteins and peptides such as GLP-1 and hGH, cytokines such as IFNα, and complex proteins such as FVII and FIX. This enhancement in pharmacokinetic properties has generally translated into improved pharmacodynamic effects. Ongoing clinical studies of albumin fusion proteins will help further define the clinical utility of this approach and its ability to permit less frequent dosing while retaining clinical efficacy.

Acknowledgments

The authors would like to acknowledge the editorial assistance of Dr. Sandra Cox (Swiss Medical Press GmbH), and the valuable statistical support of Dr. Jochen Müller-Cohrs in establishing the figures.

References

1 Schmidt, S.R. (2009) Fusion-proteins as biopharmaceuticals – applications and challenges. *Curr. Opin. Drug Discov. Devel.*, **12**, 284–295.

2 Olsen, E., Duvic, M., Frankel, A., Kim, Y., Martin, A., Vonderheid, E., *et al.* (2001) Pivotal phase III trial of two dose levels of denileukin diftitox for the treatment of cutaneous T-cell lymphoma. *J. Clin. Oncol.*, **19**, 376–388.

3 Prince, H.M., Duvic, M., Martin, A., Sterry, W., Assaf, C., Sun, Y., *et al.* (2010) Phase III placebo-controlled trial of denileukin diftitox for patients with cutaneous T-cell lymphoma. *J. Clin. Oncol.*, **28**, 1870–1877.

4 Weinblatt, M.E., Kremer, J.M., Bankhurst, A.D., Bulpitt, K.J., Fleischmann, R.M., Fox, R.I., *et al.* (1999) A trial of etanercept, a recombinant tumor necrosis factor receptor:Fc fusion protein, in patients with rheumatoid arthritis receiving methotrexate. *N. Engl. J. Med.*, **340**, 253–259.

5 Andersen, J.T., and Sandlie, I. (2009) The versatile MHC class I-related FcRn protects IgG and albumin from degradation: implications for development of new diagnostics and therapeutics. *Drug Metab. Pharmacokinet.*, **24**, 318–332.

6 Chuang, V.T., Kragh-Hansen, U., and Otagiri, M. (2002) Pharmaceutical strategies utilizing recombinant human serum albumin. *Pharm. Res.*, **19**, 569–577.

7 Veronese, F.M., and Mero, A. (2008) The impact of PEGylation on biological therapies. *BioDrugs*, **22**, 315–329.

8 Harris, J.M., and Chess, R.B. (2003) Effect of PEGylation on pharmaceuticals. *Nat. Rev. Drug Discov.*, **2**, 214–221.

9 Byrne, B., Donohoe, G.G., and O'Kennedy, R. (2007) Sialic acids: carbohydrate moieties that influence the biological and physical properties of biopharmaceutical proteins and living cells. *Drug Discov. Today*, **12**, 319–326.

10 Gao, Z., Bai, G., Chen, J., Zhang, Q., Pan, P., Bai, F., *et al.* (2009) Development, characterization, and evaluation of a fusion protein of a novel glucagon-like peptide-1 (GLP-1) analog and human serum albumin in *Pichia pastoris*. *Biosci. Biotechnol. Biochem.*, **73**, 688–694.

11 Subramanian, G.M., Fiscella, M., Lamousé-Smith, A., Zeuzem, S., and McHutchison, J.G. (2007) Albinterferon α-2b: a genetic fusion protein for the treatment of chronic hepatitis C. *Nat. Biotechnol.*, **25**, 1411–1419.

12 Kontermann, R.E. (2009) Strategies to extend plasma half-lives of recombinant antibodies. *BioDrugs*, **23**, 93–109.

13 Osborn, B.L., Olsen, H.S., Nardelli, B., Murray, J.H., Zhou, J.X., Garcia, A., *et al.* (2002) Pharmacokinetic and pharmacodynamic studies of a human serum albumin-interferon-alpha fusion protein in cynomolgus monkeys. *J. Pharmacol. Exp. Ther.*, **303**, 540–548.

14 Osborn, B.L., Sekut, L., Corcoran, M., Poortman, C., Sturm, B., Chen, G., *et al.* (2002) Albutropin: a growth hormone-albumin fusion with improved pharmacokinetics and pharmacodynamics in rats and monkeys. *Eur. J. Pharmacol.*, **456**, 149–158.

15 Müller, D., Karle, A., Meissburger, B., Höfig, I., Stork, R., and Kontermann, R.E. (2007) Improved pharmacokinetics of recombinant bispecific antibody molecules by fusion to human serum albumin. *J. Biol. Chem.*, **282**, 12650–12660.

16 Evans, L., Hughes, M., Waters, J., Cameron, J., Dodsworth, N., Tooth, D., Greenfield, A., *et al.* (2010) The production, characterization and enhanced pharmacokinetics of scFv-albumin fusions expressed in *Saccharomyces cerevisiae*. *Protein Expr. Purif.*, **73**, 113–124.

17 Huang, Y.S., Chen, Z., Chen, Y.Q., Ma, G.C., Shan, J.F., Liu, W., *et al.* (2008) Preparation and characterization of a novel exendin-4 human serum albumin fusion protein expressed in *Pichia pastoris*. *J. Pept. Sci.*, **14**, 588–595.

18 Weimer, T., Wormsbächer, W., Kronthaler, U., Lang, W., Liebing, U., and Schulte, S. (2008) Prolonged in-vivo half-life of factor VIIa by fusion to albumin. *Thromb. Haemost.*, **99**, 659–667.

19 Metzner, H.J., Weimer, T., Kronthaler, U., Lang, W., and Schulte, S. (2009) Genetic fusion to albumin improves the pharmacokinetic properties of factor IX. *Thromb. Haemost.*, **102**, 634–644.

20 Dou, W.-F., Lei, J.-Y., Zhang, L.-F., Xu, Z.-H., Chen, Y., and Jin, J. (2008) Expression, purification, and characterization of recombinant human serum albumin fusion protein with two human glucagon-like peptide-1 mutants in *Pichia pastoria*. *Protein Expr. Purif.*, **61**, 45–49.

21 Chen, J.-H., Zhang, X.-G., Jiang, Y.-T., Yan, L.-Y., Tang, L.O., Yin, Y.-W., Cheng, D.-S., *et al.* (2010) Bioactivity and pharmacokinetics of two human serum albumin-thymosin α1-fusion proteins, rHSA-Tα1 and rHSA-L-Tα1, expressed in recombinant *Pichia pastoris*. *Cancer Immunol. Immunother.*, **59**, 1335–1345.

22 Chen, J., Bai, G., Cao, Y., Gao, Z., Zhang, Q., Zhu, Y., and Yang, W. (2007) One-step purification of a fusion protein of glucagon-like peptide-1 and human serum albumin expressed in *Pichia pastoris* by an immunomagnetic separation technique. *Biosci. Biotechnol. Biochem.*, **71**, 2655–2662.

23 Daly, R., and Hearn, M.T. (2005) Expression of heterologous proteins in *Pichia pastoris*: a useful experimental tool in protein engineering and production. *J. Mol. Recognit.*, **18**, 119–138.

24 Meloun, B., Morávek, L., and Kostka, V. (1975) Complete amino acid sequence of human serum albumin. *FEBS Lett.*, **58**, 134–137.

25 He, X.M., and Carter, D.C. (1992) Atomic structure and chemistry of human serum albumin. *Nature*, **358**, 209–215.

26 Anderson, C.L., Chaudhury, C., Kim, J., Bronson, C.L., Wani, M.A., and Mohanty, S. (2006) Perspective–FcRn transports albumin: relevance to immunology and medicine. *Trends Immunol.*, **27**, 343–348.

27 Chaudhury, C., Mehnaz, S., Robinson, J.M., Hayton, W.L., Pearl, D.K., Roopenian, D.C., *et al.* (2003) The major histocompatibility complex-related Fc receptor for IgG (FcRn) binds albumin and prolongs its lifespan. *J. Exp. Med.*, **197**, 315–322.

28 Nilsson, J., Ståhl, S., Lundeberg, J., Uhlén, M., and Nygren, P.A. (1997) Affinity fusion strategies for detection, purification, and immobilization of recombinant proteins. *Protein Expr. Purif.*, **11**, 1–16.

29 Carter, J., Zhang, J., Dang, T.L., Hasegawa, H., Cheng, J.D., Gianan, I., *et al.* (2010) Fusion partners can increase the expression of recombinant interleukins via transient transfection in 2936E cells. *Protein Sci.*, **19**, 357–362.

30 Vis, A.N., van der Gaast, A., van Rhijn, B.W., Catsburg, T.K., Schmidt, C., and Mickisch, G.H. (2002) A phase II trial of methotrexate-human serum albumin (MTX-HSA) in patients with metastatic renal cell carcinoma who progressed under immunotherapy. *Cancer Chemother. Pharmacol.*, **49**, 342–345.

31 Wunder, A., Müller-Ladner, U., Stelzer, E.H., Funk, J., Neumann, E., Stehle, G., *et al.* (2003) Albumin-based drug delivery as novel therapeutic approach for rheumatoid arthritis. *J. Immunol.*, **170**, 4793–4801.

32 Chemmanur, A.T., and Wu, G.Y. (2006) Drug evaluation: Albuferon-α–an antiviral interferon-alpha/albumin fusion protein. *Curr. Opin. Investig. Drugs.*, **7**, 750–758.

33 Schulte, S. (2008) Use of albumin fusion technology to prolong the half-life of recombinant factor VIIa. *Thromb. Res.*, **122** (Suppl. 4), S14–S19.

34 Baker, M.P., and Jones, T.D. (2007) Identification and removal of immunogenicity in therapeutic proteins. *Curr. Opin. Drug Discov. Devel.*, **10**, 219–227.

35 De Groot, A.S., and Scott, D.W. (2007) Immunogenicity of protein therapeutics. *Trends Immunol.*, **28**, 482–490.

36 Tomkin, G.H. (2009) Albiglutide, an albumin-based fusion of glucagon-like peptide 1 for the potential treatment of type 2 diabetes. *Curr. Opin. Mol. Ther.*, **11**, 579–588.

37 Baggio, L.L., Huang, Q., Brown, T.J., and Drucker, D.J. (2004) A recombinant human glucagon-like peptide (GLP)-1-albumin protein (albugon) mimics peptidergic activation of GLP-1 receptor-dependent pathways coupled with satiety, gastrointestinal motility, and glucose homeostasis. *Diabetes*, **53**, 2492–2500.

38 Duttaroy, A., Kanakaraj, P., Osborn, B.L., Schneider, H., Pickeral, O.K., Chen, C., *et al.* (2005) Development of a long-acting insulin analog using albumin fusion technology. *Diabetes*, **54**, 251–258.

39 Marques, J.A., George, J.K., Smith, I.J., Bhakta, V., and Sheffield, W.P. (2001) A barbourin-albumin fusion protein that is slowly cleared *in vivo* retains the ability to inhibit platelet aggregation *in vitro*. *Thromb. Haemost.*, **86**, 902–908.

40 Syed, S., Schuyler, P.D., Kulczycky, M., and Sheffield, W.P. (1997) Potent antithrombin activity and delayed clearance from the circulation characterize recombinant hirudin genetically fused to albumin. *Blood*, **89**, 3243–3252.

41 Hagedorn, I., Schmidbauer, S., Pleines, I., Kleinschnitz, C., Kronthaler, U., Stoll, G., *et al.* (2010) Factor XIIa inhibitor recombinant human albumin Infestin-4 abolishes occlusive arterial thrombus formation without affecting bleeding. *Circulation*, **121**, 1510–1517.

42 Wang, W., Ou, Y., and Shi, Y. (2004) AlbuBNP, a recombinant B-type natriuretic peptide and human serum albumin fusion hormone, as a long-term therapy of congestive heart failure. *Pharm. Res.*, **21**, 2105–2111.

43 Chen, H.H., Matrin, F.L., Cataliotti, A., Schirger, J.A., and Burnett, J.C. (2007) Abstract 1382: AlbuBNP (Cardeva), a novel recombinant human B-type natriuretic peptide serum albumin fusion protein has prolonged renal enhancing properties when compared to human BNP. *Circulation*, **116**, 284.

44 Ikuta, S., Chuang, V.T.G., Ishima, Y., Nakajou, K., Furukawa, M., Watanabe, H., *et al.* (2010) Albumin fusion of thioredoxin–the production and evaluation of its biological activity for potential therapeutic applications. *J. Control. Release*, **147**, 17–23.

45 Nelson, D.R., Rustgi, V., Balan, V., Sulkowski, M.S., Davis, G.L., Muir, A.J., *et al.* (2009) Safety and antiviral activity of albinterferon alfa-2b in prior interferon nonresponders with chronic hepatitis C. *Clin. Gastroenterol. Hepatol.*, **7**, 212–218.

46 Zeuzem, S., Sulkowski, M.S., Lawitz, E.J., Rustgi, V.K., Rodriguez-Torres, M., Bacon, B.R., *et al.* (2010) Albinterferon Alfa-2b was not inferior to pegylated interferon-α in a randomized trial of patients with chronic hepatitis C virus genotype 1. *Gastroenterology*, **139**, 1257–1266.

47 Nelson, D.R., Benhamou, Y., Chuang, W.L., Lawitz, E.J., Rodriguez-Torres, M., Flisiak, R., *et al.* (2010) Albinterferon Alfa-2b was not inferior to pegylated interferon-α in a randomized trial of patients with chronic hepatitis C virus genotype 2 or 3. *Gastroenterology*, **139**, 1267–1276.

48 Müller, D., Schneider, B., Pfizenmaier, K., and Wajant, H. (2010) Superior serum half life of albumin tagged TNF ligands. *Biochem. Biophys. Res. Commun.*, **396**, 793–799.

49 Yao, Z., Dai, W., Perry, J., Brechbiel, M.W., and Sung, C. (2004) Effect of albumin fusion on the biodistribution of interleukin-2. *Cancer Immunol. Immunother.*, **53**, 404–410.

50 Melder, R.J., Osborn, B.L., Riccobene, T., Kanakaraj, P., Wei, P., Chen, G., *et al.* (2005) Pharmacokinetics and *in vitro* and *in vivo* anti-tumor response of an interleukin-2-human serum albumin fusion protein in mice. *Cancer Immunol. Immunother.*, **54**, 535–547.

51 Halpern, W., Riccobene, T.A., Agostini, H., Baker, K., Stolow, D., Gu, M.L., *et al.* (2002) Albugranin, a recombinant human granulocyte colony stimulating factor (G-CSF) genetically fused to recombinant human albumin induces prolonged myelopoietic effects in mice and monkeys. *Pharm. Res*, **19**, 1720–1729.

52 Joung, C.H., Shin, J.Y., Koo, J.K., Lim, J.J., Wang, J.S., Lee, S.J., *et al.* (2009) Production and characterization of long-acting recombinant human albumin-EPO fusion protein expressed in CHO cell. *Protein Expr. Purif.*, **68**, 137–145.

53 Sheffield, W.P., Mamdani, A., Hortelano, G., Gataiance, S., Eltringham-Smith, L., Begbie, M.E., *et al.* (2004) Effects of genetic fusion of factor IX to albumin on *in vivo* clearance in mice and rabbits. *Br. J. Haematol.*, **126**, 565–573.

54 Huang, Y.J., Lundy, P.M., Lazaris, A., Huang, Y., Baldassarre, H., Wang, B., *et al.* (2008) Substantially improved pharmacokinetics of recombinant human butyrylcholinesterase by fusion to human serum albumin. *BMC Biotechnol.*, **8**, 50.

55 Brimijoin, S., Gao, Y., Anker, J.J., Gliddon, L.A., Lafleur, D., Shah, R., *et al.* (2008) A cocaine hydrolase engineered from human butyrylcholinesterase selectively blocks cocaine toxicity and reinstatement of drug seeking in rats. *Neuropsychopharmacology*, **33**, 2715–2725.

56 Gao, Y., LaFleur, D., Shah, R., Zhao, Q., Singh, M., and Brimijoin, S. (2008) An albumin-butyrylcholinesterase for cocaine toxicity and addiction: catalytic and pharmacokinetic properties. *Chem. Biol. Interact.*, **175**, 83–87.

57 Matthews, J.E., Stewart, M.W., De Boever, E.H., Dobbins, R.L., Hodge, R.J., Walker, S.E., *et al.* (2008) Pharmacodynamics, pharmacokinetics, safety, and tolerability of albiglutide, a long-acting glucagon-like peptide-1 mimetic, in patients with type 2 diabetes. *J. Clin. Endocrinol. Metab.*, **93**, 4810–4817.

58 Rosenstock, J., Reusch, J., Bush, M., Yang, F., and Stewart, M. (2009) Potential of albiglutide, a long-acting GLP-1 receptor agonist, in type 2 diabetes: a randomized controlled trial exploring weekly, biweekly, and monthly dosing. *Diabetes Care*, **32**, 1880–1886.

59 Byetta Prescribing Information. (2010) Amylin Pharmaceuticals, Inc., September 2010, http://pi.lilly.com/us/byetta-pi.pdf (accessed 1 October 2010).

60 Elbrønd, B., Jakobsen, G., Larsen, S., Agersø, H., Jensen, L.B., Rolan, P., *et al.* (2002) Pharmacokinetics, pharmacodynamics, safety, and tolerability of a single-dose of NN2211, a long-acting glucagon-like peptide 1 derivative, in healthy male subjects. *Diabetes Care*, **25**, 1398–1404.

61 Lee, S.H., Lee, S., Youn, Y.S., Na, D.H., Chae, S.Y., Byun, Y., *et al.* (2005) Synthesis, characterization, and pharmacokinetic studies of PEGylated glucagon-like peptide-1. *Bioconjug. Chem.*, **16**, 377–382.

62 Sheffield, W.P., Smith, I.J., Syed, S., and Bhakta, V. (2001) Prolonged *in vivo* anticoagulant activity of a hirudin-albumin fusion protein secreted from Pichia pastoris. *Blood Coagul. Fibrinolysis*, **12**, 433–443.

63 Rustgi, V.K. (2009) Albinterferon alfa-2b, a novel fusion protein of human albumin and human interferon alfa-2b, for chronic hepatitis C. *Curr. Med. Res. Opin.*, **25**, 991–1002.

64 Fried, M.W., Shiffman, M.L., Reddy, K.R., Smith, C., Marinos, G., Gonçales, F.L. Jr, *et al.* (2002) Peginterferon alfa-2a plus ribavirin for chronic hepatitis C virus infection. *N. Engl. J. Med.*, **347**, 975–982.

65 Manns, M.P., McHutchison, J.G., Gordon, S.C., Rustgi, V.K., Shiffman, M., Reindollar, R., *et al.* (2001) Peginterferon alfa-2b plus ribavirin compared with interferon alfa-2b plus ribavirin for initial treatment of chronic hepatitis C: a randomised trial. *Lancet*, **358**, 958–965.

66 Balan, V., Nelson, D.R., Sulkowski, M.S., Everson, G.T., Lambiase, L.R., Wiesner, R.H., *et al.* (2006) A Phase 1/11 study evaluating escalating doses of recombinant human albumin-interferon-α fusion protein in chronic hepatitis C patients who have failed previous interferon-α-based therapy. *Antivir. Ther.*, **11**, 35–45.

67 Yeh, P., Landais, D., Lemaître, M., Maury, I., Crenne, J.Y., Becquart, J., *et al.* (1992) Design of yeast-secreted albumin derivatives for human therapy: biological and antiviral properties of a serum albumin-CD4 genetic conjugate. *Proc. Natl. Acad. Sci. U.S.A.*, **89**, 1904–1908.

68 Bain, V.G., Kaita, K.D., Yoshida, E.M., Swain, M.G., Heathcote, E.J., Neumann, A.U., *et al.* (2006) A phase 2 study to evaluate the antiviral activity, safety, and pharmacokinetics of recombinant human albumin-interferon alfa fusion protein in genotype 1 chronic hepatitis C patients. *J. Hepatol.*, **44**, 671–678.

69 Perry, C.M., and Jarvis, B. (2001) Peginterferon-alpha-2a (40 kD): a review of its use in the management of chronic hepatitis C. *Drugs*, **61**, 2263–2288.

70 Glue, P., Fang, J.W., Rouzier-Panis, R., Raffanel, C., Sabo, R., Gupta, S.K., *et al.* (2000) Pegylated interferon-alpha2b: pharmacokinetics, pharmacodynamics, safety, and preliminary efficacy data. Hepatitis C Intervention Therapy Group. *Clin. Pharmacol. Ther.*, **68**, 556–567.

71 Zeuzem, S., Yoshida, E.M., Benhamou, Y., Pianko, S., Bain, V.G., Shouval, D., *et al.* (2008) Albinterferon alfa-2b dosed every two or four weeks in interferon-naïve patients with genotype 1 chronic hepatitis C. *Hepatology*, **48**, 407–417.

72 Bain, V.G., Kaita, K.D., Marotta, P., Yoshida, E.M., Swain, M.G., Bailey, R.J., *et al.* (2008) Safety and antiviral activity of albinterferon alfa-2b dosed every four weeks in genotype 2/3 chronic hepatitis C patients. *Clin. Gastroenterol. Hepatol.*, **6**, 701–706.

73 European Medicines Agency (2010) Withdrawn Applications: Joulferon, http://.ema.europa.eu/ema/index.jsp?curl=pages/medicines/human/medicines/002166/wapp/Initial_authorisation/human_wapp_000096.jsp&murl=menus/medicines/medicines.jsp&mid=WC0b01ac058001d128&source=homeMedSearch&category=human (accessed 23 September 2010).

74 Zhao, H.L., Xue, C., Wang, Y., Li, X.Y., Xiong, X.H., Yao, X.Q., *et al.* (2007) Circumventing the heterogeneity and instability of human serum albumin-interferon-α2b fusion protein by altering its orientation. *J. Biotechnol.*, **131**, 245–252.

75 Zhao, H.L., Xue, C., Wang, Y., Sun, B., Yao, X.Q., and Liu, Z.M. (2009) Elimination of the free sulfhydryl group in the human serum albumin (HSA) moiety of human interferon-α2b and HSA fusion protein increases its stability against mechanical and thermal stresses. *Eur. J. Pharm. Biopharm.*, **72**, 405–411.

76 White, G.C. 2nd, Beebe, A., and Nielsen, B. (1997) Recombinant factor IX. *Thromb. Haemost.*, **78**, 261–265.

77 Ewenstein, B.M., Joist, J.H., Shapiro, A.D., Hofstra, T.C., Leissinger, C.A., Seremetis, S.V., *et al.* (2002) Pharmacokinetic analysis of plasma-derived and recombinant F IX concentrates in previously treated patients with moderate or severe hemophilia B. *Transfusion*, **42**, 190–197.

78 Poon, M.C. (2006) Pharmacokinetics of factors IX, recombinant human activated factor VII and factor VIII. *Haemophilia*, **12** (Suppl. 4), 61–69.

79 Björkman, S., Shapiro, A.D., and Berntorp, E. (2001) Pharmacokinetics of recombinant factor IX in relation to age of the patient: implications for dosing in prophylaxis. *Haemophilia*, **7**, 133–139.

80 World Federation of Hemophilia (2005) *Guidelines for the Management of Hemophilia*, World Federation of Hemophilia, Montreal, http://.wfh.org/2/docs/Publications/Diagnosis_and_Treatment/Guidelines_Mng_Hemophilia.pdf (accessed 25 May 2010).

81 Hedner, U., and Ingerslev, J. (1998) Clinical use of recombinant FVIIa (rFVIIa). *Transfus. Sci.*, **19**, 163–176.

82 Lusher, J.M., Roberts, H.R., Davignon, G., Joist, J.H., Smith, H., Shapiro, A., *et al.* (1998) A randomized, double-blind comparison of two dosage levels of recombinant factor VIIa in the treatment of joint, muscle and mucocutaneous haemorrhages in persons with haemophilia A and B, with and without inhibitors. rFVIIa Study Group. *Haemophilia*, **4**, 790–798.

83 Broze, G.J., and Majerus, P.W. (1980) Purification and properties of human coagulation factor VII. *J. Biol. Chem.*, **255**, 1242–1247.

84 Erhardtsen, E. (2000) Pharmacokinetics of recombinant activated factor VII (rFVIIa). *Semin. Thromb. Hemost.*, **26**, 385–391.

85 Key, N.S., Aledort, L.M., Beardsley, D., Cooper, H.A., Davignon, G., Ewenstein, B.M., *et al.* (1998) Home treatment of mild to moderate bleeding episodes using recombinant factor VIIa (Novoseven) in haemophiliacs with inhibitors. *Thromb. Haemost.*, **80**, 912–918.

86 Shapiro, A.D., Gilchrist, G.S., Hoots, W.K., Cooper, H.A., and Gastineau, D.A. (1998) Prospective, randomised trial of two doses of rFVIIa (NovoSeven) in haemophilia patients with inhibitors undergoing surgery. *Thromb. Haemost.*, **80**, 773–778.

87 Schulte, S., Weimer, T., Wormsbächer, W., Kronthaler, U., Groener, A., Lang, W., *et al.* (2007) Prolonged in-vivo half-life of FVIIa by fusion to albumin. *Blood*, 110. [Abstr. 3142].

88 Kronthaler, U., Schmidbauer, S., Liebing, U., Lang, W., Metzner, H.J., Weimer, T., *et al.* (2009) Prolonged half-life of recombinant factor VIIa fusion protein – single dose study in rabbits. XXII International Society on Thrombosis and Haemostasis Congress, Boston July 11–16, 2009. [Abstr. P-TH-561], http://.isth2009.com/poster_presentations_thursday.html (accessed 25 May 2010).

89 Di Scipio, R.G., Kurachi, K., and Davie, E.W. (1978) Activation of human factor IX (Christmas factor). *J. Clin. Invest.*, **61**, 1528–1538.

90 Horn, C., Steuber, H., Weimer, T., Wormsbächer, W., Liebing, U., and Kuo, W.-P. (2010) Concept and structure model of factor VIIa albumin fusion proteins. *Haemophilia.*, **16** (Suppl. 4), 37. [Abstr. 08P19].

91 Schulte, S. (2009) Half-life extension through albumin fusion technologies. *Thromb. Res.*, **124** (Suppl. 2), S6–S8.

92 Horn, C., Steuber, H., Metzner, H.J., Kuo, W.-P., Weimer, T., and Schulte, S. (2010) Concept and structure model of factor IX albumin fusion proteins. *Haemophilia*, **16** (Suppl. 4), 36. [Abstr. 08P18].

93 Haffner, D., Schaefer, F., Girard, J., Ritz, E., and Mehis, O. (1994) Metabolic clearance of recombinant human growth hormone in health and chronic renal failure. *J. Clin. Invest.*, **93**, 1163–1171.

94 Wilcox, A., Christenson, A.M., Spitznagel, T.M., and Krishnamurthy, R. (2003) Albumin fusion proteins: a novel sustained release option for protein therapeutics. *AAPS J.*, **5** (Suppl. 1), 982.

95 Zollner, S., Weimer, T., Schmidbauer, S., Raquet, E., Mueller-Cohrs, J., and Schulte, S. (2010) Pharmacokinetics of a recombinant albumin-fused, human coagulation factor VIIa (rVIIa-FP) exhibiting prolonged serum half-life in different animal species. *Haemophilia*, **16** (Suppl. 4), 33. [Abstr. 07P24].

96 Metzner, H., Weimer, T., Raquet, E., Nolte, W.M., Pragst, I., and Zollner, S. (2010) Prolonged serum half-life of a recombinant, albumin-fused, human coagulation factor IX (rIX-FP) in different animal species. *Haemophilia*, **16** (Suppl. 4), 40. [Abstr. 08P41].

13

AlbudAb™ Technology Platform—Versatile Albumin Binding Domains for the Development of Therapeutics with Tunable Half-Lives

Christopher Herring and Oliver Schon

13.1
Introduction

Most proteins and peptides are rapidly cleared from circulation through one or a combination of up to three routes: (i) *renal filtration* for those smaller than 55 kDa, (ii) *receptor mediated clearance*, also known as target mediated disposition (TMD), or (iii) *proteolysis* followed by renal filtration. Although biopharmaceutical drugs with short half-lives such as interferons (e.g., Roferon®, Biopharm Research and Development, GlaxoSmithKline) and erythropoietin (e.g., Amgen's Epogen®, GlaxoSmithKline) have been successfully developed and have had significant impact on patients' lives, these are increasingly being replaced by more advantageous engineered versions of these drugs (e.g., Pegasys) which exhibit longer half-life *in vivo*. Increased persistence in serum enables less frequent dosing to achieve steady-state serum concentrations within the therapeutic window, thereby improving drug efficacy, drug tolerability and ultimately patient compliance [1].

Successful engineering of any half-life extension modality to a peptide/protein drug often requires it to be a proteolytically stable before targeting the main clearance routes. Current half-life extension platform technologies (summarized in Table 13.1) can be divided into those that:

i)   **Increase hydrodynamic volume** to avoid renal clearance (such as PEGylation, HESylation, etc.; reviewed in Chapters 3–7).

ii) **Harness serum residence time to long lived abundant serum proteins** such as human serum albumin (HSA) or immunoglobulins (IgGs) (Affibodies, Fc fusions, etc.). Both HSA and IgGs have exceptionally long half-lives (up to 19 days and 21 days in human, respectively), due to a combination of their size and an active recycling mechanisms driven by the neonatal Fc receptor, FcRn, which prevents endosomal degradation following pinocytosis (also reviewed elsewhere in this book).

iii) **Modify or reduce receptor-mediated clearance** such that the therapeutic still binds or agonizes its target, but is recycled rather than cleared after binding

Therapeutic Proteins: Strategies to Modulate Their Plasma Half-Lives, First Edition. Edited by Roland Kontermann.
© 2012 Wiley-VCH Verlag GmbH & Co. KGaA. Published 2012 by Wiley-VCH Verlag GmbH & Co. KGaA.

Table 13.1 Various strategies to improve half-lives of therapeutic proteins.

Strategy	Who	Principle	Production		Mechanism		Size [kDa]
			Genetic fusion	Chem. conjugation	>hydrodynamic size	FcRn-mediated	
IgGs/Fcs	Various	*Harnessing half-life into natural life cycle of molecules*	✓	✓	✓	✓	50–150
PEGylation	Various	*Coupling of polyethylene glycol*		✓	✓		20–40 plus
Biodegradable/recombinant tails	Various	*Fusion/coupling of unstructured polypeptid/polyglycans and fatty acid chains*	✓	✓	✓		>50; varying
Alternative scaffolds	Various	*Multimerisation*	✓		✓		varying
Albumin binding–covalent	CoGenesys, GSK, Conjuchem, HGS	*Coupling of therapeutic to SA*		✓	✓	✓	66
Albumin binding–transient	Domantis; Ablynx; Affibody	*Protein binding to SA*	✓		✓	✓	12–15
	NExpedite Genentech	*Peptide binding to SA*	✓		✓	✓	~1
	Novo Nordisk	*FA binding to SA*		✓	✓	✓	?

FA, fatty acid; SA, serum albumin. Covered are only a few properties of each platform technology excluding degradability, manufacture and costs for example. This list is not exhaustive.

[2, 3]. This effect can provide additive benefits to the other half-life extension approaches (i) or)ii) above.

The ideal half-life extension platform would enable a fused therapeutic to mimic or exceed the serum residence time of long lived proteins such as IgG and serum albumin. A minimally sized, stable molecule that can be produced in a single step would be ideal. As we discuss below, a single domain protein transiently binding serum albumin can directly exploit mechanisms (i) and (ii) above, and–with suitable engineering–also mechanism (iii). Thus, this approach might have great potential to develop improved therapeutics.

13.2
The Domain Antibody

Domain antibodies (dAb)s are derived from the isolated single variable domain of an IgG (shown in Figure 13.1) and are approximately 1/10th of the size of a conventional antibody (12–15 kDa *vs.* 150 kDa).

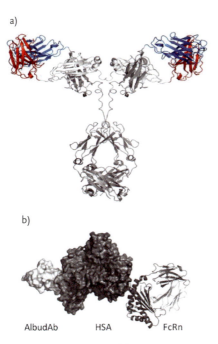

Figure 13.1 Structural representation of (a) an IgG1 and (b) HSA-FcRn complex (modeled after Kenanova *et al.* [30]). In (a) the variable domains are highlighted in blue (heavy chain or V$_H$) and red (here: light kappa chain) or V$_L$), respectively. These are the antigen-binding monoclonal antibody (mAB) fragments constituting domain antibodies or "dAbs". (b) trimeric complex model of a single domain antibody binding serum albumin (termed *AlbudAb*™; co-complex solved in house) and FcRn. The latter is responsible for the recycling of serum albumin *in vivo*. The binding sites on serum albumin for both proteins are distinctly different [30, 33].

They first were identified when it was realized that much of the binding affinity of an antibody could be obtained from either the V_H or V_L domain alone, requiring no contribution from its partner [4]. Similar single domain antibody fragments have since been described in sharks and *camelidae* which express heavy chain antibodies (VHHs) only as part of their natural immune system [5]. Early work on domain antibodies revealed the potential of a small stable protein domain that could be selected efficiently by phage display technology [6] or other *in vitro* selection approaches to bind monovalently to a variety of antigen types, including soluble as well as cell-bound targets and receptors.

Based on these potentially useful properties of domain antibodies the AlbudAb™ Technology Platform was developed. An AlbudAb is a domain antibody selected to bind serum albumin with moderate to high affinity [7]. Upon administration, an AlbudAb binds rapidly to serum albumin in the tissues and vasculature. The resulting non-covalent complex has an effective size of ~81 kDa (15 kDa plus 66 kDa), thus avoiding renal filtration, and imparting the long half-life of serum albumin to the AlbudAb [8]. Importantly, those desirable *in vivo* pharmacokinetic and biodistribution properties can also be extended to a short-lived therapeutic molecule, peptide or protein covalently attached to an AlbudAb. As a consequence, dosing regimen, efficacy and potentially toxicity profiles of such therapeutic drugs could be significantly improved.

13.3
Key Considerations for AlbudAb™-Based Molecules

Considering the above, AlbudAbs have been identified based on the following key properties:

a) **High affinity** binding to serum albumin, such that its pharmacokinetics approaches or equals that of serum albumin, also imparting these properties to fusion partners.

b) **Cross-reactivity to a range of species' serum albumins**, thus enabling the use of the same molecule throughout development. This eases dose- and pharmacokinetic predictions in other species and ultimately in human.

c) **Versatility**, allowing a wide variety of fusion options from amino/carboxy-terminal and internal conjugation sites to different molecule classes altogether, by retaining the potency of the fusion partner and the affinity of the AlbudAb to albumin.

d) **High stability** in terms of biological, biophysical and biochemical properties. Although the stability and manufacturability of AlbudAb–drug fusions may greatly depend on the nature of the therapeutic moiety, a well expressed, stable AlbudAb would contribute fewer potential liabilities to the final molecule.

e) **An affinity and binding site on serum albumin that is not modulated by endogenous or exogenous ligands** of serum albumin (e.g., fatty acids, bilirubin, etc.

or drugs such as ibuprofen, warfarin, diazepam) as this could cause adverse metabolic or drug-drug interactions.

f) **No predicted immunogenicity**, at least in humans.

In practice, selecting for and engineering in all of those properties in one single molecule proved to be very challenging, and we have developed a suite of different base AlbudAbs to maximize developmental options. In this chapter, some of the challenges, opportunities and progress in developing AlbudAb-based molecules as potential therapeutics will be discussed.

13.4
Challenges of Albumin as a Target

Despite a high structural and overall primary sequence homology between serum albumin from different species (approximately 80% between human and other species), a sequence alignment of serum albumin domain 2 (Figure 13.2a) reveals that this homology decreases significantly for surface exposed residues. This is one reason that historically, drug binding studies used different species depending on which binding site was investigated [9]. The exquisite specificity of antibodies could therefore mean that a different AlbudAb would be required for each model species. This contrasts with approaches such as PEGylation which yield PK improvements that scale linearly through species [10].

These surface differences posed a significant challenge for the selection of base AlbudAbs with a broad and relevant cross-species reactivity. A successful approach for isolating AlbudAb leads was to perform crossover phage selections, where a first round selection on HSA was followed by two consecutive rounds on mouse (MSA) or rat serum albumin (RSA), respectively, as the second antigen at lower concentrations (Liu *et al.*, 2010, unpublished data). This approach yielded a high proportion of promiscuous serum albumin binders with occasionally very broad cross-species reactivity (see Figure 13.2b), which were taken forward for further analysis. Schon *et al.* (unpublished data) found different AlbudAb lineages (related sequences, Figure 13.2c) with different patterns of species cross-reactivity. Presumably crossover selections resulted in clones that targeted (i) conserved epitopes on serum albumin (SA) or (ii) bound to areas with residue substitutions not critical or detrimental for binding.

Alanine scanning epitope mapping on HSA was used to further map individual residues that were crucial for cross-reactive binding and to footprint the epitope (Schon *et al.*, manuscript in preparation). For some AlbudAb leads this data helped to explain the pattern of cross-reactivity. For example, the AlbudAb lead DOM7h-11-3 bound human and cynomolgus SA with good affinity however failed to bind rodent SA with significant affinity. Alanine scanning experiments demonstrated that methionine 353 in HSA contributed to high affinity binding, and the substitution to threonine in rat and mouse greatly reduced binding affinity (Figure 13.3c). Insertions/deletions in marmoset and guinea pig SA add to the complexity. X-ray

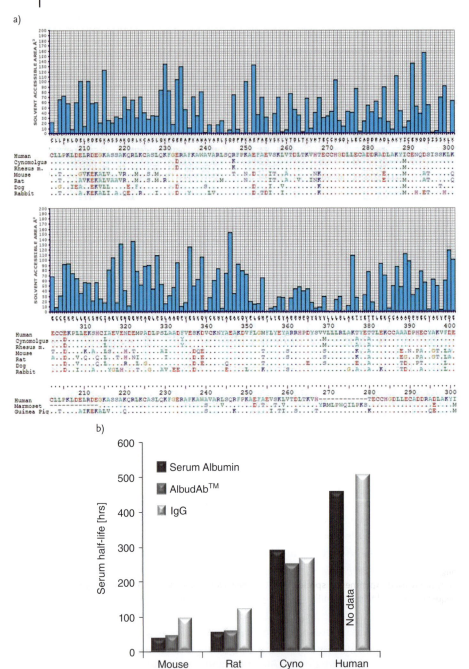

a)

b)

c)

	AlbudAb™/Species	Human	Cyno	Marmoset	Mouse	Rat	Dog	Ferret	Guinea Pig	Rabbit
	DOM7h-14-10	5.2	6.4	7370	0.7	3.6	26	133	52	NB
V$_k$ frame	DOM7h-14-119	4.9	5.7	6890	20	4.1	37	119	49	1350
work	DOM7h-11-3	48	425	-	1430	3000	-	-	-	-
	DOM7h-11-15	1	2.6	10	6.9	20	62	29	174	NB
	DOM7h-11-90	4.3	12.1	48	26	78	440	30	1750	NB
V$_H$ frame	DOM7r-31-104	88	31	4900	59	6	8.2	14	1630	1990
work	DOM7r-92-100	3.5	10	7.4	61	54	450	317	NB	136

crystallography of two AlbudAb-lead/human serum albumin complexes (data not shown) confirmed a tight packing of dAb and SA with contributions from numerous residues within the paratope to the overall binding.

Other authors have reported limited cross-species reactivity for VHH [11, 12] after immunization with a single serum albumin antigen, or via phage selections of peptides [34] and Affibody-based [13] anti-serum albumins against only one antigen. This could be seen as a fortuitous finding as neither author was systematically targeting wide cross-species reactivities.

13.5
Interactions of Albumin with AlbudAbs[TM]

AlbudAbs isolated from the phage-displayed repertoire of domain antibodies were selected for further characterization primarily on the basis of *in vitro* affinity to serum albumin as estimated by surface plasmon resonance (SPR) followed by pharmacokinetic analysis. Serum albumin has three domains, which share a similar disulphide bonded backbone structure, and pack to form a heart shaped molecule (Figure 13.1b). Initial epitope mapping of these primary leads was performed using recombinant HSA domains 1, 2 and 3 produced in *Pichia pastoris* [14] coated onto CM5 BIAcore™ chips via lysine ε-amino groups. During phage selections, randomly biotinylated soluble as well as passively coated full length serum albumin was used, and both approaches yielded predominantly domain 2 binders ([8] and unpublished). Given that the degree of homology between domains 1, 2 and 3 across the species used in selections is similar, the reason for this domain 2 bias is not clear. This is however a fortuitous finding. Domain 1 has a free-cysteine residue at residue 34, as has been used for derivitization and conjugation by others (Conjuchem for example), but can be occupied *in vivo* by various thiols, etc. [15]. Thus, this might represent a heterogeneous antigen population *in vivo* for high affinity ligands (such as AlbudAbs) in different patient populations potentially impacting on half-life and efficacy.

←

Figure 13.2 (a) Amino acid sequence alignment of serum albumin domain 2 from commonly used experimental species. All sequences were obtained from the NCBI sequence database. Bars represent the accessible surface area (ASA) for each residue (HSA crystal structure; 2XVV; RCSB: Protein Data Bank–www.pdb.org), which was used together with the crystal structure of HSA to identify target residues for alanine scanning mutagenesis based on their ASA, size and charge. Below: Sequence section exemplifying insertions/deletions in marmoset and Guinea pig SA. (b) Serum half-life of selected base AlbudAb™ leads compared to reported half-lives of SA and IgG in different species. (c) Cross-species reactivity of a selection of base AlbudAbs™ to core species' SA as determined by SPR. Note the cross-reactivity clustering for same sequence lineages. No significant cross-reactivity was obtained for rabbit. Values given are as KDs in [nM]. NB = no binding, -: not tested. Detailed analysis in Schon *et al.*, manuscript in preparation.

Domain 3 contains the interaction site for FcRn. Using wild-type and FcRn knockout mice, Stork *et al.* [16] have shown that a serum albumin binding antibody fragment prolonged half-life both, by avoiding renal clearance and "piggy-backing" on serum albumin through the FcRn-mediated endosomal recovery pathway. Thus, blocking domain 3-FcRn interactions would prevent recycling of albumin via this route.

With domain 2 binding AlbudAbs, we were able to create a 3-way complex *in vitro* of HSA, AlbudAb and FcRn extracellular domain (expressed in HEK293 cells), which was stable at both, pH 7 and pH 5.0 (Schon *et al.*, manuscript in preparation). Based on this data and the pharmacokinetics of domain 2 binding AlbudAbs, it has been proposed that AlbudAbs also take advantage of this recycling mechanism.

13.6
Bio-Analytical Characterization of AlbudAb™ Leads

For a serum half-life ($T_{1/2}$) extension platform technology to be successful, it needs to be versatile and amenable to extending the half-life of a variety of active compounds, so that it could be used in a similar way to PEG and other clinically proven technologies. To date, AlbudAbs have been fused to a wide range of molecule types, as is outlined in Figure 13.3. Others have developed similar molecules based on VHHs [11, 12, 17]. In addition to biological and biochemical characteristics, general biophysical properties such as oxidation, deamidation and aggregation propensities impact on the production, formulation, shelf-life and ultimately immunogenicity of biopharmaceuticals. We have sought to develop biophysically robust and stable base AlbudAbs as high quality building blocks to maximize the utility of the platform. In this section, some properties of the platform will be elaborated on in more detail.

13.6.1
Versatility

AlbudAbs offer multiple strategies to covalently link a therapeutic moiety. In the most classical approach, payloads such proteins and peptides are genetically fused to either the N- or C-termini of the polypeptide. Alternatively, synthetic or recombinant peptides and NCEs can be conjugated via chemical coupling to a natural or engineered non-critical surface-exposed scaffold residue [18].

For a genetic fusion, both N- or C-terminal fusions to the AlbudAb are usually explored to identify which orientation ensures retention of the therapeutic's potency as well as to minimize the effect on affinity of the AlbudAb.

The chemical nature and target interaction of fusion partners require the AlbudAbs' ability to accept different linkers (for genetic as well as chemical conjugation) to position each therapeutic in its optimal orientation to the target. For natural linkers, this could include linker length, steric rigidity or residue composition,

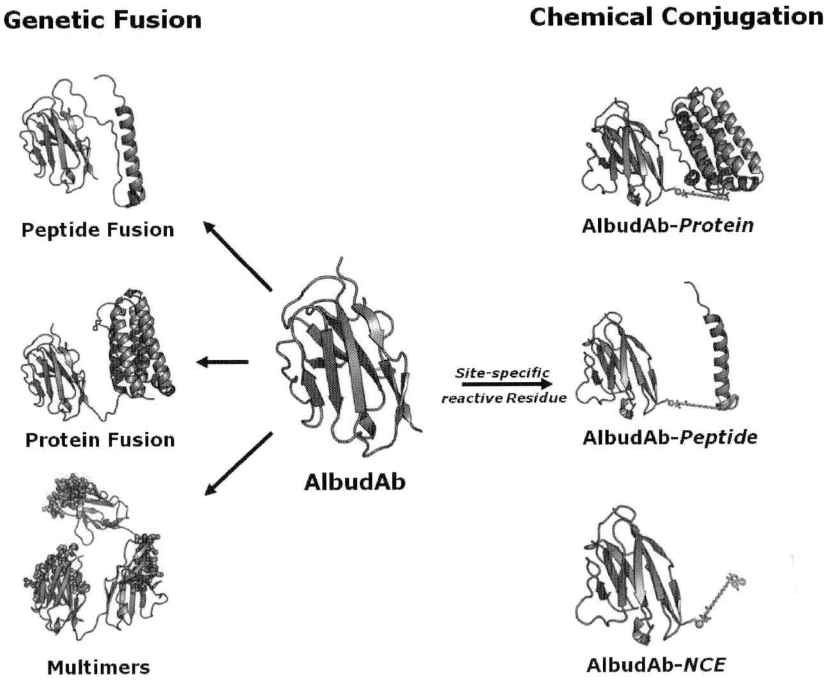

Genetic Fusion   **Chemical Conjugation**

Figure 13.3   Schematic representation of
different AlbudAb™-based formats as used in
various therapeutic programs. Chemical
conjugation is achieved via a reactive cysteine
residue (represented by small sphere) and a
maleimide-reactive synthetic linker between
AlbudAb and therapeutic moiety.
Final format structures are modeled.
Figure prepared by H. Arulanantham
using Pymol.

whereas chemical linkers are very varied in nature indeed, with the main body most often consisting of ethylene glycol repeats. In an ideal case a generic set of linkers/spacers can be identified to streamline discovery and development stages.

13.6.2
Affinity to Serum Albumin and Potency

Upon fusion of AlbudAb leads to therapeutic agents or other domain antibodies, the binding affinities to their targets of both partners of the now bifunctional molecule can be impacted to varying degrees. Both, Holt *et al.* [8] and Walker *et al.* [19] noted an optimal orientation for their AlbudAb fusions, with the affinity to HSA, target potency and molecule stability being affected. This could result from the orientation of the therapeutic to the AlbudAb as well as the connecting linker resulting in steric hindrance and a decreased diffusion coefficient (*D*) of the now larger AlbudAb-therapeutic. Walker *et al.* showed that the impact on fusion partner potency was lower than seen with a direct HSA fusion or PEGylated therapeutic, probably driven by the smaller size and non-covalent linkage in the AlbudAb fusion reducing steric hindrance in receptor binding. The impact on affinity and potency has to be determined individually for each therapeutic fusion. A typical decrease in affinity to serum albumin ranges from 2–10 fold compared to the un-conjugated AlbudAb. Figure 13.2b shows the serum residence times as obtained for some base AlbudAb leads in various animal models.

13.6.3
Solution State

The solution state of biopharmaceuticals can impact not only their purification and formulation, but also their activity. One of the key benefits of domain antibodies is that they are monomeric and monovalent, compared with the conventional IgG-based therapeutics which exhibit bivalency within their architecture. This offers the opportunity to neutralize cell surface receptors without the potential of induced agonism due to crosslinking. Therefore, our selection strategies for AlbudAbs included screening for monomer propensity at high concentration in order to be able to make long-lived monomeric AlbudAb fusion proteins of antagonistic dAbs or other proteins such as IL-1ra [8]. However, many therapeutic proteins are not natively monomeric, such as interferon alpha, and dimerization may be required for activity [20]. Walker *et al.* [19] used size exclusion chromatography-multi-angle laser light scattering (SEC-MALLS) analysis to demonstrate the strong dimerization propensity of an interferon alpha AlbudAb, which was driven by the interferon moiety. Thus the AlbudAb is a flexible enough fusion partner that it can be used with both monomeric and dimeric partners, retaining the potency of the fusion partner, and albumin binding ability.

13.6.4
Thermal Stability and Aggregation Resistance

High thermal stability of proteins has been found to be predictive of good performance under stress testing and downstream processing (DSP). As many biological proteins unfold irreversibly upon heat denaturation, only the apparent thermal stability (apparent T_m) can be determined; however such data are still indicative of the compact folding of a protein, and hence are useful as a ranking tool for a set of leads with comparable other properties. Single domain antibodies naturally exhibit relatively high thermal stability (into the 60 s°C to 70 s°C [21]), which makes them in most cases suitable for development. However – in some cases – additional resistance to aggregation upon stress or long-term storage is desirable.

We and others have adapted phage display selections to mimic stress conditions and therefore facilitate the isolation of stress-resistant candidates. Published strategies include acid selections [22] and reversible heat unfolding during selections [21]. Using the latter, the proteins undergo reversible unfolding on the phage tip and only those that escape aggregation will propagate in the next infection step. Depending on the stress selection, this can result in outputs with increased stability, proteolytic resistance or potency. Other display and selection systems can achieve similar results, but the robustness of the phage particles enables a broader range of stress conditions to be imposed (e.g., [21]).

Of interest is the observation that AlbudAbs show a significantly increased proteolytic and thermal stability when complexed with albumin, which should be beneficial for stability *in vivo*. Albumin has a long history in protein formulation as a stability enhancing excipient [23]; however the requirement for stoichiometric amounts of albumin in an AlbudAb formulation suggests a rather infrequent utility.

13.7
Production of AlbudAb™ Fusions

The single domain architecture of domain antibodies means that they can be efficiently expressed in a range of systems, either insolubly or secreted in *E. coli*, as well as solubly in yeast and mammalian systems (reviewed in [7]). The similarly structured camelid VHHs have additionally been expressed in plants (e.g., [24]). This versatility and simplicity of the scaffold means that AlbudAbs can be an efficient modular building block for a variety of molecule formats (Figure 13.3). This contrasts with some larger fusion partners such as albumin whose complex disulphide bonding restricts them to eukaryotic expression hosts.

Thus, the choice of expression system can be dictated by the fusion partner as well as speed and cost considerations. Most work with AlbudAb fusions has used secreted periplasmic expression in *E. coli* [7], (Holt *et al*, manuscript in preparation). Coppieters *et al.* [11] also used secreted expression in *E. coli* to obtain their trimeric, bivalent α-TNFα/α-MSA VHH. Where a fusion partner requires glycosylation for activity or is not efficiently expressed in *E. coli*, a mammalian system (or appropriately engineered eukaryote) can be used. Walker *et al.* [19] for example used mammalian expression to produce interferon-alpha AlbudAbs as the most adequate route due to product proteolysis occurring in *E. coli* systems.

As described above, the AlbudAb is amenable to chemical conjugation of synthetic moieties. Our preferred approach for chemical conjugation to date has been to engineer a free cysteine residue at either the carboxy-terminus or via an engineered internal cysteine residue at a site remote from the CDRs but within the germline V-gene scaffold. Engineering of framework residues requires careful consideration, as the molecule's stability and residue reactivity can impact its affinity as well as the activity/potency of the final conjugate. Gomez *et al.* [25] reported similar findings when engineering free cysteine groups for conjugation into IgGs. Different cysteine positions gave different expression levels and conjugation efficiencies. So far, maleimide-cysteine chemistry is the preferred method for dAbs due to its high coupling efficiency and final product quality, although other conjugation chemistries may have advantages and are well reviewed in Hermanson [18].

13.8
Purification of AlbudAbs™

Purification of AlbudAbs at laboratory scale is simplified due to their generic V_H and V_K scaffolds that naturally bind immobilized bacterial superantigens protein A or L, respectively. This provides a generic ligand for affinity chromatography purification and analysis during the early stages of development. This protocol has proven to be tolerant of a large number of AlbudAb fusion proteins with no need for purification tags or scouting of optimum purification conditions. As has been established for mAb manufacture, Protein-A based matrices have the potential to give very good clearance levels of host cell proteins (HCPs) and other contaminants, are scalable and potentially useable to make clinical grade material. Being high affinity albumin binders, affinity chromatography on serum albumin matrices has also been used as a robust generic purification method; however, the use of protein ligands for purification at scale may be expensive, non-feasible or not FDA approved. Mixed-mode resins work well for primary purifications once optimized and polishing steps such as conventional cation or anion exchange or hydrophobic interaction chromatography (HIC) yield final, high purity product.

The small size of an AlbudAb means that characterization of the target products is easy, particularly with small molecule conjugation as relatively simple mass spectrometry-based approaches can be used to analyze derivitisation stoichiometry and site specificity. This contrasts somewhat with larger and potentially more heterogeneous polymers such as PEG as well as monoclonal antibodies.

13.9
Biodistribution of AlbudAbs™

The biodistribution of AlbudAbs has been assessed in healthy animals by two approaches, one using a tritiated label, and another using an I^{123}/I^{125} dual labeling approach.

Using quantitative whole body autoradiography (QWBA), Holt *et al.* [8] examined the biodistribution of tritiated MSA and a MSA-specific AlbudAb 12 hours post injection (Figure 13.4a). The distribution of both albumin and AlbudAb in a

Figure 13.4 (a) Quantitative whole body autoradiography (QWBA) 1 h after administration of (i) MSA-binding AlbudAb™ i.v., (ii) s.c. and (iii) MSA alone (2.5 mg/kg each). The biodistribution of the AlbudAb is similar to that of the serum albumin. Key organs have been labeled to aid orientation. (b) Tissue distribution of rat serum albumin (top) and base AlbudAb (bottom) in a selection of tissues and brain sections in rat after bolus injection of 5 mg/kg I^{125}-labeled material. Shown is the ratio of labeled protein in tissue *vs.* blood. 6–12% of labeled RSA/AlbudAb in tested organs, whereas only 2.5–5‰ of labeled material in the brain. Readings were corrected for blood contamination in tissues and brain sections. Brain 1: Cerebellum. Brain 2: Medulla oblongata, Pons, Superior and Inferior colliculus. Brain 3: Hypothalamus, Hyppocampus, Thalamus, Septum.

a)

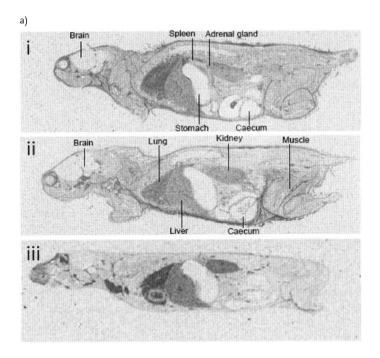

b)

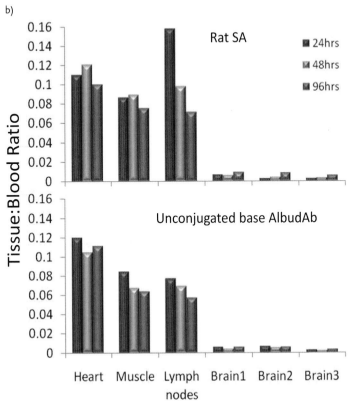

sagital section appeared virtually identical, with almost no radioactivity in the brain, but albumin and dAb widely distributed throughout the body. In an extensive unpublished study in rat, Schon *et al.* investigated the biodistribution of an AlbudAb and rat serum albumin in the brain as well as 17 other tissues. Again, for serum albumin and the AlbudAb, a broad distribution was observed, with no appreciable accumulation of either molecule in various brain sections with brain:blood ratios as low as 2.5–5‰ (Figure 13.4b). This is in good agreement with reports of an active exclusion mechanism of serum albumin from the brain [26].

Several recent publications have demonstrated increased serum albumin concentrations in tumors [27] and inflamed tissues and joints [28] driven via the leaky vasculature found in such disease settings. It is possible to use this differential albumin distribution to drive therapeutics to the site of action. Coppieters *et al.* [11] have shown accumulation of an α-TNF-α-MSA trimeric camelid VHH in inflamed mouse joints via a proposed enhanced permeability and retention effect. This effect has also been reported for PEGylated *vs.* non-PEGylated molecules [29].

Thus, a therapeutically relevant increase in concentration at the site of action may be achievable in certain tissues and diseases. Similarly, some drugs have beneficial systemic therapeutic effects, but an adverse side effect profile that is driven by central nervous system (CNS) penetration. An AlbudAb fusion could be exploited to maximize systemic exposure, while at the same time minimizing CNS penetration and potential side effects. Drugs modulating exposure in this fashion are currently in the clinic, for example Nektar's PEGylated opioid receptor antagonist NKTR-181.

The distribution of AlbudAb fusions can also be driven by receptor concentration gradients in tissues. Here, the small size and non-covalent interaction with albumin aids rapid and deep tissue penetration, as has been seen for small antibody fragments [30, 31].

13.9.1
Pharmacokinetics and Efficacy of AlbudAb™ Fusions

Various groups working on serum half-life extension platforms utilizing transient binding to serum albumin have shown significant reduction in clearance rates, and concomitant increases of half-lives of the fusion proteins compared to the unmodified therapeutic (see also Chapters 14 and 15). Holt *et al.* [8] used interleukin-1 receptor antagonist (IL-1ra) alone and as an amino-terminal genetic AlbudAb fusion in mice. The unmodified IL-1ra was renally excreted rapidly with a calculated half-life of only 2 minutes. As an AlbudAb fusion, the molecule remained in circulation significantly longer ($T_{1/2}$ of 4.3 hours or 130 fold improvement). Coppieters *et al.* [11] used a bivalent α-TNFα VHH alone and as an α-MSA VHH fusion. Again, the serum half life improved significantly from 54 min to 2.2 days (58-fold improvement).

Walker *et al.* [19] used IFN-α2b in fusions with various AlbudAbs of the same sequence lineage yet different affinities to SA covering 3 logs of affinity range ([19] and unpublished). Thus, any difference in PK was derived solely from affinity effects and was not influenced by epitope or lineage-specific properties. The PK parameters of these AlbudAb fusions, as determined in rats for those proteins, compare favorably in terms of their long serum half life, low clearance rate and bioavailability compared to HSA-IFN and IFN alone (Figure 13.5b). They found a linear correlation between albumin affinity and serum half-life in rat (Figure 13.6). This observation has been repeated in various other models and species and should ultimately enable the prediction of serum residence times in man. Using this correlation, the design of AlbudAb therapeutics for various indications requiring different dosing regimen should be possible.

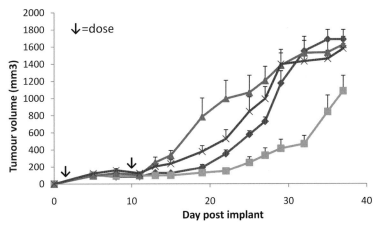

	IFN-α2b-DOM7h-14	IFN-α2b-DOM7h-14-10	HSA-IFN-α2b	IFN-α2b standard
$T_{1/2}$ elim (h)	22.6	30.9	14.2	1.2
Cmax (ug/mL)	34.7	38.8	60.1	25.3
AUC$_{0-\infty}$ (h ug/mL)	737.5	750.5	689.2	18.8
Clearance (mL/h/kg)	2.6	2.7	3.1	135.2
Bioavailability (%)	47.4	45.8	19.9	29.2

Figure 13.5 *In vivo* efficacy and PK summary table of IFN-a2b fusion proteins. (a) Tumor volume in 518A2 xenografted mice injected with 8.75 mg/kg IFN-a2b-DOM7h-14 (closed rectangles), IFN-α2b-DOM7h-14-10 (closed squares), AlbudAb™ alone (closed triangles) or 24 mg/kg HSA-IFN-α2b (crosses) at days 1 and 8 post-tumor implant. Tumor volume was measured at the indicated times, with *n* = 10 at each time point. (b) Pharmacokinetics of interferon fusions following intravenous administration in rats. IFN-a2b-DOM7h-14-10 has a 10-fold higher affinity to MSA compared to IFN-a2b-DOM7h-14 (13 *vs.* 150 nM). *N* = 3 per compound.

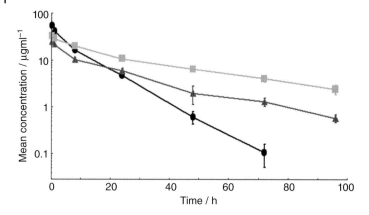

Figure 13.6 Pharmacokinetic profiles of IFN-α2b/AlbudAb™ fusions after i.v. bolus administration of 2 mg kg⁻¹ fusion protein in rats. AlbudAbs are from the same sequence lineage but affinity matured: low (circles) to medium (triangles; fourfold improvement) to high (squares; 50-fold improvement over parent) affinity to RSA. An increase of area under the curve (AUC) and $T_{1/2}$ with increased affinity to albumin was observed.

Improved efficacy of therapeutic AlbudAb fusions over the unconjugated parent protein has been demonstrated in various disease models. Holt *et al.* [8], showed in a collagen induced arthritis (CIA) mouse model that IL-1ra AlbudAbs showed a significant, yet concentration-and dose dependent decrease in the arthritic score. Walker *et al.* ([19] and unpublished) compared the efficacy of two IFN-AlbudAb fusions with different affinities to MSA with an HSA-IFN molecule in a SCID mice xenografted with human 515A2 tumors. Mice were dosed twice with equimolar amounts of therapeutics post tumor implantation.

The plot of tumor volume over time (Figure 13.5A) shows that the low affinity AlbudAb fusion compares well with HSA-IFN. However, the affinity matured AlbudAb fusion at the same dosing shows a significantly slower tumor growth. From these data, one could conclude that improved pharmacokinetics translates into improved pharmacodynamics. Coppieters *et al.* [11], Roovers *et al.* [12] and Tijink *et al.* [17] all have shown similar efficacy gains using various albumin binding VHH models.

As described elsewhere (Holt and others), the half-life of IL-1ra/AlbudAb fusions was not as long as expected from data of the same AlbudAb alone. This observation was also made for other therapeutic fusions targeting cell surface receptors, indicating that some other mechanism specific to the IL-1ra part may have mediated the apparent clearance of the fusion from the bloodstream.

Coffey *et al.* [32] and others have observed a similar phenomenon for IgGs to certain targets in the past. It has since been described as 'target-mediated disposition' (TMD) or 'receptor-mediated clearance'. Upon binding to the receptor, the complex is internalized for e.g. signal transduction before targeted for lysosomal degradation. This additional clearance mechanism will affect the overall serum concentration and consequently the half-life of such therapeutic significantly.

13.10
Summary and Conclusion

Serum half-life extension of biopharmaceuticals has been widely and successfully used either by harnessing to long-lived formats such as IgGs and Fc fusions or by chemical conjugation of natural, but rapidly excreted therapeutics to synthetic polymers such as PEG. Recently, a new generation of more refined technologies have emerged. These include biodegradable polymers, alternative engineered scaffolds and a range of albumin-binding technologies.

Here, we described the AlbudAb platform technology in more detail. The ease of production and versatility, tunability of serum residence time and possible routes of administration makes the AlbudAb technology platform a viable competitor to established formats and a vehicle for therapeutics to new indications, which have been off-target for IgGs so far. The broad cross-reactivity gives the AlbudAb platform a similar diversity of use to PEGylation, and its transient, non-covalent complexation with serum albumin has shown better tissue penetrative properties compared to other, larger molecular scaffolds.

In conclusion, one can say that the AlbudAb platform benefits from the wide systemic distribution of serum albumin to the point where albumin accumulates/infiltrates inflamed or diseased tissues carrying with it the AlbudAb and therefore increases the local concentration of the therapeutic at the site of action. On the other hand, serum albumin is excluded from the brain and we can exploit this to limit exposure of drugs to the brain and consequently minimize CNS side effects.

Not all therapeutic targets or diseases require the longest possible serum half-life. A strong differentiation of any albumin-binding serum half-life extension technology is the ability to tune its residence time. Numerous researchers have described a good correlation between affinity to albumin and serum half-life. This is a very good opportunity to spread out equally well into acute as well as chronic indications using therapeutic molecules, including peptides, proteins and even small molecules (NCEs). Due to their size and robustness, dAbs can not only be administered *intravenously,* as conventional antibodies but also *subcutaneously* and work is currently being conducted on inhaled as well as topical delivery. This allows targeting new diseases, but should also increase patient compliance, a major factor in the clinic.

Although no clinical data has yet been reported, published data to date on AlbudAbs appear consistent with infrequent dosing intervals and in the target range for therapeutics of this type, although this will be driven by the fusion partner and further work will have to be conducted. Despite the good progress made to date, these new technologies are faced with numerous challenges, and none currently have products with marketing authorization. Only continued scientific and clinical development can de-risk not only the next therapeutic, but the whole platform principle it is using.

AlbudAb™, AlbudAbs™, Domain Antibody, Domain Antibodies, dAb and dAbs are trademarks of the GlaxoSmithKline group of companies.

References

1 Ridruejo, E., Adrover, R., Cocozzella, D., Fernández, N., and Reggiardo, M.V. (2010) Efficacy, tolerability and safety in the treatment of chronic hepatitis C with combination of PEG-Interferon–Ribavirin in daily practice. *Ann. Hepatol.*, **9** (1), 46–51.

2 Igawa, T., Ishii, S., Tachibana, T., Maeda, A., Higuchi, Y., Shimaoka, S., Moriyama, C., Watanabe, T., Takubo, R., Doi, Y., Wakabayashi, T., Hayasaka, A., Kadono, S., Miyazaki, T., Haraya, K., Sekimori, Y., Kojima, T., Nabuchi, Y., Aso, Y., Kawabe, Y., and Hattori, K. (2010) Antibody recycling by engineered pH-dependent antigen binding improves the duration of antigen neutralization. *Nat. Biotechnol.*, **28** (11), 1203–1207.

3 Sarkar, C.A., and Lauffenburger, D.A. (2003) Cell-level pharmacokinetic model of granulocyte colony-stimulating factor: implications for ligand lifetime and potency *in vivo. Mol. Pharmacol.*, **63** (1), 147–158.

4 Ward, E.S., Güssow, D., Griffiths, A.D., Jones, P.T., and Winter, G. (1989) Binding activities of a repertoire of single immunoglobulin variable domains secreted from *Escherichia coli. Nature*, **341** (6242), 544–546.

5 Muyldermans, S. (2001) Single domain camel antibodies: current status. *J. Biotechnol.*, **74** (4), 277–302. Review.

6 Davies, J., and Riechmann, L. (1996) Affinity improvement of single antibody VH domains: residues in all three hypervariable regions affect antigen binding. *Immunotechnology*, **2** (3), 169–179.

7 Holt, L.J., Herring, C., Jespers, L.S., Woolven, B.P., and Tomlinson, I.M. (2003) Domain antibodies: proteins for therapy. *Trends Biotechnol.*, **21** (11), 484–490.

8 Holt, L.J., Basran, A., Jones, K., Chorlton, J., Jespers, L.S., Brewis, N.D., and Tomlinson, I.M. (2008) Anti-serum albumin domain antibodies for extending the half-lives of short lived drugs. *Protein Eng. Des. Sel.*, **21** (5), 283–288.

9 Kosa, T., Maruyama, T., and Otagiri, M. (1997) Species differences of serum albumins: I. Drug binding sites. *Pharm. Res.*, **11**, 1607–1612.

10 Kurtzhals, P., Havelund, S., Jonassen, I., Kiehr, B., Larsen, U.D., Ribel, U., and Markussen, J. (1995) Albumin binding of insulins acylated with fatty acids: characterization of the ligand-protein interaction and correlation between binding affinity and timing of the insulin effect *in vivo. Biochem. J.*, **312**, 725–731.

11 Coppieters, K., Dreier, T., Silence, K., de Haard, H., Lauwereys, M., Casteels, P., Beirnaert, E., Jonckheere, H., Van de Wiele, C., Staelens, L., Hostens, J., Revets, H., Remaut, E., Elewaut, D., and Rottiers, P. (2006) Formatted anti-tumor necrosis factor alpha VHH proteins derived from camelids show superior potency and targeting to inflamed joints in a murine model of collagen-induced arthritis. *Arthritis Rheum.*, **54** (6), 1856–1866.

12 Roovers, R.C., Laeremans, T., Huang, L., De Taeye, S., Verkleij, A.J., Revets, H., de Haard, H.J., and Henegouwen, P.M. (2007) Efficient inhibition of EGFR signaling and of tumour growth by antagonistic anti-EFGR Nanobodies. *Cancer Immunol. Immunother.*, **56** (3), 303–317.

13 Jonsson, A., Dogan, J., Herne, N., Abrahmsén, L., and Nygren, P.A. (2008) Engineering of a femtomolar affinity binding protein to human serum albumin. *Protein Eng. Des. Sel.*, **21** (8), 515–527.

14 Dockal, M., Carter, D.C., and Rüker, F. (1999) The three recombinant domains of human serum albumin. Structural characterization and ligand binding properties. *J. Biol. Chem.*, **274** (41), 29303–29310.

15 Peters, T. Jr (1995) *All about Albumin; Biochemistry, Genetics and Medical Applications*, Elsevier Inc.

16 Stork, R., Campigna, E., Robert, B., Müller, D., and Kontermann, R.E. (2009) Biodistribution of a bispecific single-chain diabody and its half-life extended derivatives. *J. Biol. Chem*, **284** (38), 25612–25619.

17 Tijink, B.M., Laeremans, T., Budde, M., Stigter-van Walsum, M., Dreier, T., de Haard, H.J., Leemans, C.R., and van Dongen, G.A. (2008) Improved tumor targeting of anti-epidermal growth factor receptor nanobodies through albumin binding: taking advantage of modular nanobody technology. *Mol. Cancer Ther.*, **7** (8), 2288–2297.

18 Hermanson, G.T. (2008) *Bioconjugate Techniques*, 2nd edn, Academic Press, Inc.

19 Walker, A., Dunlevy, G., Rycroft, D., Topley, P., Holt, L.J., Herbert, T., Davies, M., Cook, F., Holmes, S., Jespers, L., and Herring, C. (2010) Anti-serum albumin domain antibodies in the development of highly potent, efficacious and long-acting interferon. *Protein Eng. Des. Sel.*, **23** (4), 271–278.

20 Radhakrishnan, R., Walter, L.J., Hruza, A., Reichert, P., Trotta, P.P., Nagabhushan, T.L., and Walter, M.R. (1996) Zinc mediated dimer of human interferon-alpha 2b revealed by X-ray crystallography. *Structure*, **4** (12), 1453–1463.

21 Jespers, L., Schon, O., Famm, K., and Winter, G. (2004) Aggregation-resistant domain antibodies selected on phage by heat denaturation. *Nat. Biotechnol.*, **9**, 1161–1165.

22 Famm, K., Hansen, L., Christ, D., and Winter, G. (2008) Thermodynamically stable aggregation-resistant antibody domains through directed evolution. *J. Mol. Biol.*, **376** (4), 926–931.

23 Hussain, R., and Siligardi, G. (2010) Novel drug delivery system for lipophilic therapeutics of small molecule, peptide-based and protein drugs. *Chirality*, **22**, E44–E46.

24 Winichayakul, S., Pernthaner, A., Scott, R., Vlaming, R., and Roberts, N. (2009) Head-to-tail fusions of camelid antibodies can be expressed in planta and bind in rumen fluid. *Biotechnol. Appl. Biochem.*, **53** (Pt 2), 111–122.

25 Gomez, N., Vinson, A.R., Ouyang, J., Nguyen, M.D., Chen, X.N., Sharma, V.K., and Yuk, I.H. (2010) Triple light chain antibodies: factors that influence its formation in cell culture. *Biotechnol. Bioeng.*, **105** (4), 748–760.

26 Kim, K.S., Wass, C.A., and Cross, A.S. (1997) Blood–brain barrier permeability during the development of experimental bacterial meningitis in the rat. *Exp. Neurol.*, **145**, 253–257.

27 Lichtenbeld, H.C., Yuan, F., Michel, C.C., and Jain, R.K. (1996) Perfusion of single tumor microvessels: application to vascular permeability measurement. *Microcirculation*, **4**, 349–357.

28 Wunder, A., Müller-Ladner, U., Stelzer, E.H., Funk, J., Neumann, E., Stehle, G., Pap, T., Sinn, H., Gay, S., and Fiehn, C. (2003) Albumin-based drug delivery as novel therapeutic approach for rheumatoid arthritis. *J. Immunol.*, **170** (9), 4793–4801.

29 Veronese, F.M., and Mero, A. (2008) The impact of PEGylation on biological therapies. *BioDrugs*, **22** (5), 315–329. Review.

30 Kenanova, V.E., Olafsen, T., Salazar, F.B., Williams, L.E., Knowles, S., and Wu, A.M. (2010) Tuning the serum persistence of human serum albumin domain III: diabody fusion proteins. *Protein Eng. Des. Sel.*, **23**, 789–798.

31 Schmidt, M.M., and Wittrup, K.D. (2009) A modeling analysis of the effects of molecular size and binding affinity on tumor targeting. *Mol. Cancer Ther.*, **8** (10), 2861–2871.

32 Coffey, G.P., Fox, J.A., Pippig, S., Palmieri, S., Reitz, B., Gonzales, M., Bakshi, A., Padilla-Eagar, J., and Fielder, P.J. (2005) Tissue distribution and receptor-mediated clearance of anti-CD11a antibody in mice. *Drug Metab. Dispos.*, **33** (5), 623–629.

33 Chaudhury, C., Brooks, C.L., Carter, D.C., Robinson, J.M., and Anderson, C.L. (2006) Albumin binding to FcRn: distinct from the FcRn-IgG interaction. *Biochemistry*, **45** (15), 4983–4990.

34 Dennis, M.S., Zhang, M., Meng, Y.G., Kadkhodayan, M., Kirchhofer, D., Combs, D., and Damico, L.A. (2002) Albumin binding as a general strategy for improving the pharmacokinetics of proteins. *J. Biol. Chem.*, **277** (38), 35035–35043.

14

Half-Life Extension by Binding to Albumin through an Albumin Binding Domain

Fredrik Y. Frejd

14.1
Introduction

Serum half-life extension of small therapeutic proteins has become an established method to increase potency and improve patient compliance by allowing for fewer administrations. Several technologies are available, with PEGylation and Fc fusion being the ones most established clinically. Fc fusion is interesting as it utilizes the natural FcRn-mediated recycling mechanism described earlier in this book (Chapter 9). Another protein that takes advantage of FcRn-mediated recycling is serum albumin. In contrast to antibodies, albumin is also present at high concentration in the interstitial compartment and has a larger volume of distribution. Proteins that take advantage of albumin to increase their half-life will therefore have a different tissue distribution profile than Fc-fused proteins. Albumin fused protein therapeutics have been tested clinically, described earlier in this book (Chapter 13). Depending on the nature of the therapeutic protein, its fusion or coupling to serum albumin can complicate the manufacturing process, especially if the protein as such is produced recombinantly in, for example, *E. coli*. Preclinical development can also be affected by the fact that human serum albumin (HSA) does not always cross-react with the neonatal receptor of other species, murine FcRn, for example, binds very weakly to HSA [1]. An alternative strategy is to use albumin binding molecules that can be genetically fused or conjugated to the therapeutic protein and that will confer albumin binding to the therapeutic. This principle was first described by Nygren and coworkers for the half-life extension of soluble CD4 (sCD4) using an albumin binding domain of streptococcal protein G (SpG) [2]. Albumin association was later explored using antibody fragments [3], domain antibodies [4, 5] and peptides [6, 7] among others. In this chapter, the use of albumin binding SpG derivatives will be described, with a special focus on a 46-amino acid residue albumin binding motif commonly denoted albumin binding domain (ABD).

Therapeutic Proteins: Strategies to Modulate Their Plasma Half-Lives, First Edition. Edited by Roland Kontermann.
© 2012 Wiley-VCH Verlag GmbH & Co. KGaA. Published 2012 by Wiley-VCH Verlag GmbH & Co. KGaA.

14.2
Albumin Binding Domains and Engineered Derivatives

Serum protein binding is a natural property in many gram-positive bacteria [8–10], and is involved in host escape mechanisms of the bacteria [11]. Streptococcal protein G (SpG) is a bi-functional receptor present on the surface of certain strains of streptococci and is capable of binding to both immunoglobulin (IgG) and serum albumin of various species [12, 13]. The protein has several structurally and functionally different domains [14, 15]. More precisely, three Ig-binding domains and three serum ABDs [16–18] accounts for the bispecific binding ability (Figure 14.1). The highest affinity of the three homologous ABDs is found for serum albumin

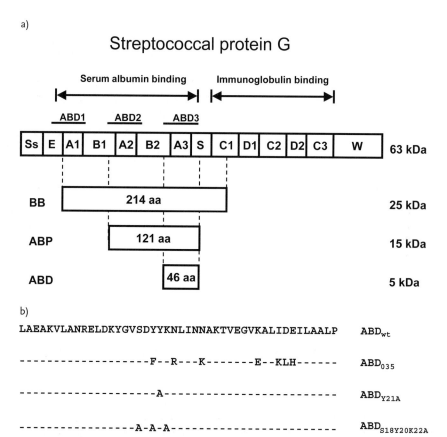

Figure 14.1 (a) Schematic representation of streptococcal protein G (SpG) and examples of various albumin-binding SpG-derived domains described in the text. (Reproduced with permission from [21]. Copyright © (1997) Elsevier. (b) Sequences mentioned in the text of wild-type G148 ABD3 (ABD$_{wt}$), one high affinity (ABD$_{035}$) and two low affinity ABD variants (ABD$_{Y20A}$ and ABD$_{S18Y20K22A}$), respectively.

from rat, man and mouse, whereas the binding to serum albumin from horse, hen, cow, rabbit, and sheep, as well as to ovalbumin, is low [13]. Early studies using various fragments of SpG, e.g. BB (25 kDa) [19], and ABP (15 kDa) [20] have been done (Figure 14.1a). A 46-amino acid motif was defined as the ABD and represents the smallest albumin binding unit in SpG. Other bacterial ABDs than the ones in protein G have also been identified, some of which are structurally similar to the ones of protein G. Examples of proteins containing such ABDs are the PAB, PPL, MAG and ZAG proteins [22]. Studies of structure and function of different ABDs have been carried out and reported, e.g. by Johansson and co-workers [23]. The third ABD of protein G from *Streptococcus* G148, (ABD3, G148-GA3), has been extensively studied and shown to be a three helical bundle using NMR spectroscopy [24, 25]. It is a 5 kDa protein that is chemically and thermally stable, and is independent from disulfide bridges, bound ligands, crosslinks or metal ions. ABD3 (hereinafter denoted ABD) binds with low nanomolar affinity to human, monkey, rat, and mouse serum albumin [24], and its binding surface to human serum albumin has been mapped by alanine scan analysis and been found to reside mainly in helix two [26]. Structural data of HSA in complex with an ABD homolog (the GA domain of protein PAB of *Finegoldia magna*) revealed that helix two and three of ABD interacts with domain II of HSA [27]. This is spacially separated from the binding site of FcRn, proposed to reside in domain III of albumin [28, 29]. Binding to albumin is not reduced at low pH (pH 6.0) and the presence of ABD does not interfere with the pH-dependent binding of HSA to FcRn *in vitro* [1]. ABD has been successfully fused to a number of proteins, retaining the functional activity both of the albumin binding part and the fused part, as will be discussed below in examples describing half-life extension. To investigate the impact of very high affinity binding to albumin, an affinity matured variant of ABD has recently been developed by Jonsson and coworkers [30]. ABD3 was used as a starting point to construct a library of domain variants. Fifteen of the 46 amino acids in the protein were subject to constrained randomization and displayed on bacteriophage for selection. A first set of binders showed improved binding affinities with K_D in the low picomolar affinity range. Sequence features of individual variants were combined to create a second generation set of seven binders. One of the resulting binders, denoted ABD035, had relatively high T_m, excellent thermal refolding properties and an increased affinity for albumin (Figure 14.1b). Initial SPR experiments suggested a very high affinity for human serum albumin with K_D between 50–500 fM [30]. The affinity was later measured in solution, using a Kinexa device, and a K_D of 120 fM was obtained. The affinity was improved also for albumin of other species including rat, mouse and cynomolgus. When used as fusion partner for single-chain diabody (scDb) constructs, the affinity matured binder increased the affinity of the trispecific fusion protein for HSA with a factor 80 and for MSA with a factor 12 [31].

The albumin binding region of SpG strain 148 (G148) has been experimentally shown to contain a few T-cell epitopes [32], and the affinity maturation process to obtain ABD035 did not address this issue. In a recent protein engineering effort, the ABD035 has been subject to a deimmunisation program, removing T-cell epitopes while maintaining stability, solubility, and high affinity. *In vitro* immunogenicity studies of the new molecule, denoted ABD094, are described below.

14.3
Albumin Binding Domains and Half-Life Extension *In Vivo*

A number of reports show that albumin binding derivatives of SpG can be used to increase the plasma half-life of a fusion partner, outlined in Figure 14.2. Pioneering approaches from the 1990s included fusion to larger albumin binding fragments, BA or BABA (BB), derived from SpG (Figure 14.1). In 1991, Nygren and coworkers reported on the use of BB to extend the half-life of a soluble form of CD4 [2]. CD4 is an important component of the infection mechanism of HIV, being a target receptor for the virus envelope protein gp120. By providing an excess of half-life extended CD4, the authors reasoned that the infectious action of viral gp120 would be diluted or even blocked. A fusion construct containing an N-terminal BB followed by two domains of the extracellular part of CD4 (E1 and E2) followed by a C-terminal BB region was produced and shown to bind recombinant gp120. In a kinetics study in mice, the BB-stabilized CD4 was compared with an IgG1 Fc- stabilized CD4 or CD4 or BB alone. It turned out that the half-life of the BB-stabilized protein or BB alone ($T_{1/2}$ 15–24 h) was similar or slightly better than the Fc-stabilized counterpart, and much longer than CD4 alone. The long half-life of BB was confirmed in macaques, where it had a half-life of 16 days.

In another early study, Makrides *et al.* wanted to enhance the effect of soluble complement receptor 1 (sCR1) [33]. CR1 has the capacity to dissociate the C3 and C5 convertases, and could have a role in reducing complement mediated tissue injury. A long circulating sCR1 could thus have a therapeutic effect *in vivo*. The authors fused the SpG fragments B2A3 to sCR1 to form sCR1-BA, or SpG domains

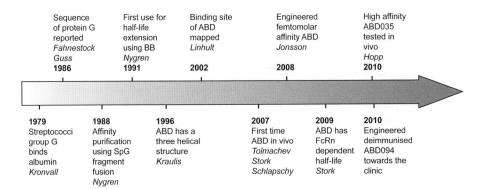

Figure 14.2 Historical outline of the development of using ABD for half-life extension. Please note that the years indicated are the years of publication, except for unpublished data mentioned in the text.

B1A2B2A3 to form more avid RSA-binding sCR1-BABA (also called BB, Figure 14.1). A kinetic study in rats showed that the half-life was modestly increased by 2.9- and 1.6-fold for BA and BABA fusions respectively, compared with sCR alone. The circulatory half-life for the more avid BB was not better than for the monomeric albumin binding, suggesting that one ABD could be enough for enhanced half-life.

As the SpG derivatives are of bacterial origin, Nygren and coworkers suggested the use of a much smaller but still albumin binding part of the receptor (<6 kDa, denoted ABD) to minimize potential recognition by the immune system [2]. ABD was initially used as an affinity tag (reviewed in [21]), for example, in fusion with Affibody molecules in a phage display library [34]. König and Skerra later used ABD as an affinity handle fused to antibody fragments such as Fv or Fab, to facilitate capture in a sandwich ELISA format [35]. Later, the same group used ABD to increase the half-life of a HER2-binding Fab fragment derived from the clinically used monoclonal trastuzumab (Herceptin®) [36]. Terminal half-lives in mice were increased from 2.1 h for the Fab-fragment alone to 20.9 h for the ABD fusion. In comparison, the half-life of the Fab fragment fused to a glycine rich homo-amino-acid polymer (HAP), was increased to 5.7 h. Similar results were described by Kontermann and colleagues, using ABD to extend the half-life of a bispecific scDb developed in their laboratory. The aim was to achieve a long lived T-cell recruiting antitumor agent by constructing an ABD fused scDb specific for CD3 and for the tumor antigen CEA which is overexpressed in colon cancer. The trispecific fusion protein bound CEA-expressing tumor cell lines as well as CD3-positive PBMCs in the presence of HSA or 50% mouse serum, and could activate human effector cells as shown in an IL-2 release assay [37]. Presence of HSA did not reduce cell binding, but reduced the effector cell activation by a factor of four (EC$_{50}$ 6 nM compared with 1.5 nM). *In vivo*, the half-life of the ABD-fused scDb was increased almost fivefold, from 5.6 to 27.6 h in mice. This can be compared with an albumin-fused scDb that had a comparable half-life of 25.0 h in the same model. In fact, ABD seems to follow the same kinetics and distribution properties as serum albumin itself, as shown in a recent study by Andersen *et al.* [1]. Here, an ABD fusion protein and rat serum albumin (RSA) were radiolabeled using two different radionuclides and then coinjected in rats. By using the different emission spectra of the nuclides, very accurate comparative data were obtained. The central distribution volumes of ABD and RSA were nearly identical (13.1 and 13.5 mL respectively) and in accordance with rat blood volume. Also the biodistribution profile and blood terminal half-life were very similar, suggesting that ABD does indeed benefit from the FcRn-mediated recycling mechanism of albumin. Another study supporting that ABD benefits from FcRn mediated long half-life was recently described, using the scDb-construct introduced above [38]. Here, *in vivo* plasma kinetics of ABD-fused scDb were compared in normal and FcRn heavy chain knockout mice. The terminal half-life in wild type mice was 53 h, while the half-life in the knock-out mice was only 24.8 h, confirming both that the long half-life of ABD does indeed depend on the FcRn-dependent recycling mechanism of albumin *in vivo*, and that ABD-binding to albumin does not dramatically affect the *in vivo* interaction with

FcRn [38]. In contrast, the half-life of a PEGylated scDb was not altered in the FcRn-heavy chain knock-out mice. When the tissue distribution profiles of scDb modified with either PEG or ABD were compared in tumor-bearing mice, the total uptake in blood of scDb-PEG was somewhat higher compared with scDb-ABD (1.2 times) [38]. Interestingly, the tumor uptake of scDb-ABD was 1.7 times higher than for scDb-PEG, suggesting that for this construct, albumin association by ABD confers not only long half-life but also an improved tumor uptake, compared with PEG. This could be due to better extravasation and tumor penetration properties due to the smaller hydrodynamic radius of scDb-ABD bound to HSA (4.8 nm) compared with scDb-PEG (7.9 nm) [39], but other properties due to the nature of albumin itself should not be excluded. In fact, there is more albumin in the interstitium than in the plasma, and endothelial cells express receptors that mediate albumin transport [40, 41]. Albumin has an intrinsic capability to extravasate and accumulate in both tumoral [42] and inflammatory lesions [43], and it has been argued that albumin is a promising carrier for various drugs (reviewed in [44]).

Another even smaller ABD-fused tumor-targeting construct has been described. It derives from a HER2-specific Affibody molecule [45] which belongs to a class of small (6.5 kDa) Protein A-derived antibody mimetics (reviewed in [46]). The HER2-specific Affibody molecule was originally developed for imaging purposes and recently clinically tested for imaging of metastatic breast cancer [47]. Its rapid kinetics was ideal for imaging purposes, but resulted in a high kidney uptake which was incompatible with radionuclide therapy that would be a logical extension of the successful tumor targeting obtained. Therefore, Nilsson and co-workers decided to convert the imaging tracer to a candidate for targeted radiotherapy by fusion to ABD [48]. The albumin association did not compromise the tumor targeting properties (Figure 14.3). Compared with the non-fused Affibody molecule, kidney uptake was decreased 25-fold and the tumor uptake increased three- to fivefold. This, combined with retained high specificity for the tumor, allowed for successful radioimmunotherapy using the beta-emitting nuclide ^{177}Lu. Treatment of mice bearing HER2-overexpressing microxenografts completely prevented tumor formation, in contrast to radioactivity coupled to an ABD-fused construct against an irrelevant antigen [48].

It is very difficult to compare tumor uptake of molecules targeting different receptors. Thus, the higher tumor uptake of ABD-fused HER2-specific Affibody molecule [48], compared with the ABD-fused CEA-specific scDb [38], could be due to its smaller hydrodynamic radius and hence better tumor penetration properties, but could equally well depend on differences in antigen expression density on the tumor cells, size of the tumor, and degree of tumor vascularization. Normal organs are however easier to compare, and here the blood values of the two molecules as well as most other organs have approximately the same degree of uptake, suggesting that albumin is the major determinant for tissue distribution. However, one exception suggests the importance of reducing the free fraction of the ABD fusion protein in order to take full advantage of the distribution properties of albumin. The ABD-fused Affibody molecule shows an increased kidney uptake compared

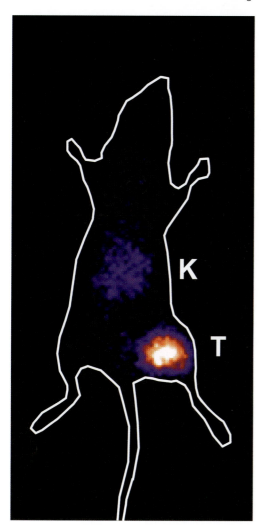

Figure 14.3 Imaging of the distribution of a ABD-fused HER2-targeting Affibody molecule in mice bearing HER2-expressing SKOV-3 xenografts. Planar gamma-camera images were accquired 52 h after administration of radioactively labeled (^{177}Lu) ABD fusion protein. Accumulation in tumors (T, right femur) was clearly visualized. Some accumulation can be seen in the kidneys (K). Adapted from [48], printed with kind permission from AACR.

with ABD-fused scDb. This could be due to the smaller apparent size of the ABD–Affibody molecule compared with the ABD–scDb, making the albumin-ABD fusion protein complex smaller. More likely however, the increased kidney uptake is associated with the affinity of ABD for albumin, where the nanomolar K_D of ABD to MSA allows for a small free fraction of the ABD fusion protein, not associated with albumin in the filtering unit of the kidneys (the glomeruli). This free fraction results in an increased kidney uptake of the much smaller free ABD–Affibody fusion compared with the relatively large free ABD-fused scDb. It has been postulated that the fraction of unbound fusion protein can have a rather high impact on the clearance rate, as was shown for a Fab fused to an albumin binding peptide [7], though in a lower affinity interval. In the case of the ABD-fused

Affibody molecule, increase of the affinity for albumin from low nanomolar K_D to sub-nanomolar K_D by fusing Affibody molecule to the high affinity ABD035 instead of the original ABD, decreased the kidney uptake by a factor of four, without increasing the half-life in the blood (Affibody AB, unpublished data). This would support the free fraction hypothesis, though not for plasma residence time where even low nanomolar affinity is sufficient for long half-life, but rather for kidney uptake in this case. If this is a general observation, it could also have an impact for other processes where albumin is involved, including FcRn-mediated recycling.

Kontermann and colleagues have been investigating the role of affinity and valency of ABD on the half-life of the scDb protein described above. A low affinity ABD fusion variant (ABD$_{Y20A}$) with a K_D of 634 nM for MSA, and a high affinity ABD fusion variant (ABD$_{035}$) with a K_D for MSA of 1.8 nM were created (Figure 14.1b) and compared with wild type ABD (K_D of 21.4 nM for MSA) fused to the scDb [31]. Both the low affinity ABD fusion and original affinity ABD fusion did indeed have a shorter terminal half-life (28.4 and 36.4 h respectively) compared with the high affinity ABD fusion (47.5 h), which had a half-life similar to a scFv–Fc fusion used as control (2 days). However, the low affinity ABD fusion did not provide an intermediate low half-life when compared with the non-fused scDb with an half-life of 5.6 h. Given the very high concentration of albumin in plasma, the affinity of ABD would likely have to be further lowered to obtain an intermediate half-life. The very low affinity S18Y20K22/A ABD mutant described in Linhult *et al.* [26] has been tested, but had too low affinity (estimated K_D of 0.5 mM) to show any significant half-life extension effect of an Affibody molecule as fusion partner (Affibody AB, unpublished data). This was in spite of the fact that the fusion protein was affinity-purified on an immobilized albumin column. Thus, in terms of half-life, intermediate affinity for the albumin is sufficient, whereas to avoid kidney uptake and fully exploit the tissue distribution profile of albumin, a high affinity is probably warranted. Simultaneous binding of two albumins by having two ABD-molecules in one fusion protein does not seem to be a solution for increasing the functional half-life, as shown by Hopp *et al.* [31].

Also the half-life of nontumor targeting proteins has been extended by ABD. The granulocyte colony stimulating factor (G-CSF) is a cytokine used to stimulate production of white blood cells in cancer patients with neutropenia. It has a short circulatory half-life and a PEGylated variant (Neulasta) has been marketed for improved efficacy. An ABD-fused G-CSF showed full *in vitro* activity in an NFS-60 proliferation assay in the presence of HSA, and the plasma half-life of G-CSF–ABD was increased tenfold compared with G-CSF alone (Neupogen) in a pharmacokinetic study in rats. In a pharmacodynamic study in rats, the fusion protein showed the same enhanced efficacy profile as the marketed PEGylated G-CSF (Neulasta), both in terms of total increase of granulocytes and in duration of response. Neupogen-induced granulocyte increase reached baseline level at 48 h, whereas G-CSF–ABD reached baseline at 120 h, increasing the area under curve by a factor of four (Affibody AB, unpublished data).

14.4
Albumin Binding Domains and Immunogenicity

When considering the prolonged exposure of a protein for therapeutic use, immunogenicity is a key issue. ABD is a bacterial protein domain, and as such could be expected to stimulate an immune response. On the other hand, it is derived from a protein that evolved to mediate the escape of the bacteria from the immune system. Indeed, strong adjuvants are needed to raise an immune response against BB, and repeated immunizations in rabbits using PBS only induced very low or nondetectable antibody responses [49]. However, by using strong adjuvants such as Freunds adjuvant, ISCOMs or aluminum hydroxide (Alhydrogel®), immune responses have been obtained and studied [19, 32, 49, 50]. It was hypothesized that the prolonged half-life of BB fusions, along with the bacterial origin of BB itself, could make BB useful as a carrier-protein in vaccine development, and some effort has been made in this direction. BB has been fused to antigens such as the repeated structures of the *P. falciparum* malaria antigen Pf155/RESA, denoted M3 [19] or to the G2Na fragment of RSV (BBG2Na) [50] The BBG2Na vaccine candidate has even been tested in healthy volunteers and was found to be safe, and well-tolerated [51]. If co-administered with a strong adjuvant, fusion of BB to the antigen will indeed increase the immune response to the antigen and also result in antibodies directed at BB [19, 32, 50]. It is hence possible to induce antibodies recognizing BB and indeed, T-cell epitopes have been found and verified [32]. One T-cell epitope that was identified by a Pepscan analysis, and which was reacting with sera from both Balb/c mice and human volunteers immunized with BBG2Na [32], resides in the minimal ABD. It should however be kept in mind that all these studies have been carried out using strong adjuvants to achieve an immunization effect.

Few studies have been performed investigating the effect of BB or ABD without adjuvant. In the early example by Sjölander and co-workers mentioned above [49], it was stated that strong adjuvants are needed to induce immune responses recognizing BB. Our own experience with ABD-fused proteins is coherent with this. We have investigated what effect ABD will add when fused to an Affibody molecule, since it will provide a longer exposure to the immune system. Most Affibody molecules do not induce immune responses [52]. Thus, a model Affibody molecule was selected for use in this investigation based on its ability to induce an immune response in animals. Outbred NMRI mice were injected s.c. with 20 µg dimeric Affibody molecule fused to ABD. Following an immunization schedule that could raise antibodies to the non-fused dimeric Affibody molecule alone, injections were given six times over a period of 21 days. The first five injections were administered every third day and the sixth at day 21. The mice were bled every week, in total five times over a period of 34 days. Serum was prepared and analyzed for antibodies, and interestingly, no specific antibodies against the Affibody moiety could be detected. The experiment was repeated in another mouse strain (CD1) with identical results (Affibody AB, unpublished results).

The same combination of proteins was tested in rats in a long term study. Female Sprague–Dawely rats were injected s.c. with 100 μg of the ABD-fused dimeric Affibody molecule. The proteins were administered four times during the first 21 days and then once monthly until the end of study which lasted for 273 days. Serum samples were obtained weekly the first month and then once monthly, approximately 15 days after the latest injection. In spite of the long duration of the study, as well as an administration protocol designed to trigger the immune system, very low levels of antibodies could be detected in only a fraction of the animals (Affibody AB, unpublished results). This suggests that the immune response to SpG-derived fusion proteins can be very different depending on whether adjuvant is used or not. It also seems that if extended half-life through albumin association has any effect on the immune response, it reduces the response rather than increasing it. However, given the previous use of BB as a carrier for vaccines, and the fact that there is at least one T-cell epitope in ABD [32], and since it has been concluded that the larger BB is a T cell-dependent antigen [50], it would nevertheless be attractive to reduce the potential immunogenicity of the protein. Thus, to further improve ABD towards use of half-life extension of biopharmaceuticals, efforts have been made to deimmunize the protein. Based on available literature on BB/ABD and open source *in silico* T-cell epitope prediction programs, the high affinity ABD035 [30] was used as starting point to produce a series of improved ABD mutants. These mutants were characterized with respect to retained high albumin affinity, solubility, thermal stability as well as expression yield, and the two variants with best properties were subjected to a 52 human donor PBMC T-cell proliferation assay (Algonomics/Lonza) and comparing them with ABD wild-type (SpG-ABD3). Compared with buffer control, only PBMCs from two individuals responded to ABD094, whereas ABD001 elicited response in samples from ten individuals. It should be noted that two individuals also responded to rHSA alone and 51 individuals to the positive control (KLH). The stimulation indices (SI) for all 52 donors were not significantly different for ABD094 compared with rHSA (0.99 and 1.0 respectively). In contrast, the SI for ABD001 was significantly higher (1.33). Thus, apparently, the number of T-cell epitopes were substantially decreased as no significant immunogenicity could be detected with ABD094, showing that this construct has very low immunogenic potential (Affibody AB, unpublished results).

14.5
Bispecific Albumin Binding Domains for Novel Target Binding and Long Half-Life

A recent interesting development of ABD is to add a new binding specificity to the ABD-molecule while retaining the native albumin binding capacity. Based on their previous mutational analysis of the albumin binding epitope of ABD [26], Hober and colleagues randomized 11 amino acid residues that should not interfere with albumin binding [53]. As scaffold, an in house ABD* that was stabilized to withstand alkaline conditions by replacement of four asparagines in the structure [54] was chosen. The library was displayed on bacteriophage and subjected to selections. The initial aim was to isolate a dual affinity purification handle for fusion proteins, and the protein A derivative, domain Z was chosen as target. Initial binders were characterized and a lead binder could be successfully affinity-purified both on HSA–sepharose and from crude lysate on the commercially available MabSelect SuRe matrix containing a modified domain Z and used for purification of monoclonal antibodies [53]. Next, the group investigated the approach for creating dual ABD binders intended for therapeutic applications. Using the same library, binders were selected against human tumor necrosis factor-alpha (TNF) [55]. First-generation binders were identified, but although several were bispecific, a relatively high affinity for TNF seemed to limit the affinity for albumin and vice versa. Thus, a new affinity maturation library was designed and a bacterial-based cell display system was used for isolation of improved binders. As cell display enables the use of fluorescence-activated cell sorting (FACS), selections could be performed against both targets simultaneously by labeling TNF and HSA with different fluorophores and indeed, bispecific binders with high-affinity against both targets could now be isolated. Thus, ABD can be used not only as a fusion domain for extending the plasma half-life of proteins, but can also be used as a target-binding domain itself, with intrinsic long half-life.

14.6
Conclusion

ABD is a small albumin binding domain that can be used to enhance the *in vivo* performance of biotherapeutic drugs. ABD does not interfere with the binding of albumin to FcRn, and when fused to a therapeutic protein, ABD will confer the albumin binding property to the resulting fusion protein, hence prolonging its plasma half-life. In addition, the tissue distribution profile will be altered, and different from the one of antibodies, mainly by having a better exposure to the interstitial compartment. ABD has been shown to extend the half-life of a number of biomolecules, including Fab fragments, scDbs and Affibody molecules. ABD compares well with other half-life extension technologies and had similar or better half-life than glycosylation, PEGylation, Fc fusion or serum albumin fusion in comparative scDb studies by Kontermann and colleagues. A HER2-specific Affi-body molecule retained its tumor targeting ability *in vivo* when fused to ABD and the cytokine G-CSF obtained the same enhanced efficacy profile as the marketed PEGylated G-CSF (Neulasta) when fused to ABD.

Importantly, ABD does not induce an immune response neither after multiple administrations in mice, nor in a 273-day chronic multiple administration study in rats, and an engineered, further deimmunized, high affinity ABD094 is now being developed towards clinical testing. It can thus be concluded that ABD provides an interesting alternative technology for half-life extension of biopharmaceuticals.

Acknowledgments

The author would like to thank professor Per-Åke Nygren for constructive comments on the manuscript

References

1 Andersen, J.T., Pehrson, R., Tolmachev, V., Daba Bekele, M., Abrahmsén, L., Sandlie, I., and Ekblad, C. (2011) Extending half-life by indirect targeting of the neonatal Fc receptor (FcRn) using a minimal albumin binding domain (ABD). *J. Biol. Chem.*, **286**, 5234–5241. Epub 2010 Dec 7, 10.1074/jbc.M110.164848

2 Nygren, P.A., Uhlen, M., Flodby, P., Andersson, R., and Wigzell, H. (1991) *In vivo* stabilization of a human recombinant CD4 derivative by fusion to a serum-albumin-binding receptor, in *Vaccines 91* (eds R.M. Chanock, H.S. Ginsberg, F. Brown, and R.A. Lerner), Cold Spring Harbor Laboratory Press, Cold Spring Harbor, NY, pp. 363–368.

3 Smith, B.J., Popplewell, A., Athwal, D., Chapman, A.P., Heywood, S., West, S.M., Carrington, B., Nesbitt, A., Lawson, A.D., Antoniw, P., Eddelston, A., and Suitters, A. (2001) Prolonged *in vivo* residence times of antibody fragments associated with albumin. *Bioconjug. Chem.*, **12** (5), 750–756.

4 Holt, L.J., Basran, A., Jones, K., Chorlton, J., Jespers, L.S., Brewis, N.D., and Tomlinson, I.M. (2008) Anti-serum albumin domain antibodies for extending the half-lives of short lived drugs. *Protein Eng. Des. Sel.*, **21** (5), 283–288. Epub 2008 Apr 2.

5 Roovers, R.C., Laeremans, T., Huang, L., De Taeye, S., Verkleij, A.J., Revets, H., de Haard, H.J., and van Bergen en Henegouwen, P.M. (2007) Efficient inhibition of EGFR signaling and of tumour growth by antagonistic anti-EFGR nanobodies. *Cancer Immunol. Immunother.*, **56** (3), 303–317.

6 Dennis, M.S., Zhang, M., Meng, Y.G., Kadkhodayan, M., Kirchhofer, D., Combs, D., and Damico, L.A. (2002) Albumin binding as a general strategy for improving the pharmacokinetics of proteins. *J. Biol. Chem.*, **277** (38), 35035–35043. Epub 2002 Jul 15.

7 Nguyen, A., Reyes, A.E. 2nd, Zhang, M., McDonald, P., Wong, W.L., Damico, L.A., and Dennis, M.S. (2006) The pharmacokinetics of an albumin-binding Fab (AB.Fab) can be modulated as a function of affinity for albumin. *Protein Eng. Des. Sel.*, **19** (7), 291–297. Epub 2006 Apr 18.

8 Kronvall, G., Simmons, A., Myhre, E.B., and Jonsson, S. (1979) Specific absorption of human serum albumin, immunoglobulin A, and immunoglobulin G with selected strains of group A and G streptococci. *Infect. Immun.* **25** (1), 1–10.

9 Myhre, E.B., and Kronvall, G. (1980) Demonstration of specific binding sites for human serum albumin in group C and G streptococci. *Infect. Immun.* **27** (1), 6–14.

10 Forsgren, A., and Sjöquist, J. (1966) "Protein A" from *S. aureus*. I. Pseudo-immune reaction with human gamma-globulin. *J. Immunol.* **97** (6), 822–827.

11 Langone, J.J. (1982) Protein A of *Staphylococcus aureus* and related immunoglobulin receptors produced by streptococci and pneumonococci. *Adv. Immunol.*, **32**, 157–252. Review.

12 Björck, L., Kastern, W., Lindahl, G., and Widebäck, K. (1987) Streptococcal protein G, expressed by streptococci or by *Escherichia coli*, has separate binding sites for human albumin and IgG. *Mol. Immunol.*, **24** (10), 1113–1122.

13 Nygren, P.A., Ljungquist, C., Trømborg, H., Nustad, K., and Uhlén, M. (1990) Species-dependent binding of serum albumins to the streptococcal receptor protein G. *Eur. J. Biochem.*, **193** (1), 143–148.

14 Fahnestock, S.R., Alexander, P., Nagle, J., and Filpula, D. (1986) Gene for an immunoglobulin-binding protein from a group G streptococcus. *J. Bacteriol.*, **167** (3), 870–880.

15 Guss, B., Eliasson, M., Olsson, A., Uhlén, M., Frej, A.K., Jörnvall, H., Flock, J.I., and Lindberg, M. (1986) Structure of the IgG-binding regions of streptococcal protein G. *EMBO J.*, **5** (7), 1567–1575.

16 Akerström, B., Nielsen, E., and Björck, L. (1987) Definition of IgG- and albumin-binding regions of streptococcal protein G. *J. Biol. Chem.*, **262** (28), 13388–13391.

17 Olsson, A., Eliasson, M., Guss, B., Nilsson, B., Hellman, U., Lindberg, M., and Uhlén, M. (1987) Structure and evolution of the repetitive gene encoding streptococcal protein G. *Eur. J. Biochem.*, **168** (2), 319–324.

18 Nygren, P.A., Eliasson, M., Abrahmsén, L., Uhlén, M., and Palmcrantz, E. (1988) Analysis and use of the serum albumin binding domains of streptococcal protein G. *J. Mol. Recognit. Apr;* **1** (2), 69–74.

19 Sjölander, A., Nygren, P.A., Stahl, S., Berzins, K., Uhlen, M., Perlmann, P., and Andersson, R. (1997) The serum albumin-binding region of streptococcal protein G: a bacterial fusion partner with carrier-related properties. *J. Immunol. Methods*, **201** (1), 115–123.

20 Nilsson, J., Larsson, M., Ståhl, S., Nygren, P.A., and Uhlén, M. (1996) Multiple affinity domains for the detection, purification and immobilization of recombinant proteins. *J. Mol. Recognit.*, **9** (5–6), 585–594.

21 Nilsson, J., Ståhl, S., Lundeberg, J., Uhlén, M., and Nygren, P.A. (1997) Affinity fusion strategies for detection, purification, and immobilization of recombinant proteins. *Protein Expr. Purif.*, **11** (1), 1–16.

22 Rozak, D.A., Alexander, P.A., He, Y., Chen, Y., Orban, J., and Bryan, P.N. (2006) Using offset recombinant polymerase chain reaction to identify functional determinants in a common family of bacterial albumin binding domains. *Biochemistry*, **45** (10), 3263–3271.

23 Johansson, M.U., de Château, M., Wikström, M., Forsén, S., Drakenberg, T., and Björck, L. (1997) Solution structure of the albumin-binding GA module: a versatile bacterial protein domain. *J. Mol. Biol.*, **266** (5), 859–865.

24 Johansson, M.U., Frick, I.M., Nilsson, H., Kraulis, P.J., Hober, S., Jonasson, P., Linhult, M., Nygren, P.A., Uhlén, M., Björck, L., Drakenberg, T., Forsén, S., and Wikström, M. (2002) Structure, specificity, and mode of interaction for bacterial albumin-binding modules. *J. Biol. Chem.*, **277** (10), 8114–8120. Epub 2001 Dec 18.

25 Kraulis, P.J., Jonasson, P., Nygren, P.A., Uhlén, M., Jendeberg, L., Nilsson, B., and Kördel, J. (1996) The serum albumin-binding domain of streptococcal protein G is a three-helical bundle: a heteronuclear NMR study. *FEBS Lett.*, **378** (2), 190–194.

26 Linhult, M., Binz, H.K., Uhlén, M., and Hober, S. (2002) Mutational analysis of the interaction between albumin-binding domain from streptococcal protein G and human serum albumin. *Protein Sci.*, **11** (2), 206–213.

27 Lejon, S., Frick, I.M., Björck, L., Wikström, M., and Svensson, S. (2004) Crystal structure and biological implications of a bacterial albumin binding module in complex with human serum albumin. *J. Biol. Chem.*, **279** (41), 42924–42928. Epub 2004 Jul 21.

28 Anderson, C.L., Chaudhury, C., Kim, J., Bronson, C.L., Wani, M.A., and Mohanty, S. (2006) Perspective – FcRn transports albumin: relevance to immunology and medicine. *Trends Immunol.*, **27** (7), 343–348. Epub 2006 May 30. Review.

29 Chaudhury, C., Brooks, C.L., Carter, D.C., Robinson, J.M., and Anderson, C.L. (2006) Albumin binding to FcRn: distinct from the FcRn-IgG interaction. *Biochemistry*, **45** (15), 4983–4990.

30 Jonsson, A., Dogan, J., Herne, N., Abrahmsén, L., and Nygren, P.A. (2008) Engineering of a femtomolar affinity binding protein to human serum albumin. *Protein Eng. Des. Sel.*, **21** (8), 515–527. Epub 2008 May 22.

31 Hopp, J., Hornig, N., Zettlitz, K.A., Schwarz, A., Fuss, N., Müller, D., and Kontermann, R.E. (2010) The effects of affinity and valency of an albumin-binding domain (ABD) on the half-life of a single-chain diabody-ABD fusion protein. *Protein Eng. Des. Sel.*, **23** (11), 827–834. Epub 2010 Sep 3.

32 Goetsch, L., Haeuw, J.F., Champion, T., Lacheny, C., N'Guyen, T., Beck, A., and Corvaia, N. (2003) Identification of B- and T-cell epitopes of BB, a carrier protein derived from the G protein of Streptococcus strain G148. *Clin. Diagn. Lab. Immunol.*, **10** (1), 125–132.

33 Makrides, S.C., Nygren, P.A., Andrews, B., Ford, P.J., Evans, K.S., Hayman, E.G., Adari, H., Uhlén, M., and Toth, C.A. (1996) Extended *in vivo* half-life of human soluble complement receptor type 1 fused to a serum albumin-binding receptor. *J. Pharmacol. Exp. Ther.*, **277** (1), 534–542.

34 Nord, K., Nilsson, J., Nilsson, B., Uhlén, M., and Nygren, P.A. (1995) A combinatorial library of an alpha-helical bacterial receptor domain. *Protein Eng.*, **8** (6), 601–608.

35 König, T., and Skerra, A. (1998) Use of an albumin-binding domain for the selective immobilisation of recombinant capture antibody fragments on ELISA plates. *J. Immunol. Methods*, **218** (1–2), 73–83.

36 Schlapschy, M., Theobald, I., Mack, H., Schottelius, M., Wester, H.J., and Skerra, A. (2007) Fusion of a recombinant antibody fragment with a homo-amino-acid polymer: effects on biophysical properties and prolonged plasma half-life. *Protein Eng. Des. Sel.* **20** (6), 273–284. Epub 2007 Jun 26.

37 Stork, R., Müller, D., and Kontermann, R.E. (2007) A novel tri-functional antibody fusion protein with improved pharmacokinetic properties generated by fusing a bispecific single-chain diabody with an albumin-binding domain from streptococcal protein G. *Protein Eng. Des. Sel.*, **20** (11), 569–576. Epub 2007 Nov 3.

38 Stork, R., Campigna, E., Robert, B., Müller, D., and Kontermann, R.E. (2009) Biodistribution of a bispecific single-chain diabody and its half-life extended derivatives. *J. Biol. Chem.*, **284** (38), 25612–25619. Epub 2009 Jul 23.

39 Stork, R., Zettlitz, K.A., Müller, D., Rether, M., Hanisch, F.G., and Kontermann, R.E. (2008) N-glycosylation as novel strategy to improve pharmacokinetic properties of bispecific single-chain diabodies. *J. Biol. Chem.*, **283** (12), 7804–7812. Epub 2008 Jan 22.

40 Goldblum, S.E., Ding, X., Funk, S.E., and Sage, E.H. (1994) SPARC (secreted protein acidic and rich in cysteine) regulates endothelial cell shape and barrier function. *Proc. Natl. Acad. Sci. U.S.A.*, **91** (8), 3448–3452.

41 Tiruppathi, C., Song, W., Bergenfeldt, M., Sass, P., and Malik, A.B. (1997) Gp60 activation mediates albumin transcytosis in endothelial cells by tyrosine kinase-dependent pathway. *J. Biol. Chem.*, **272** (41), 25968–25975.

42 Maeda, H., Wu, J., Sawa, T., Matsumura, Y., and Hori, K. (2000) Tumor vascular permeability and the EPR effect in macromolecular therapeutics: a review. *J. Control. Release.*, **65** (1–2), 271–284. Review.

43 Ballantyne, F.C., Fleck, A., and Dick, W.C. (1971) Albumin metabolism in rheumatoid arthritis. *Ann. Rheum. Dis.*, **30** (3), 265–270.

44 Kratz, F. (2008) Albumin as a drug carrier: design of prodrugs, drug conjugates and nanoparticles. *J. Control. Release*, **132** (3), 171–183. Epub 2008 May 17. Review.

45 Orlova, A., Magnusson, M., Eriksson, T.L., Nilsson, M., Larsson, B., Höidén-Guthenberg, I., Widström, C., Carlsson, J., Tolmachev, V., Ståhl, S., and Nilsson, F.Y. (2006) Tumor imaging using a picomolar affinity HER2 binding affibody molecule. *Cancer Res.*, **66** (8), 4339–4348.

46 Löfblom, J., Feldwisch, J., Tolmachev, V., Carlsson, J., Ståhl, S., and Frejd, F.Y. (2010) Affibody molecules: engineered proteins for therapeutic, diagnostic and biotechnological applications. *FEBS Lett.*, **584** (12), 2670–2680. Epub 2010 Apr 11. Review.

47 Baum, R.P., Prasad, V., Müller, D., Schuchardt, C., Orlova, A., Wennborg, A., Tolmachev, V., and Feldwisch, J. (2010) Molecular imaging of HER2-expressing

malignant tumors in breast cancer patients using synthetic 111In- or 68Ga-labeled affibody molecules. *J. Nucl. Med.*, **51** (6), 892–897. Epub 2010 May 19.

48 Tolmachev, V., Orlova, A., Pehrson, R., Galli, J., Baastrup, B., Andersson, K., Sandström, M., Rosik, D., Carlsson, J., Lundqvist, H., Wennborg, A., and Nilsson, F.Y. (2007) Radionuclide therapy of HER2-positive microxenografts using a ^{177}Lu-labeled HER2-specific Affibody molecule. *Cancer Res.*, **67** (6), 2773–2782.

49 Sjölander, A., Stahl, S., and Perlmann, P. (1993) Bacterial expression systems based on protein A and protein G designed for the production of immunogens: Applications to *Plasmodium falciparum* malara antigens. *Immunomethods*, **2**, 79–92.

50 Libon, C., Corvaïa, N., Haeuw, J.F., Nguyen, T.N., Ståhl, S., Bonnefoy, J.Y., and Andreoni, C. (1999) The serum albumin-binding region of streptococcal protein G (BB) potentiates the immunogenicity of the G130-230 RSV-A protein. *Vaccine*, **17** (5), 406–414.

51 Power, U.F., Nguyen, T.N., Rietveld, E., de Swart, R.L., Groen, J., Osterhaus, A.D., de Groot, R., Corvaia, N., Beck, A., Bouveret-Le-Cam, N., and Bonnefoy, J.Y. (2001) Safety and immunogenicity of a novel recombinant subunit respiratory syncytial virus vaccine (BBG2Na) in healthy young adults. *J. Infect. Dis.*, **184** (11), 1456–1460. Epub 2001 Nov 13.

52 Ahlgren, S., Orlova, A., Wållberg, H., Hansson, M., Sandström, M., Lewsley, R., Wennborg, A., Abrahmsén, L., Tolmachev, V., and Feldwisch, J. (2010) Targeting of HER2-expressing tumors using 111In-ABY-025, a second-generation affibody molecule with a fundamentally reengineered scaffold. *J. Nucl. Med.*, **51** (7), 1131–1138. Epub 2010 Jun 16.

53 Alm, T., Yderland, L., Nilvebrant, J., Halldin, A., and Hober, S. (2010) A small bispecific protein selected for orthogonal affinity purification. *Biotechnol. J.*, **5** (6), 605–617.

54 Gülich, S., Linhult, M., Nygren, P., Uhlén, M., and Hober, S. (2000) Stability towards alkaline conditions can be engineered into a protein ligand. *J. Biotechnol.*, **80** (2), 169–178.

55 Nilvebrant, J., Alm, T., Hober, S., and Löfblom, J. (2011) Engineering bispecificity into a single albumin binding domain. *PLoS One.* **6** (10), e25791. Epub 2011 Oct 3.

15
Half-Life Extension by Binding to Albumin through Small Molecules

Sabrina Trüssel, Joerg Scheuermann, and Dario Neri

15.1
Albumin and Albumin Binders

Albumin is the most abundant protein in plasma with a concentration of approximately $45\,mg\,mL^{-1}$. In addition to maintaining osmotic pressure in blood, albumin has an important physiological role associated with the transport of various compounds with low water solubility. This ability to tightly bind a variety of different (mainly hydrophobic) molecules is commonly exploited in pharmaceutical development, as many small organic molecules achieve acceptable blood circulation time by engaging in a binding interaction with albumin. Furthermore, small organic albumin binders could be considered as "portable moieties" for the functionalization of bioactive agents (e.g., drugs, contrast agents, therapeutic proteins), thus conferring a long blood circulatory half-life. In this chapter, two sources of albumin-binding compounds will be discussed: naturally occurring fatty acids (which have been used to develop long-acting insulin derivatives) and compounds isolated from chemical libraries, which have been used for the development of a general "Albu" tagging technology, both for small molecules and for proteins.

The considerably long circulatory half-life of albumin of approximately 19–22 days in humans has been generally explained by its size of 66.5 kDa, which excludes it from renal clearance that has a threshold of approximately 50 to 60 kDa. However, other proteins of similar size clearly display a shorter half-life so the size of albumin cannot be the sole reason for its long half-life. In fact, it has recently been discovered that albumin exhibits the same molecular mechanism as immunoglobulin (IgG), the only other protein with a half-life in the range of albumin, in order to extend its half-life. Both bind simultaneously to different binding sites of the MHC-class I-like neonatal Fc receptor (FcRn) and are thus prevented from lysosomal degradation and recycled to the circulatory system. FcRn, a broadly expressed receptor, is therefore the key homeostatic regulator of both. The interaction of albumin and IgG with FcRn is strictly pH dependent, binding at acidic pH, and release or no binding under physiological conditions. Subsequently, albumin

Therapeutic Proteins: Strategies to Modulate Their Plasma Half-Lives, First Edition. Edited by Roland Kontermann.
© 2012 Wiley-VCH Verlag GmbH & Co. KGaA. Published 2012 by Wiley-VCH Verlag GmbH & Co. KGaA.

and IgG, when taken up by cells via pinocytosis, bind to FcRn upon acidification of the endosome and are transported back to the cell surface, where they are released into the circulatory system again [1–3].

Serum albumin is synthesized in and secreted by the liver and unlike most other extracellular proteins, it is not glycosylated and displays a free cysteine (Cys34) on the surface. Its sequence is moderately well conserved over a broad series of species. A comparison of the primary sequence of mouse, rat, rabbit, cat, dog, and human serum albumin indicates >70% identity in amino acid sequence [4].

Serum albumin is a multifunctional protein. Because of its high concentration in the plasma it accounts for 80% of the colloid osmotic pressure, thereby substantially maintaining the osmotic pressure in the blood and keeping the fluids in the blood vessels as well as contributing to the buffering system in the blood. Further it serves as a transporter and depot for an impressive number of otherwise insoluble or toxic compounds, including endogenous and exogenous ligands, such as ions, fatty acids, the hormone thyroxine, several waste products, amino acids and a broad range of therapeutic agents. The majority of ligands bind to albumin in a reversible noncovalent way. However, ligands can also bind covalently to albumin, for example nitric oxide or compounds with a free thiol group capable of interacting with the free cysteine residue on the surface of the protein. In addition, certain types of drug metabolites interact irreversibly with the protein [5, 6].

These remarkable binding properties of albumin have been studied for over 40 years, revealing the presence of multiple binding sites as well as structural plasticity. Albumin consists of a single polypeptide chain with a high portion of α-helical structures and no β-sheets. The three-dimensional structure shows a heart-shaped molecule with three homologous helical domains I–III, each in turn further divided into the two sub-domains A and B. The A and B sub-domains consist of six and four α-helices, respectively, linked by flexible loops (Figure 15.1). The major orientation of the domains and the heart-shape form is likely to be the structure of all mammalian albumin molecules under physiological conditions and is maintained even in the presence of a wide variety of ligands [6, 7].

The primary endogenous ligands of serum albumin are fatty acids. Crystallographic analysis of albumin bound to medium and long-chain saturated fatty acids (with chain lengths of C_{10}, C_{12}, C_{14}, C_{16} and C_{18}) revealed seven binding sites, which are common to all these fatty acids, distributed across the protein and displaying varying affinities. Three out of these seven sites were identified to bind palmitic acid (C_{16}) with high affinity by anchoring the carboxylate group through ionic interactions at the entrance to the binding pocket. The carboxylate group of fatty acids seems to be a crucial feature for high-affinity binding though the distinct high-affinity binding sites differ with chain length and saturation of the fatty acids. Fatty acids constitute the main cargo of serum albumin and have the most binding sites. Nevertheless many other endogenous compounds are transported by albumin, invariably in one of the fatty acid binding sites. The tremendous versatility of serum albumin is exemplified by the binding site of sub-domain IB. Binding of fatty acids to this site requires a minor conformational change in order to generate the cavity that accommodates the lipid. In contrast, bilirubin, the degradation

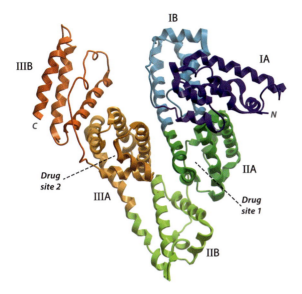

Figure 15.1 Structure of defatted human serum albumin (PDB 1e78) with secondary structure elements depicted schematically and colored by domains. Drug site 1 in domain IIA and drug site 2 in domain IIIA are indicated.

product of heme, which is transported from the spleen to the liver by albumin, binds to the same site without involving this conformational change. Thus the binding site shows a considerably different shape depending on the bound ligand [7].

In addition to the numerous endogenous ligands, albumin also associates with a broad spectrum of therapeutic agents with a lipophilic core structure and acidic or electronegative features. Studies based on the displacement of fluorescent probes revealed that most drugs bind with association constants in the range of 10^4–10^6 M^{-1} to one of two sites, called drug sites 1 and 2 (Figure 15.1). Additional binding sites of low affinity have been identified, but as the concentration of administered drugs is almost always lower than that of albumin, they are negligible for pharmaceutical considerations. Drug sites 1 and 2 consist of six α-helices in the sub-domain IIB and IIIA, respectively, forming a large apolar cavity with defined polar features. Despite this structural similarity the two binding sites show considerable differences. Drug site 1 has some polar amino acids positioned in the center of two apolar chambers which coordinate the binding of aromatic ligands with centrally located negative or electronegative features. Further, drug site 1 exhibits a rather complex architecture involving parts of the sub-domains IB, IIB and IIIA. The binding cavity thus created is of such a voluminous size and flexibility that it allows the binding of bulky compounds as well as the binding of two compounds at the same time. Smaller bound ligands often do not fill the pocket completely and likely derive some of the binding energy from contacts with bound water molecules. Drug site 2 in contrast, has one single dominant polar patch near the entrance to the binding pocket, favoring the binding of aromatic compounds with peripherally located negative or electronegative features. The binding pocket is smaller and built in a more compact way leading to less flexible binding properties. Indeed, site 2 binding is often strongly affected by stereoselectivity or minor modification of known binders. In the end, it is very hard to determine which ligands bind with high affinity to which site. Although some binders to either site are well known, both sites bind various compounds without evident preference. Table 15.1 lists some compounds and their preferential binding site, for example warfarin for site 1 and the non-steroidal anti-inflammatory drugs (NSAIDs) and diazepam for site 2. In addition, fatty acids binding to or close to the drug sites may induce conformational changes within the binding pocket, which influence the binding properties of drugs. This effect appears to be relatively small for drugs binding to the voluminous site 1. Drug site 2 though overlaps with one of the high affinity fatty acid binding sites resulting in direct competition of drugs with fatty acids for binding to the sub-domain IIIA. Site 2 drugs are therefore more likely to be displaced from albumin with rising levels of fatty acids [8–16].

Binding to albumin has a significant impact on the pharmacokinetic properties of drugs. Association with albumin can overcome solubility problems and therefore enhance the distribution and bioavailability of drug molecules with a poor solubility or the tendency to aggregate, while enhancing circulatory half-life. On the other hand, if the binding is too tight and more than 95% of the administered dose is bound to albumin, the free and active concentration of the drug is drastically reduced and the delivery to the site of action may be impeded. It is also

Table 15.1 Ligands with high-affinity to site 1 or 2 of serum albumin[a].

Site 1	K_A (M^{-1})[b]	Site 2	K_A (M^{-1})[b]
Drugs			
Warfarin	3.4×10^5	Diazepam	3.8×10^5
Azapropazone	2.8×10^5	Ketoprofen	2.5×10^6
Salicylate	1.9×10^5	Ibuprofen	2.7×10^6
Indomethacin	1.4×10^6	Diclofenac	3.3×10^6
Sulfathioazole	2.5×10^4	S(-)-Thiamylal	8.7×10^4
Furosemide	1.9×10^5	S-Naproxen	3.7×10^6
Valproate	2.8×10^5	Chlofibrate	7.6×10^5
Endogenous compounds			
Bilirubin	9.7×10^7	L-Trytophan	4.4×10^4
		Octanoate	1.6×10^6
		Indoxylsuflate	1.6×10^6

a) The table shows a selection of ligands, which have been proposed to bind to either site 1 or 2.
b) High-affinity associations constants K_A obtained at pH 7.4.

important to mention that peptide drugs are often susceptible to degradation by proteases. Binding to albumin increases the half-life and provides protection against proteolytic digestion, allowing the drugs to reach the site of action at pharmacologically relevant concentrations [17].

Because of the positive effect of albumin binding on the pharmacokinetic behavior of numerous drugs, this protein continues to play a central role in pharmaceutical development. Although the binding site of the majority of drugs has not yet been fully elucidated, the investigation of the interaction of pharmaceutical agents with albumin serves as a basis for the design and optimization of drugs with improved pharmacokinetic properties [5–7, 17, 18].

As discussed in this book, several strategies may be considered in order to influence the pharmacokinetic properties of compounds through albumin binding. This chapter focuses on the increase of the circulatory half-life of therapeutic proteins and contrast agents by means of their modification with small organic albumin-binding molecules.

The conjugation of chemical moieties to therapeutic proteins can be performed in a statistical manner at multiple reaction sites, or in a site-specific manner. In the first case, molecules are allowed to react with the primary amino groups of accessible lysine residues and of the N-terminus on the surface of the therapeutic protein of interest. Alternatively, the thiol groups of unique cysteine residues can be considered for chemical modification strategies. Such approaches are facilitated by the fact that reactive cysteine residues may be rare in certain classes of therapeutic proteins, either because other cysteine residues are engaged in disulfide bond formation or because they are absent. Random conjugation strategies result in heterogeneous products, containing a mixture of species with different molar ratios of molecule conjugated to the protein, linked at different sites and, consequently, with distinct *in vivo* pharmacokinetic, efficacy, and safety profiles. Batch-to-batch reproducibility may be difficult to ensure for the preparation of clinical-grade samples and may compromise industrial development plans. By contrast, the site-specific modification of therapeutic proteins may yield structurally defined, homogenous products, thus facilitating pharmaceutical development [19, 20].

15.2
Albumin-Binding Insulin Derivatives

The peptide hormone insulin is a key player in the energy and glucose metabolism of the body. It causes cells to take up glucose and store it as glycogen if the blood glucose level is rising. This prompts the body to use glucose as an energy source instead of fat as in the absence of insulin. Complete or partial loss of insulin levels leads to diabetes mellitus – either type 1 diabetes mellitus where no insulin is produced at all, or type 2 diabetes mellitus that is characterized by an insulin resistance and relative insulin deficiency. Type 2 patients are not treated with insulin until other medications fail to control blood glucose levels adequately.

Patients with type 1 diabetes though are completely dependent on externally delivered insulin.

Insulin consists of two peptide chains A and B, has a size of 5.8 kDa, and is produced in the pancreatic β-cells. Within the β-cells, the insulin heterodimer is triggered by its high concentration and the presence of zinc ions to assembly into hexamers. This storage form rapidly dissociates into biologically active monomers upon secretion and following dilution in the blood circulatory system. Since the advent of recombinant DNA technology in the 1980s, varying recombinant analogs of insulin have been constructed in order to mimic the endogenous insulin availability and, consequently, facilitate the metabolic control. These analogs are artificially modified insulin molecules, which basically differ in respect of their shorter or longer metabolic effect. Rapid-acting insulin analogues, such as insulin lispor (Humalog®, Lilly) and insulin aspart (NovoLog®, Novo Nordisk), are designed in a way to impede self-association leading to a high absorption and elimination rate of the monomeric molecules. In contrast, long-acting insulin analogs ideally show no pronounced peak of action but a constant low insulin level over a period of 18–26 h. Nowadays the common therapy for type 1 diabetes mellitus consists of the combination of long-acting insulin for the fasting periods and administration of a rapid-acting analog after food ingestion [21–25].

In recent years, different attempts have been made to design insulin analogs with protracted absorption from the subcutaneous tissue, to imitate the constant low insulin profile observed overnight in normal physiology. So far, long-acting insulin analogs following two different approaches have become clinically available. One approach, represented by insulin glarginine (Lantus®, Sanofi Aventis), is based on site-directed mutagenesis of insulin, resulting in isolelectric point changes which reduce solubility at physiological pH values. Insulin glarginine is injected in acidic solution of pH 4 and precipitates upon administration, forming a depot in the subcutaneous tissue from which insulin is slowly released. The second approach, represented by insulin detemir (Levemir®, Novo Nordisk), features the modification of insulin with fatty acids, allowing albumin binding and a consequent extension of the blood circulatory half-life. Insulin detemir has the threonine at position B30 removed and the myristoyl fatty acid (C_{14}) acetylated to the lysine residue at position B29, resulting in a derivative which consists of an insulin analog and a fatty acid (Figure 15.2). It is believed that binding to albumin occurs not only through the interaction with the myristoyl fatty acid, but also with the side chain of lysine B29 and with the C-terminal carboxylate of the B chain. Approximately 98% of injected insulin detemir binds to serum albumin after administration. Considering the very high concentration of albumin in blood, the fraction of albumin-binding sites occupied at therapeutic doses of insulin detemir are negligibly low and no clinically relevant interactions with other albumin-binding drugs (which would result in pharmacokinetic changes) have been reported so far. Insulin detemir represents a relevant example in which the engineering of albumin binding was successfully applied to product development in terms of improved half-life and bioavailability [21, 26–29].

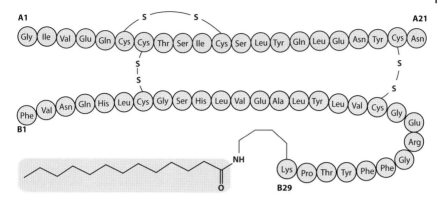

Figure 15.2 Schematic representation of Insulin detemir with chains A and B linked by two disulfide bridges. Myristic acid conjugated to the lysine residue at position B29 is shown in the gray box.

15.3
"Albu" Tagging Technology

In principle, tagging proteins and small molecules with chemical compounds in order to confer albumin binding is not limited to modification strategies featuring the use of fatty acids. Indeed, these compounds tend to be hydrophobic and may seriously compromise the solubility of the resulting conjugates. On the other hand, while many pharmaceutical agents are known to bind to serum albumin, in the majority of the cases albumin binding is lost upon chemical modification of the drug.

Recently a general class of portable albumin binding molecules was identified from selections with a DNA-encoded chemical library, that is a set of organic molecules, individually tagged with a synthetic oligonucleotide serving as amplifiable identification bar code. These biopanning experiments, performed by using human serum albumin immobilized on a solid support, led to the identification of 4-(*p*-iodophenyl)butanoic acid and related compounds as a general class of albumin binders. Adding a negative charge in the form of a carboxylate group, and introducing a readily reactive group allowed conjugation to other pharmaceutical agents without loss of binding affinity. This moiety has been named "Albu" tag and binds to albumin with a dissociation constant of 3 µM at 37 °C. Competition studies with several known ligands for drug sites 1 and 2 suggest preferential binding of the Albu tag at drug site 2 which is in line with the general findings of chemical structures of drug site 2 binders. In contrast to other existing methods to increase circulatory half-life, the Albu tagging technology is generally applicable because it neither results in a substantial gain of weight nor affects the solubility. As described subsequently, the Albu tag moiety has been used to modify contrast agents (e.g., fluorescein and Gd-DTPA) and antibodies [30, 31].

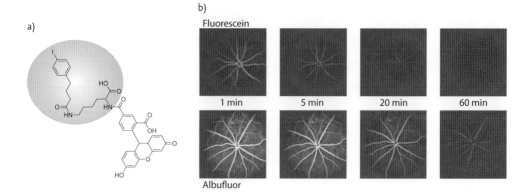

a)

b)

Fluorescein

| 1 min | 5 min | 20 min | 60 min |

Albufluor

Figure 15.3 (a) Chemical structure of "Albufluor" with the albumin-binding moiety (Albu tag) highlighted in the gray circle. (b) *In vivo* fluorescein angiography performed in mice. Images were recorded over 1 h after i.v. injection of 50 nmol fluorescein (top row) and Albufluor (bottom row).

Figure 15.3a shows the chemical structure of fluorescein modified with the Albu tag, named "Albufluor". Fluorescein is widely used in ophthalmology as a contrast agent in fluorescein angiography of the eye fundus in patients with retinal disorders. However, the short circulatory half-life of fluorescein requires injection of high doses from 200 to 500 mg per patient, which causes non-negligible side effects and impedes the comparative study of both eyes. When fluorescein was used to image vascular structure in the retina of mice, one minute after intravenous (i.v.) administration most of the fluorophore had already been cleared, yielding a reduced performance for the imaging of ocular vascular structure (Figure 15.3b, upper panel). By contrast, Albufluor exhibited a long circulation time *in vivo*, allowing a sensitive and high-resolution detection of microvascular structures up to 20 min following i.v. administration (Figure 15.3b, lower panel). This pharmacokinetic behavior is likely to be of clinical relevance, considering the fact that often only one eye can be imaged in human fluorescein angiographic procedures, due to the rapid clearance of unmodified fluorescein [30, 32].

Similarly, Albu tagging was shown to improve blood clearance profiles for gadolinium-DTPA (Magnevist®, Bayer) which represents the most widely used contrast agent in magnetic resonance imaging (MRI) procedures [33]. Magnevist modified with the Albu tag (Albuvist®) revealed superior pharmacokinetic behavior over the unmodified drug after i.v. injection into mice. MRI analysis of major blood vessels in the head of mice showed an extended half-life and decrease of extravasation, slowing down the reduction of signal intensity and resolution. This may be of clinical interest given that high-resolution MRI studies of blood vessels hold a potential advantage for the early diagnosis of atherosclerosis, a major cause of mortality [30, 34].

In full analogy to the previous examples, Albu tagging was used to extend circulatory half-life of small antibody fragments (i.e., scFv and diabody fragments). Antibody fragments, having a size below 60 kDa, are typically cleared very rapidly from circulation via the renal route shortly after i.v. administration. For certain pharmaceutical applications (e.g., clearance of toxic drugs, transient interference with the function of a target protein *in vivo*) short serum half-life values are desirable. However, in many cases a long circulation time in the blood is preferable in order to ensure antibody function *in vivo* over an extended period of time and reduce the number of administrations [35].

Albu tagging was applied to the scFv and diabody fragments of F8 human antibody, which bind specifically and with high affinity to the alternatively spliced extradomain A (EDA) of fibronectin, a marker of tumor angiogenesis [36, 37]. Figure 15.4a presents the schematic structure of F8 antibody fragments either

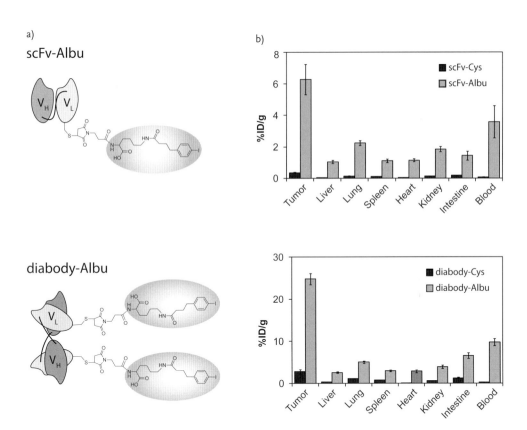

Figure 15.4 (a) Schematic representation of scFv-Cys and diabody-Cys modified with the Albu tag forming the conjugates scFv-Albu and diabody-Albu, respectively. The albumin-binding moiety (Albu tag) is highlighted in gray circles. (b) Increased tumor uptake by Albu tagging of antibody fragments resulting from biodistribution studies of radioiodinated scFv and diabody with and without the Albu tag 24 h after i.v. injection. Mean targeting results are expressed as % injected dose per gram (%ID/g) +/− standard error. The upper panel shows the values for scFv-Cys (0.4%ID/g) and scFv-Albu (6.3% ID/g). In the lower panel the values for diabody-Cys (2.8% ID/g) and diabody-Albu (24.8%ID/g) are depicted.

forming a scFv or, if a short linker is applied to connect V_H and V_L, a non-covalent homodimeric diabody. Cysteine residues added to the C-termini of these antibody fragments could be selectively and quantitatively modified using a derivative of the Albu tag with a maleimide modification. The pharmacokinetic behavior and tumor targeting properties of the antibody fragments with and without the Albu tag were studied by quantitative biodistribution analysis. For this purpose, radio-iodinated protein preparations were intravenously injected in mice bearing sub-cutaneous teratocarcinoma tumors and the amount of antibody fragments present in the tumor, organs and blood was determined after 24 h. Figure 15.4b indicates the impact of albumin binding for small antibody fragments: while the non modi-fied scFv and diabody exhibited rapid clearance and reduced uptake in the tumor, the corresponding Albu tagged antibodies showed approximately 10-fold higher tumor uptake. Additional pharmacokinetic studies of the Albu tagged diabody performed over three days revealed half-live values of approximately 1000 min corresponding to an approximately 35-fold reduction of the blood clearance rate.

Albu tagging technology features two major benefits for small antibody frag-ments and other pharmaceutical agents. First, circulatory half-lives are increased by the induced binding to albumin through the Albu tag and consequently the accumulation of the antibody at the tumor site. Additionally, the influence of the Albu tag with its low molecular weight on the size of the antibody fragments is negligible, whereby the advantage of the small size, such as a potentially more homogenous distribution in the targeted tissue, is maintained.

15.4
Concluding Remarks

The modification of small antibody fragments with the chemical Albu tag consid-erably extended serum half-lives, leading to higher accumulation at the site of disease and reduction in the number of administrations. Albu tagging technology represents a potent strategy for the extension of serum half-lives of therapeutic proteins and contrast agents of the blood-pool. In addition, the albumin binding mediated by the Albu tag may be of interest for neurotoxic drugs to prevent cross-ing of the blood–brain barrier and the generation of produgs with long circulatory half-lives. Albu tags are suitable for the chemical modification of drugs, peptides and proteins. They are small in size and the corresponding conjugates can be characterized using standard analytical techniques.

References

1 Popov, S., Hubbard, J.G., Kim, J., Ober, B., Ghetie, V., and Ward, E.S. (1996) The stoichiometry and affinity of the interaction of murine Fc fragments with the MHC class I-related receptor, FcRn. *Mol Immunol.*, **33** (6), 521–530.

2 Ghetie, V., and Ward, E.S. (2002) Transcytosis and catabolism of antibody. *Immunol. Res.*, **25** (2), 97–113.

3 Kim, J., Bronson, C.L., Hayton, W.L., *et al.* (2006) Albumin turnover: FcRn-mediated recycling saves as much albumin from degradation as the liver produces. *Am. J. Physiol. Gastrointest. Liver Physiol.*, **290** (2), G352–G360.

4 Peters, T. (1996) *All about Albumin: Biochemistry, Genetics and Medical Applications*, Academic Press, San Diego.

5 Andersen, J.T., and Sandlie, I. (2009) The versatile MHC class I-related FcRn protects IgG and albumin from degradation: implications for development of new diagnostics and therapeutics. *Drug Metab. Pharmacokinet.*, **24** (4), 318–332.

6 Kragh-Hansen, U., Chuang, V.T., and Otagiri, M. (2002) Practical aspects of the ligand-binding and enzymatic properties of human serum albumin. *Biol. Pharm. Bull.*, **25** (6), 695–704.

7 Curry, S. (2009) Lessons from the crystallographic analysis of small molecule binding to human serum albumin. *Drug Metab. Pharmacokinet.*, **24** (4), 342–357.

8 Kragh-Hansen, U. (1988) Evidence for a large and flexible region of human serum albumin possessing high affinity binding sites for salicylate, warfarin, and other ligands. *Mol. Pharmacol.*, **34** (2), 160–171.

9 Takamura, N., Haruta, A., Kodama, H., *et al.* (1996) Mode of interaction of loop diuretics with human serum albumin and characterization of binding site. *Pharm. Res.*, **13** (7), 1015–1019.

10 Petersen, C.E., Ha, C.E., Harohalli, K., Feix, J.B., and Bhagavan, N.V. (2000) A dynamic model for bilirubin binding to human serum albumin. *J. Biol. Chem.*, **275** (28), 20985–20995.

11 Chuang, V.T., Kuniyasu, A., Nakayama, H., Matsushita, Y., Hirono, S., and Otagiri, M. (1999) Helix 6 of subdomain III A of human serum albumin is the region primarily photolabeled by ketoprofen, an arylpropionic acid NSAID containing a benzophenone moiety. *Biochim. Biophys. Acta*, **1434** (1), 18–30.

12 Sakai, T., Takadate, A., and Otagiri, M. (1995) Characterization of binding site of uremic toxins on human serum albumin. *Biol. Pharm. Bull.*, **18** (12), 1755–1761.

13 Yamasaki, K., Rahman, M.H., Tsutsumi, Y., *et al.* (2000) Circular dichroism simulation shows a site-II-to-site-I displacement of human serum albumin-bound diclofenac by ibuprofen. *AAPS PharmSciTech*, **1** (2), E12.

14 Sueyasu, M., Fujito, K., Shuto, H., Mizokoshi, T., Kataoka, Y., and Oishi, R. (2000) Protein binding and the metabolism of thiamylal enantiomers *in vitro*. *Anesth. Analg.*, **91** (3), 736–740.

15 Kragh-Hansen, U. (1991) Octanoate binding to the indole- and benzodiazepine-binding region of human serum albumin. *Biochem. J.*, **273** (Pt 3), 641–644.

16 Irikura, M., Takadate, A., Goya, S., and Otagiri, M. (1991) 7-Alkylaminocoumarin-4-acetic acids as fluorescent probe for studies of drug-binding sites on human serum albumin. *Chem. Pharm. Bull.*, **39** (3), 724–728.

17 Ghuman, J., Zunszain, P.A., Petitpas, I., Bhattacharya, A.A., Otagiri, M., and Curry, S. (2005) Structural basis of the drug-binding specificity of human serum albumin. *J. Mol. Biol.*, **353** (1), 38–52.

18 Sudlow, G., Birkett, D.J., and Wade, D.N. (1975) The characterization of two specific drug binding sites on human serum albumin. *Mol. Pharmacol.*, **11** (6), 824–832.

19 Junutula, J.R., Flagella, K.M., Graham, R.A., *et al.* (2010) Engineered thio-trastuzumab-DM1 conjugate with an improved therapeutic index to target HER2-positive breast cancer. *Clin. Cancer Res.*, **16** (19), 4769–4778.

20 Kontermann, R.E. (2009) Strategies to extend plasma half-lives of recombinant antibodies. *BioDrugs*, **23** (2), 93–109.

21 Havelund, S., Plum, A., Ribel, U., *et al.* (2004) The mechanism of protraction of insulin detemir, a long-acting, acylated analog of human insulin. *Pharm. Res.*, **21** (8), 1498–1504.

22 Brems, D.N., Alter, L.A., Beckage, M.J., *et al.* (1992) Altering the association properties of insulin by amino acid replacement. *Protein Eng.*, **5** (6), 527–533.

23 Holleman, F., and Hoekstra, J.B. (1997) Insulin lispro. *N. Engl. J. Med.*, **337** (3), 176–183.

24 Simpson, K.L., and Spencer, C.M. (1999) Insulin aspart. *Drugs*, **57** (5), 759–765. discussion 766–757.

25 Plank, J., Wutte, A., Brunner, G., *et al.* (2002) A direct comparison of insulin aspart and insulin lispro in patients with type 1 diabetes. *Diabetes Care*, **25** (11), 2053–2057.

26 Owens, D.R., Coates, P.A., Luzio, S.D., Tinbergen, J.P., and Kurzhals, R. (2000) Pharmacokinetics of 125I-labeled insulin glargine (HOE 901) in healthy men: comparison with NPH insulin and the influence of different subcutaneous injection sites. *Diabetes Care*, **23** (6), 813–819.

27 Vigneri, R., Squatrito, S., and Sciacca, L. (2010) Insulin and its analogs: actions via insulin and IGF receptors. *Acta Diabetol.*, **47** (4), 271–278.

28 Bolli, G.B., and Owens, D.R. (2000) Insulin glargine. *Lancet*, **356** (9228), 443–445.

29 Markussen, J., Havelund, S., Kurtzhals, P., *et al.* (1996) Soluble, fatty acid acylated insulins bind to albumin and show protracted action in pigs. *Diabetologia*, **39** (3), 281–288.

30 Dumelin, C.E., Trussel, S., Buller, F., *et al.* (2008) A portable albumin binder from a DNA-encoded chemical library. *Angew. Chem. Int. Ed. Engl.*, **47** (17), 3196–3201.

31 Trussel, S., Dumelin, C., Frey, K., Villa, A., Buller, F., and Neri, D. (2009) New strategy for the extension of the serum half-life of antibody fragments. *Bioconjug. Chem.*, **20** (12), 2286–2292.

32 Kwiterovich, K.A., Maguire, M.G., Murphy, R.P., *et al.* (1991) Frequency of adverse systemic reactions after fluorescein angiography. Results of a prospective study. *Ophthalmology*, **98** (7), 1139–1142.

33 de Roos, A., Doornbos, J., Baleriaux, D., Bloem, H.L., and Falke, T.H. (1988) Clinical applications of gadolinium-DTPA in MRI. *Magn. Reson. Annu.*, 113–145.

34 Goyen, M. (2008) Gadofosveset-enhanced magnetic resonance angiography. *Vasc. Health Risk Manag.*, **4** (1), 1–9.

35 Holliger, P., and Hoogenboom, H. (1998) Antibodies come back from the brink. *Nat. Biotechnol.*, **16** (11), 1015–1016.

36 Rybak, J.N., Roesli, C., Kaspar, M., Villa, A., and Neri, D. (2007) The extra-domain A of fibronectin is a vascular marker of solid tumors and metastases. *Cancer Res.*, **67** (22), 10948–10957.

37 Villa, A., Trachsel, E., Kaspar, M., *et al.* (2008) A high-affinity human monoclonal antibody specific to the alternatively spliced EDA domain of fibronectin efficiently targets tumor neo-vasculature *in vivo*. *Int. J. Cancer*, **122** (11), 2405–2413.

Part Four
Half-Life Extension with Pharmaceutical Formulations

16
Half-Life Extension with Pharmaceutical Formulations: Liposomes

Astrid Hartung and Gerd Bendas

16.1
Rationale

Liposomes, small sized phospholipid vesicles, show great promise as selective drug delivery systems for numerous medical applications. Due to their bilayer composition, liposomes have the ability to entrap hydrophilic drugs into the aqueous interior or to incorporate hydrophobic compounds into the core of the surrounding bilayer. The decoration of the liposomal membrane with surface-grafted hydrophilic polymers, that is, polyethylene glycol (PEG) strongly decreases the clearance by macrophages or the reticuloendothelial system after systemic applications, and thus tremendously increases their circulation half-life. These long-circulating carriers, also referred to as sterically stabilized liposomes (SSL), have successfully been applied to numerous drug targeting approaches, and a number of liposomal formulations have been approved for therapy, that is, as carriers for cytostatic agents.

In view of this, PEGylated liposomes appear promising to prolong the circulation half-life of therapeutically relevant proteins. The incorporation of proteins, such as interleukin-2 (IL-2) or tumor necrosis factor α (TNF-α) into liposomes has been found to extend the circulation half-life or to protect the proteins from degradation.

A new platform technology is based on the noncovalent, but highly specific binding of proteins to the grafted PEG chains, which substantially enhances the pharmacodynamic characteristics of the proteins. This has been demonstrated in preclinical studies or clinical trials including liposomes associated with coagulation factor VIII or VIIa for the treatment of hemophilia A.

This chapter focuses on approaches to improve the pharmacokinetic or pharmacodynamic properties of therapeutically relevant proteins by PEGylated liposomes, based on these two strategies.

Therapeutic Proteins: Strategies to Modulate Their Plasma Half-Lives, First Edition. Edited by Roland Kontermann.
© 2012 Wiley-VCH Verlag GmbH & Co. KGaA. Published 2012 by Wiley-VCH Verlag GmbH & Co. KGaA.

16.2
Prospects of Liposomes as Drug Carriers and Their Pharmacokinetic Properties

Since the early 1970s, when the use of liposomes as drug carriers was proposed, liposomes have been extensively exploited for this application [1]. Liposomes are colloidal particles composed of lipid bilayer membranes of self-assembled amphiphiles (mostly phospholipids) surrounding discrete aqueous compartments [3]. Liposomes are characterized by their particle size, number of lamellae, lipid composition and thus, surface charge; all of these factors dictate their properties and stability. Unilamellar vesicles in a size range around 100 nm are most often used for pharmaceutical formulations.

Hydrophilic drugs can be entrapped in the aqueous interior of the liposomes, while hydrophobic compounds can be incorporated into the bilayer region (Figure 16.1). Referring to this, liposomes offer several advantages over other drug carries, such as biocompatibility, low immunogenicity and toxicity, while protecting the entrapped drug from degradation. A wide range of physical characteristics of liposomes can be modified to systematically affect the pharmacokinetic and pharmacodynamic properties of entrapped drugs. Furthermore, liposomes possess the feasibility of industrial large-scale production.

Nevertheless, despite these favorable characteristics, the use of liposomes as drug delivery systems in systemic applications has been seriously limited by their rapid clearance from the circulation caused by their uptake into the mononuclear phagocyte system [3–5], resulting in circulation half-life of less than 2 h.

Much longer circulation of liposomes could be achieved when the concept of steric stabilization of liposomes was introduced in the early 1990s. The incorporation of bulky, highly flexible and hydrophilic surface grafted moieties reduced the interaction of liposomes with serum proteins and opsonins dramatically, thus

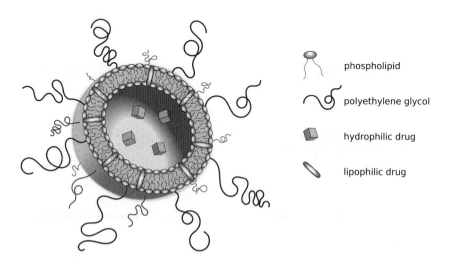

phospholipid

polyethylene glycol

hydrophilic drug

lipophilic drug

Figure 16.1 Schematic view of a PEGylated liposome as drug carrier system.

tremendously prolonging the systemic circulation [6, 7]. Polyethylene glycol (PEG) appeared as the optimal choice for this sterically stabilizing effect. The protective PEG shield around the liposomes has been optimized with respect to the molecular weight and concentration of the polymer in a great number of studies [8]. Nowadays, a certain concentration (mostly 5 mol%) of PEG_{2000}-coupled phospholipids displays the standard for long-circulating liposomes, also referred as "PEGylated liposomes".

PEGylated liposomes have a circulation half-life of about 20 h in mice [7], 29 h in dogs [9] and 45 h in human [10, 11]. This prolonged systemic circulation is the essential prerequisite for targeting liposomes to other cells and tissues than the macrophage system. Liposomes tend to accumulate especially at sites of leaky vasculature, which is found at inflammatory regions or sites of tumor growth [12, 13]. This represents the basis for a great number of preclinical and clinical approaches to the targeting of tumor tissues with liposomal cytostatics, which leads to impressive reduction of unwanted side effects and increases in the drug index. Several formulations, that is, carrying the cytostatics doxorubicin or daunorubicin have successfully been approved for therapy and meanwhile been established in the market [14].

In contrast to these "passive" targeting approaches, the coupling of certain structures onto the liposomal surface for a selective biological recognition and "active" targeting of cells and tissues encompass a very active research field. Carbohydrates, peptides, proteins, and antibodies/antibody fragments have been described for these target strategies. Most commonly, these targeting devices were coupled covalently onto reactive linker lipids, that is, at the liposomal surface (Type I targeted liposomes). For a successful combination of steric stabilization with sufficient accessibility of the coupled homing devices, the coupling onto the terminal ends of the grafted PEG chains appears to be most promising (Type II targeted liposomes) (Figure 16.2).

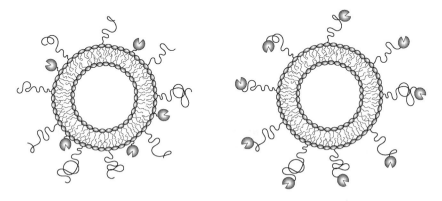

Type I targeted liposomes   Type II targeted liposomes

Figure 16.2 Schematic drawing of targeted liposomes. Ligand coupling directly to the liposomal surface results in Type I targeted liposomes. If the ligand is attached to the distal end of a PEG chain, the liposomes are referred to as Type II targeted liposomes.

Since details towards these approaches are beyond the scope of this article, the reader is referred to several review in this field [15–18].

In summary, PEGylated liposomes are attractive carriers for numerous targeting approaches because of their long systemic circulation. However, several further factors have to be considered critically. The controlled release of the drug payload from the liposomes often restricts the therapeutic benefit, furthermore, the fate of the liposomes at the target site, that is, the cellular uptake and intracellular trafficking are important factors.

16.3
Entrapment of Therapeutically Relevant Proteins in PEGylated Liposomes

Combining liposomes and proteins is traditionally associated with the coupling of proteins as homing devices onto the outer surface of liposomes for active drug targeting purposes. As in the case of antibody(-fragment)–coupled liposomes, so called immunoliposomes, the protein is not the active therapeutic component [15]. The covalent coupling of proteins onto the liposomal surface helps to accumulate drug loaded liposomes specifically at the target site and thus to reduce drug doses and negative side-effects, but this approach will not be stressed in this chapter.

In this section we focus on the systemic use of proteins as therapeutic agents themselves, which often comes along with several challenges like systemic toxicity, short blood residence time or inactivation due to blocking antibodies or proteolytic enzymes. One approach to circumvent these problems is to incorporate highly active proteins into liposomes. The kind of association is controlled by their hydrophilic/hydrophobic characteristics, but proteins are most often incorporated within the aqueous interior. This permits systematic modification of the pharmacokinetics and biodistribution of the proteins and besides, affects the targeting or overcoming cellular resistance.

The following section introduces recent approaches and developments regarding TNF-α or IL-2 carrying liposomes.

Tumor necrosis factor α (TNF-α) is a macrophage/monocyte-derived cytokine with several biological functions including immunomodulatory and antitumor activities. As it combines direct cytotoxic effects with stimulation of the host immune antitumor response and induces hemorrhagic necrosis of tumor tissue, it is a promising agent for the therapy of cancer. Impressive antineoplastic effects were found in *in vitro* experiments. Unfortunately, systemic administration is limited due to severe toxicity and short plasma half-life in the range of minutes.

Several *in vitro* and *in vivo* studies including liposomal TNF-α have been published to date. There is strong evidence that liposome-encapsulated TNF-α retains immunomodulatory activity but significantly reduces toxic inflammatory effects *in vivo* when compared with soluble TNF-α alone or co-injected with buffer-loaded liposomes [19–20]. Furthermore, Morishige *et al.* reported that it was possible to overcome TNF-α resistance *in vitro* by incorporation of TNF-α in immunoliposomes [21].

Kedar *et al.* focused on the biodistribution and pharmacokinetic properties of TNF-α liposomes in mice [19]. They found out that plasma half-life and area under the curve (AUC) of the entrapped cytokine were 10–15 times greater than those of free TNF-α (259 vs. 25 min; 21.8×10^6 vs. 1.4×10^6 U × mL min^{-1}). Thus, an important requirement for systemic TNF-α therapy could be fulfilled. Interestingly, 75% of the cytokine activity in plasma 24 h after injection could be traced back to liposome-associated TNF-α. Therefore, pharmacokinetics of lipids and active drug were analyzed separately. The findings suggest that TNF-α liposomes remain intact in the plasma for around 4 h after which the cytokine is gradually released and eliminated while lipid half-life is prolonged. The sustained release could be an explanation for the diminished systemic toxicity. Unfortunately, with respect to biodistribution of TNF-α bearing liposomes, the authors observed strong liposome accumulation in nonrelevant organs like liver and spleen. This phenomenon is often described, as large conventional liposomes were rapidly taken up by the reticuloendothelial system (RES).

In a next step, the behavior of TNF-α bearing sterically stabilized liposomes (SSL–TNF-α) was investigated in a tumor rat model [22]. In accordance with earlier results, a 33-fold prolonged blood mean residence time (MRT) and 14-fold increase in the AUC was stated for SSL–TNF-α when compared with free TNF-α. Regarding tumor targeting, the authors were enthusiastic to report a significant accumulation of SSL–TNF-α in the tumor tissue in the experimental time range, while soluble TNF-α was cleared rapidly. However, there were no differences in the initial peak concentrations. Despite the use of long-circulating PEGylated liposomes, significant amounts of liposomes were detected in mononuclear phagocyte-rich organs like liver and spleen. This might be due to the tumor model used.

As TNF-α was reported to enhance the efficacy of antineoplastic radiotherapy, Kim *et al.* studied the influence of radiation in combination with TNF-α PEGylated liposomes or free TNF-α on tumor growth in a tumor mice model [23]. In contrast to radiation alone, neither free TNF-α nor SSL–TNF-α alone were able to inhibit tumor growth significantly. Nevertheless, best results were gained when tumor-bearing mice were treated with a combination of radiation and SSL–TNF-α. This led to significant reduced tumor progression compared with radiation treatment with or without free TNF-α. Besides, the authors suggest that the application of TNF-α in addition to radiotherapy might have a positive effect on certain leukocyte populations decreased by radiation. Although the upregulation was more pronounced with SSL–TNF-α, both free and encapsulated TNF-α triggered lymphocyte activation and increased natural killer cell numbers in blood and spleen.

The aim of a study by ten Hagen *et al.* [24] was to determine if various doses of SSL–TNF-α (15–200 µg kg^{-1}) could improve the antitumor activity of liposomal doxorubicin (Doxil®) in soft tissue sarcoma-bearing rats. Therefore, these therapy regimes were compared with blank or treatment with Doxil® alone. As endpoints tumor response and systemic toxicity were evaluated. The authors demonstrated that the addition of SSL–TNF-α to liposomal doxorubicin resulted in a strongly decreased tumor progression. Even a dose of 200 μg kg^{-1} SSL–TNF-α did not show relevant toxicity, whereas administration of the same amount of free TNF-α ended up in serious side effects. Thus, the authors concluded that combined administration of Doxil® and SSL–TNF-α may be an option in the therapy of solid tumors.

In contrast to the studies presented before, Mori *et al.* as well as Savva *et al.* tried to modify the pharmacokinetic parameters and systemic toxicity of TNF-α not by incorporation of the cytokine into liposomes, but by protein coupling on the outer surface of PEGylated liposomes [25, 26]. Based on Type I targeted liposomes, as depicted in Figure 16.2, Mori *et al.* formulated recombinant TNF-α (rTNF-α) bearing liposomes with the cytokine attached directly to the liposomal surface [25]. Conjugation of rTNF-α to PEGylated liposomes resulted in a significantly reduced protein clearance from blood *in vivo* compared with either free rTNF-α or rTNF-α bound to non-PEGylated liposomes. Investigating uncoupled liposomes, it could be shown that PEGylation dramatically shielded liposomes from uptake by the RES, thus resulting in increased blood circulation times. Nevertheless, upon protein coupling substantial amounts of rTNF-α were recovered from liver and spleen after administration of either liposomal formulation. This phenomenon was more distinct using non-PEGylated liposomes and was correlated to increasing protein/lipid ratios. Thus, the authors hypothesized that elevated uptake of rTNF-α liposomes into liver and spleen might be caused by direct ligand–receptor interactions in these organs. However, this was not consistent with the outcome of an *in vitro* cytotoxicity assay revealing significantly reduced biological activity of rTNF-α conjugated to PEGylated liposomes. This was attributed to limited protein-receptor interactions due to sterical hindrance by PEG polymers present on the liposomal surface.

In view of these findings Savva *et al.* developed long-circulating rTNF-α liposomes with the protein coupled to the distal ends of the surface grafted PEG chains (Type II targeted liposomes; Figure 16.2) to improve ligand–receptor interactions [26]. It was shown in the past that conjugation of proteins or antibodies to liposomal PEG terminals could provide both improved plasma half-life and specific receptor targeting [27, 28]. Surprisingly, all Type II TNF-α-liposomes were rapidly cleared from the blood stream in just about 30 min, although as little as 0.13% of the PEG polymers were conjugated to rTNF-α. In contrast, 73% of the administered protein-free liposomes were still circulating in the blood after 4.5 h. Thus, the authors conclude that ligand coupling to the surface of sterically stabilized liposomes to improve pharmacokinetic parameters is not a suitable strategy for all therapeutic proteins.

In addition to TNF-α and other mediators, the therapeutic use of interleukin-2 (IL-2) as immunostimulator takes on a certain role in the therapy of tumors, for example, metastatic renal cell carcinoma or malignant melanoma [29, 30]. IL-2, a 15 kDa hydrophobic cytokine produced by activated T-lymphocytes, induces a complex signaling cascade augmenting the immune response: among others, proliferation and differentiation of T- and B-lymphocytes is stimulated, tumor cell killing by activated macrophages is initiated and cytotoxic cells like natural killer cells (NK-cells) or lymphokine-activated killer cells (LAK-cells) were mobilized. Thus, the administration of recombinant IL-2 in tumor patients enhances the body's natural combat of cancer. Nevertheless, due to its short plasma half-life [31] very high doses are needed to reach therapeutic blood levels. As life-threatening side effects can occur after intravenous infusion of high-dose IL-2 the effective use is limited. Therefore, new formulations have to be developed facilitating an efficient and safe application of IL-2.

The first *in vivo* experiments including non-PEGylated IL-2-liposomes revealed significant immunostimulatory and antitumor effects in comparison with free IL-2 following local or regional administration, for example, on rat hepatoma [32, 33], on murine pulmonary sarcoma [34], on murine renal cell carcinoma [35] or, combined with adoptively transferred anti-CD3$^+$ IL-2-stimulated T cells, in mice bearing hepatic metastasis [36].

Pharmacokinetic characteristics and biodistribution of IL-2-liposomes were examined as a function of different application routes and compared with free IL-2 by Anderson *et al.* [37]. They reported retarded clearance and depot effects for all investigated application routes, namely intravenous (i.v.), subcutaneous (s.c.), intraperitoneal (i.p.) or intrathoracic administration, when IL-2 was incorporated into liposomes. After i.v. bolus injection into rats, plasma half-life of liposomal IL-2 was twice as long as detected for the free cytokine (102 min vs. 41 min). 36% of the injected IL-2 dose were still found in the blood circulation after 2 h in rats treated with IL-2-liposomes compared with 6.5% in rats given free IL-2. Organ distribution was profoundly different for both formulations: In contrast to soluble IL-2, significant amounts were recovered from the liver and spleen for both IL-2- and BSA-carrying liposomes, thus indicating that biodistribution pattern is determined by means of the liposomal vehicle.

In order to circumvent liposomal uptake into the RES and thereby improving the systemic treatment of neoplastic diseases outside the mononuclear phagocyte system (MPS), Kedar *et al.* investigated the delivery of IL-2 by small sterically stabilized liposomes (SSL–IL-2) *in vivo* and compared it to soluble IL-2, IL-2 encapsulated in large conventional liposomes and to PEGylated IL-2 (PEG-IL-2) [38]. As expected, following i.v. application plasma half-life and AUC of SSL–IL-2 were 10–30 times higher than those of free IL-2, but in the same range as those of PEG-IL-2. Interestingly, there was no need for IL-2 release from the liposomes to mediate biological activity. Considering the amphipathic properties of IL-2, the authors suggest that the cytokine is embedded into the phospholipid membrane with the receptor binding domain exposed on the outer liposomal surface. This thesis could be supported by the finding that SSL–IL-2, but not "empty" liposomes, bound to IL-2-receptor bearing T cells *in vitro*. Nevertheless, *in vivo* IL-2 separated from its carrier continuously with a half-life of around 40 min, although PEGylated liposomes were stable in the plasma for more than 4 h. Due to the use of PEGylated liposomes only little liposome uptake into the RES could be observed.

In a next step, Kedar *et al.* focused on tumor regression and systemic immunological acitivation by SSL–IL-2 in mice [39]. They could show that i.v. administration of SSL–IL-2 significantly elevated the amount of leukocytes in the blood and spleen and induced LAK-cell activity in the spleen in comparison to free IL-2. Treatment of tumor-bearing mice pretreated with the cytostatic cyclophosphamide resulted in two- to sixfold prolonged survival when SSL–IL-2 was applied instead of IL-2. Moreover, effective therapy with SSL–IL-2 required less frequent injections and an around 50% lower IL-2 total amount. This might be advantageous with respect to a simplified treatment protocol and lower systemic toxicity. Comparing liposomal IL-2 to PEG-IL-2, PEG-IL-2 was a more potent immunomodulator in healthy mice, but not superior in the therapy of tumor-mice. However, in contrast to SSL–IL-2 administration of PEG-IL-2 often caused severe side effects (up to 40% mortality depending on the experimental settings). Therefore, the authors conclude that IL-2 incorporated into PEGylated liposomes is the most promising candidate for the therapeutic use in tumor patients.

In a comparative study different liposomal formulations were analyzed with respect to their immunomodulatory capacity [40]. IL-2 was either encapsulated in large multilamellar vesicles with a mean diameter of 0.75–1.5 μm (MLV-IL-2), or small unilamellar long-circulating liposomes were applied as cytokine carriers as described before (SSL–IL-2). It was found that all liposomal IL-2 preparations were superior to free IL-2 regarding their biologic activity *in vitro* as well as *in vivo*. In accordance with previous results, plasma half-lives were 7 and 17 times greater for MLV-IL-2 and SSL–IL-2, respectively, following i.v. injection in comparison to the soluble cytokine. An i.p. application lead to two- to fourfold prolonged blood circulation times for the liposome preparations. With respect to the application route, SSL–IL-2 was more effective in triggering enhanced immune system activation using i.v. administration instead of intraperitoneal or subcutaneous injections. However, in spite of the predominant pharmacokinetic characteristics SSL–IL-2 were not as potent as MLV-IL-2 after i.p. or s.c. application. In search of an explanation, the authors suggest an increased production of hematopoietic and immunopotentiating cytokines upon MLV-IL-2 uptake by phagocytic cells. This hypothesis was supported by previous studies reporting unspecific stimulation of antigen-presenting macrophages or reduced production of macrophage-derived inhibitory messenger substances mediated by empty liposomes [41, 42]. Furthermore, Kedar *et al.* indicate that pharmacokinetic parameters were calculated from IL-2 plasma levels, so that potential liposomal depots in various hematopoietic or lymphoid tissues were not comprised.

An additional study assessed the antitumor activity of liposomal doxorubicin (Doxil®) and subsequent systemic i.v. or regional i.p. application of liposome-associated IL-2 in tumor-bearing mice [43]. Although all combination therapies were superior to Doxil® alone, the greatest effects were gained depending on the choice of IL-2 formulation and application route: free IL-2 or SSL–IL-2 were most efficient in the i.v. tumor model, but with regard to a regional i.p. administration MLV-IL-2 showed the best synergistic antitumor effect.

These results confirmed previous findings that liposomal IL-2 is a very appealing option in the therapy of cancer with great potential for further optimization efforts.

Collectively, liposome entrapped cytokines combine many favorable features as an immunomodulatory or anticancer drug for the clinical use, like striking immune response activation, lack of toxicity, convincing pharmacokinetic properties, and simplified application protocols, as well as easy preparation, high entrapment rates and long-term stability of the formulations. Nevertheless, the results demonstrate that for an effective therapy not only sufficient plasma stability but also liposome composition and route of administration are of remarkable importance.

16.4
Noncovalent Complex Formation of Proteins and PEGylated Liposomes

A unique approach to the significant extension of the biological activity of recombinant coagulation factor VIII (rFVIII) in combination with PEGylated liposomes compared with free rFVIII was described by Baru *et al.* in 2005 in *in vitro* and *in vivo* assays [44]. This not only opened new perspectives for the treatment of hemophilia A, but also raised the question whether complex formation with PEGylated liposomes is a general way for improving the pharmacokinetic properties of proteins, and maybe avoiding the problems of restricted release from the liposomal interior (Figure 16.3).

Hemophilia A is a recessive, X-linked genetic bleeding disorder caused by the lack of coagulation factor VIII. FVIII is a glycoprotein of about 330 kDa, which possesses a key function in the contact activation clotting pathway. The severity of

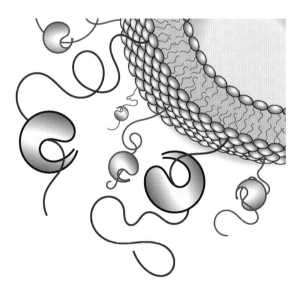

Figure 16.3 Noncovalent binding of factor VIII to PEGylated liposomes. Although the exact binding mechanism of factor VIII to the liposomal surface is not fully elucidated, it is assumed that a certain amino acid sequence of the protein binds to the surface grafted PEG chains in a specific, noncovalent manner with high affinity.

hemophilia A becomes manifest in spontaneous bleeding episodes. Therapy of hemophilia A is based on replacement of FVIII by i.v. infusions of purified plasma-derived or recombinant FVIII. Since human FVIII has a relatively short half-life in the bloodstream of about 8 to 12 h, repeated infusions every two to three days are required for the prophylactic treatment of patients.

A novel concept could be based on the finding of Baru *et al.* [44]. They described a high affinity binding in the low nanomolar range of rFVIII to PEGylated liposomes (POPC/DSPE-PEG2000; 97/3 molar ratio) with strong consequences for the biological activity. Although the molecular mechanisms remained unclear at this time, they demonstrated the specificity of the protein association with the PEG moieties at the liposomal surface, since no binding could be detected on non-PEGylated liposomes or micelles. Surface bound FVIII retained the full biological activity *in vitro* for binding vWF or inducing blood clotting in suitable assays in comparison to free FVIII. Furthermore, in hemophilic mice PEG-LipFVIII demonstrated favorable pharmocokinetic properties compared with free FVIII. A slower clearance of liposomal FVIII vs. free FVIII became evident in statistical significant differences in circulation half-life (10.0 h vs. 6.5 h) and twofold higher levels 7–48 h after injection. These changes in pharmacokinetic properties were associated with increases in and prolongations of the hemostatic activity of PEG-LipFVIII vs. free FVIII in mice bleeding models. These promising data strike a new path for the treatment of hemophilia A patients.

A first clinical trial on PEG-LipFVIII (rFVIII; Kogenate®, Bayer HealthCare) in combination with PEGylated POPC liposomes, also referred as BAY 79-4980, included adult patients with severe hemophilia A. Subjects received a single prophylactic dose (25 IU kg^{-1} or 35 IU kg^{-1}) of either PEG-LipFVIII or free rFVIII and the time between infusion and the next spontaneous bleeding was recorded [45]. The results of this trial demonstrated significant prolongation of the bleeding free interval with PEG-LipFVIII vs. free FVIII at both dose groups. The mean number of days without spontaneous bleeding was 13.3 ± 4.8 for PEG-LipFVIII vs. 7.2 days ± 1.7 days after infusion of the standard FVIII dose of 35 IU kg^{-1}. No clinically relevant adverse effects were reported for PEG-LipFVIII. These initial promising results on enhanced clinical efficacy suggested a once weekly infusion of PEG-LipFVIII for prophylactic therapy of hemophilia A patients.

A double blind, randomized clinical trial I study on the safety, bioavailability, pharmacokinetics and pharmacodynamics of a single administration of either low or high dose PEGLipFVIII compared with standard FVIII in patients with hemophilia A was reported [46]. In contrast to the former hypotheses, no differences in the pharmacokinetics between the free and the liposome-bound FVIII were found. The authors concluded that other factors than a longer circulation half-life are the reason for the increased protection from bleeding. They postulated that the liposomal carriers tend to interact with platelets and thus, locally accumulate higher concentrations of FVIII than in case of the free protein.

Another clinical trial I study could show that the infusion rate has no impact on the activity of PEG-LipFVIII. In contrast to the former studies, a bolus infusion of the liposomal product, as normally used in the free FVIII therapy, was applied.

This was well tolerated and could not explain any differences in the hemostatic efficacy of the liposomal product [47].

In search of the molecular mechanisms of the higher hemostatic activity of the liposomal FVIII, another randomized clinical crossover study was enrolled using various quantities of PEGylated liposomes (4.2; 12.6; 22.1 mg kg^{-1}) in combination with a fixed dose of FVIII at 35 IU kg^{-1} [45]. The data suggested that the number of bleeding-free days increased non-significantly with the liposomal doses, reaching a plateau at higher doses, but did not further contribute to the understanding of the mechanisms.

Dayan *et al.* could show that, similarly to the studies using rFVIII for liposomal association, plasma-derived purified FVIII displayed identical results [48]. The authors also concluded that the moderately improved pharmacokinetics in mice (18% increase in AUC, 22% increase in MRT) cannot solely explain the extended hematostatic activity of the PEGLipFVIII vs. the free protein. They also postulated a specific interaction of liposomes with platelets.

Pan *et al.* shed new light on the mechanisms behind the described phenomena [49]. First, they confirmed that the higher activity of liposomal FVIII is dependent on a complex formation between liposomes and protein before administration, which argues for the strength and specificity of the binding. Referring to the mechanisms of the more than twofold increased hemostatic efficacy *in vivo*, which was not accounted by the weakly increased half-life of 13%, the authors focused on several pathways and consequences of the cell–liposome interaction. They demonstrated, in mice experiments, that the liposomal system mainly act through a sensitization of the platelets and induction of procoagulant microparticles that sequester FVIII.

In 2009, a clinical phase II trial (Liplong study) was designed and enrolled to evaluate the efficacy and safety of a once-weekly prophylaxis treatment of hemophilia A patients with PEG-LipFVIII (BAY 79-4980; 35 IU kg^{-1}) compared with thrice-weekly prophylaxis with standard rFVIII (25 IU kg^{-1}) [50]. Surprisingly, in 2010 Bayer Schering Pharma announced that on basis of an interim analysis of the phase II trial data they have concluded that the study will not be able to achieve the predetermined efficacy endpoint. Although no safety concerns were raised, the company has decided to discontinue the study.

Despite of this nonfavorable outcome of the clinical development of PEGLip-FVIII, the general idea to increase the pharmacokinetic properties of proteins by complexation with PEGylated liposomes remains very attractive. Recent data shed new light on the basis of protein/PEG complex formation. Yatuv *et al.* demonstrated that the PEG complexation is based on a certain consensus sequence of eight amino acids in proteins (S/T-X-L/V-I/Q/S-S/T/Q-X-X-E) [51]. Consequently, other proteins than FVIII can be applied, which justifies this approach as a platform technology.

One of the suitable proteins is coagulation factor VIIa. FVIIa has been shown to induce hemostasis in the absence of FVIII [52]. Therefore, the prophylactic application of recombinant FVIIa is an attractive therapeutic option for hemophilia patients with inhibitors for FVIII and thus has been approved for therapy in many countries. Recombinant FVIIa has a relatively short half-life of about 2.3 h, which in similarity to the FVIII therapy, requires repeated dosing or high bolus injections. Again the therapeutic need for longer acting FVIIa preparations is evident.

Yatuv *et al.* showed that the PEGLip approach is well suitable and that the complexation of rFVIIa with PEGylated liposomes is highly specific, noncovalent and of high affinity (0.4 nM) [53]. In agreement to the findings with FVIII, the complexation has no impact on changed coagulant activity of FVIIa vs. the free rFVIIa. Pharmacokinetic experiments in rats confirmed that PEGLipFVIIa has a prolonges half-life of 1.49 h vs. 1.06 h of the free protein and 1.4 times higher AUC. PEGLip-FVIIa did not only show a prolonged biological activity, but also an enhanced hemostatic activity at lower doses. Based on these preliminary data the authors proposed a similar hypothesis of amplified platelet interaction of the liposomes as in the case of FVIII.

These studies were followed by a phase I/II clinical trial which evaluated the safety and efficacy of PEGLipFVIIa in hemophilia A patients with inhibitors to FVIII [50]. A significantly improved hemostatic efficacy in different readouts became evident for up to 5 h postinfusion of the PEGLipFVIIa vs. the standard FVIIa, although both did not significantly differ in the pharmacokinetic parameters. Further studies are needed to clarify the mechanisms for these finding and to evaluate whether similar postulations on platelet interactions with the liposomal carriers as in case of PEGLipFVIII are justified.

Based on the findings of a specific protein amino acid sequence responsible for PEG complexation, a recent study reported on the complexation of colony-stimulating factor (G-CSF) with PEGylated liposomes and improved biological activity in mice [51]. The authors described an extended circulation time of the liposome-complexed G-CSF vs. the free protein after both subcutaneous and intravenous application and a joined significant increased efficacy in the mobilization of stems cells from the bone marrow. Although further studies are needed to evaluate the potency of this novel approach, the studies demonstrate the practicability and potency of the PEGLip technology beyond hemophilia treatment.

Concluding, the noncovalent complexation of certain proteins with PEGylated liposomes appears as a very attractive approach to increase the biological efficacy of therapeutically relevant proteins in various medical application. Although the liposomal complexation cannot be considered as a classical way to enhance the circulation half-life of proteins, the increase in their activity via improved pharmocodynamic characteristics is a great benefit for therapy.

16.5
Conclusions

Due to their long circulation characteristics after systemic application, PEGylated liposomes appear as an interesting and vital option to affect the circulation half-life of therapeutically relevant, highly active proteins, either by incorporation into, or noncovalent attachment to the liposomes. Although none of these formulations has obtained approval for therapeutic treatments yet, promising preclinical data or clinical trials suggest the potential of these approaches.

References

1 Gregoriadis, G., and Ryman, B.E. (1972) Fate of protein-containing liposomes injected into rats. An approach to the treatment of storage diseases. *Eur. J. Biochem.*, **24**, 485.

2 Weissmann, G. and Sessa, G. (1968) Phospholipid spherules (liposomes) as a model for biological membranes. *J. Lipid Res.* **9**, (3), 310–318.

3 Freise, J., Müller, W.H., Brölsch, C., and Schmidt, F.W. (1980) "*In vivo*" distribution of liposomes between parenchymal and nonparenchymal cells in rat liver. *Biomedicine*, **32**, 118.

4 Yatvin, M.B., and Lelkes, P.I. (1982) Clinical prospects for liposomes. *Med. Phys.*, **9**, 149.

5 Senior, J.H. (1987) Fate and behavior of liposomes *in vivo*: a review of controlling factors. *Crit. Rev. Ther. Drug Carrier Syst.*, **3**, 123–193.

6 Woodle, M.C., and Lasic, D.D. (1992) Sterically stabilized liposomes. *Biochim. Biophys. Acta*, **1113**, 171–199.

7 Allen, T.M. (1994) The use of glycolipids and hydrophilic polymers in avoiding rapid uptake of liposomes by the mononuclear phagocyte system. *Adv. Drug Deliv. Rev.*, **13**, 285.

8 Torchilin, V.P., Omelyanenko, V.G., Papisov, M.I., Bogdanov, A.A., Trubetskoy, V.S., Herron, J.N., *et al.* (1994) Poly(ethylene glycol) on the liposome surface: on the mechanism of polymer-coated liposome longevity. *Biochim. Biophys. Acta*, **1195**, 11–20.

9 Gabizon, A.A., Barenholz, Y., and Bialer, M. (1993) Prolongation of the circulation time of doxorubicin encapsulated in liposomes containing a polyethylene glycol-derivatized phospholipid: pharmacokinetic studies in rodents and dogs. *Pharm. Res.*, **10**, 703–708.

10 Gabizon, A., Catane, R., Uziely, B., Kaufman, B., Safra, T., Cohen, R., *et al.* (1994) Prolonged circulation time and enhanced accumulation in malignant exudates of doxorubicin encapsulated in polyethylene-glycol coated liposomes. *Cancer Res.*, **54**, 987–992.

11 Scherphof, G.L., Morselt, H., and Allen, T.M. (1994) Intrahepatic distribution of long-circulating liposomes containing polyethylene glycol) distearoyl phosphatidylethanolamine. *J. Liposome Res.*, **4**, 213.

12 Ishida, O., Maruyama, K., Sasaki, K., and Iwatsuru, M. (1999) Size-dependent extravasation and interstitial localization of polyethyleneglycol liposomes in solid tumor-bearing mice. *Int. J. Pharm.*, **190**, 49–56.

13 Yuan, F., Leunig, M., Huang, S.K., Berk, D.A., Papahadjopoulos, D., and Jain, R.K. (1994) Microvascular permeability and interstitial penetration of sterically stabilized (stealth) liposomes in a human tumor xenograft. *Cancer Res.*, **54**, 3352–3356.

14 Soloman, R., and Gabizon, A.A. (2008) Clinical pharmacology of liposomal anthracyclines: focus on pegylated liposomal Doxorubicin. *Clin. Lymphoma Myeloma*, **8**, 21–32.

15 Cheng, W.W., and Allen, T.M. (2010) The use of single chain Fv as targeting agents for immunoliposomes: an update on immunoliposomal drugs for cancer treatment. *Expert Opin. Drug Deliv.*, **7**, 461–478.

16 Park, J.W., Benz, C.C., and Martin, F.J. (2004) Future directions of liposome- and immunoliposome-based cancer therapeutics. *Semin. Oncol.*, **31**, 196–205.

17 Schnyder, A., and Huwyler, J. (2005) Drug transport to brain with targeted liposomes. *NeuroRx.*, **2**, 99–107.

18 Torchilin, V.P. (2010) Passive and active drug targeting: drug delivery to tumors as an example. *Handb. Exp. Pharmacol.*, **197**, 3–53.

19 Kedar, E., Palgi, O., Golod, G., Babai, I., and Barenholz, Y. (1997) Delivery of cytokines by liposomes. III. Liposome-encapsulated GM-CSF and TNF-alpha show improved pharmacokinetics and biological activity and reduced toxicity in mice. *J. Immunother.*, **20**, 180–193.

20 Debs, R.J., Fuchs, H.J., Philip, R., Brunette, E.N., Düzgüneş, N., Shellito, J.E., *et al.* (1990) Immunomodulatory and toxic effects of free and liposome-encapsulated tumor necrosis factor alpha in rats. *Cancer Res.*, **50**, 375–380.

21 Morishige, H., Ohkuma, T., and Kaji, A. (1993) *In vitro* cytostatic effect of TNF (tumor necrosis factor) entrapped in immunoliposomes on cells normally insensitive to TNF. *Biochim. Biophys. Acta*, **1151**, 59–68.

22 van der Veen, A.H., Eggermont, A.M., Seynhaeve, A.L., van Tiel, S.T., and ten Hagen, T.L. (1998) Biodistribution and tumor localization of stealth liposomal tumor necrosis factor-alpha in soft tissue sarcoma bearing rats. *Int. J. Cancer*, **77**, 901–906.

23 Kim, D.W., Andres, M.L., Li, J., Kajioka, E.H., Miller, G.M., Seynhaeve, A.L., *et al.* (2001) Liposome-encapsulated tumor necrosis factor-alpha enhances the effects of radiation against human colon tumor xenografts. *J. Interferon Cytokine Res.*, **21**, 885–897.

24 Hagen, T.L.M., Seynhaeve, A.L.B., van Tiel, S.T., Ruiter, D.J., and Eggermont, A.M.M. (2002) Pegylated liposomal tumor necrosis factor-alpha results in reduced toxicity and synergistic antitumor activity after systemic administration in combination with liposomal doxorubicin (Doxil) in soft tissue sarcoma-bearing rats. *Int. J. Cancer*, **97**, 115–120.

25 Mori, A., Duda, E., and Huang, L. (1996) Recombinant human tumor necrosis factor-[alpha] covalently conjugated to long-circulating liposomes. *Int. J. Pharm.*, **131**, 57–66.

26 Savva, M., Duda, E., and Huang, L. (1999) A genetically modified recombinant tumor necrosis factor-alpha conjugated to the distal terminals of liposomal surface grafted polyethyleneglycol chains. *Int. J. Pharm.*, **184**, 45–51.

27 Blume, G., Cevc, G., Crommelin, M.D., Bakker-Woudenberg, I.A., Kluft, C., and Storm, G. (1993) Specific targeting with poly(ethylene glycol)-modified liposomes: coupling of homing devices to the ends of the polymeric chains combines effective target binding with long circulation times. *Biochim. Biophys. Acta*, **1149**, 180–184.

28 Maruyama, K., Takizawa, T., Yuda, T., Kennel, S.J., Huang, L., and Iwatsuru, M. (1995) Targetability of novel immunoliposomes modified with amphipathic poly(ethylene glycol)s conjugated at their distal terminals to monoclonal antibodies. *Biochim. Biophys. Acta*, **1234**, 74–80.

29 McDermott, D.F. (2009) Immunotherapy of metastatic renal cell carcinoma. *Cancer*, **115**, 2298–2305.

30 Bhatia, S., Tykodi, S.S., and Thompson, J.A. (2009) Treatment of metastatic melanoma: an overview. *Oncology (Williston Park)*, **23**, 488–496.

31 Konrad, M.W., Hemstreet, G., Hersh, E.M., Mansell, P.W., Mertelsmann, R., Kolitz, J.E., *et al.* (1990) Pharmacokinetics of recombinant interleukin 2 in humans. *Cancer Res.*, **50**, 2009–2017.

32 Konno, H., Yamashita, A., Tadakuma, T., and Sakaguchi, S. (1991) Inhibition of growth of rat hepatoma by local injection of liposomes containing recombinant interleukin-2. Antitumor effect of IL-2 liposome. *Biotherapy*, **3**, 211–218.

33 Konno, H., Maruo, Y., Matin, A.F., Tanaka, T., Nakamura, S., Baba, S., *et al.* (1992) Effect of liposomal interleukin-2 on ascites-forming rat hepatoma. *J. Surg. Oncol.*, **51**, 33–37.

34 Anderson, P.M., Katsanis, E., Leonard, A.S., Schow, D., Loeffler, C.M., Goldstein, M.B., *et al.* (1990) Increased local antitumor effects of interleukin 2 liposomes in mice with MCA-106

sarcoma pulmonary metastases. *Cancer Res.*, **50**, 1853–1856.

35 Oya, M. (1994) Antitumor effect of interleukin-2 entrapped in liposomes on murine renal cell carcinoma. *Keio J. Med.*, **43**, 37–44.

36 Loeffler, C.M., Platt, J.L., Anderson, P.M., Katsanis, E., Ochoa, J.B., Urba, W.J., *et al.* (1991) Antitumor effects of interleukin 2 liposomes and anti-CD3-stimulated T-cells against murine MCA-38 hepatic metastasis. *Cancer Res.*, **51**, 2127–2132.

37 Anderson, P.M., Katsanis, E., Sencer, S.F., Hasz, D., Ochoa, A.C., and Bostrom, B. (1992) Depot characteristics and biodistribution of interleukin-2 liposomes: importance of route of administration. *J. Immunother.*, **12**, 19–31.

38 Kedar, E., Rutkowski, Y., Braun, E., Emanuel, N., and Barenholz, Y. (1994) Delivery of cytokines by liposomes. I. Preparation and characterization of interleukin-2 encapsulated in long-circulating sterically stabilized liposomes. *J. Immunother. Emphasis Tumor Immunol.*, **16**, 47–59.

39 Kedar, E., Braun, E., Rutkowski, Y., Emanuel, N., and Barenholz, Y. (1994) Delivery of cytokines by liposomes. II. Interleukin-2 encapsulated in long-circulating sterically stabilized liposomes: immunomodulatory and anti-tumor activity in mice. *J. Immunother. Emphasis Tumor Immunol.*, **16**, 115–124.

40 Kedar, E., Gur, H., Babai, I., Samira, S., Even-Chen, S., and Barenholz, Y. (2000) Delivery of cytokines by liposomes: hematopoietic and immunomodulatory activity of interleukin-2 encapsulated in conventional liposomes and in long-circulating liposomes. *J. Immunother.*, **23**, 131–145.

41 Sencer, S.F., Rich, M.L., Katsanis, E., Ochoa, A.C., and Anderson, P.M. (1991) Anti-tumor vaccine adjuvant effects of IL-2 liposomes in mice immunized against MCA-102 sarcoma. *Eur. Cytokine Netw.*, **2**, 311–318.

42 de Haan, A., Groen, G., Prop, J., van Rooijen, N., and Wilschut, J. (1996) Mucosal immunoadjuvant activity of liposomes: role of alveolar macrophages. *Immunology*, **89**, 488–493.

43 Cabanes, A., Even-Chen, S., Zimberoff, J., Barenholz, Y., Kedar, E., and Gabizon, A. (1999) Enhancement of antitumor activity of polyethylene glycol-coated liposomal doxorubicin with soluble and liposomal interleukin 2, Clin. *Cancer Res.*, **5**, 687–693.

44 Baru, M., Carmel-Goren, L., Barenholz, Y., Dayan, I., Ostropolets, S., Slepoy, I., *et al.* (2005) Factor VIII efficient and specific non-covalent binding to PEGylated liposomes enables prolongation of its circulation time and haemostatic efficacy. *Thromb. Haemost.*, **93**, 1061–1068.

45 Spira, J., Plyushch, O.P., Andreeva, T.A., and Khametova, R.N. (2008) Evaluation of liposomal dose in recombinant factor VIII reconstituted with pegylated liposomes for the treatment of patients with severe haemophilia A. *Thromb. Haemost.*, **100**, 429–434.

46 Powell, J.S., Nugent, D.J., Harrison, J.A., Soni, A., Luk, A., Stass, H., *et al.* (2008) Safety and pharmacokinetics of a recombinant factor VIII with pegylated liposomes in severe hemophilia A. *J. Thromb. Haemost.*, **6**, 277–283.

47 Martinowitz, U., Luboshitz, J., Bashari, D., Ravid, B., Gorina, E., Regan, L., *et al.* (2009) Stability, efficacy, and safety of continuously infused sucrose-formulated recombinant factor VIII (rFVIII-FS) during surgery in patients with severe haemophilia. *Haemophilia*, **15**, 676–685.

48 Dayan, I., Robinson, M., and Baru, M. (2009) Enhancement of haemostatic efficacy of plasma-derived FVIII by formulation with PEGylated liposomes. *Haemophilia*, **15**, 1006–1013.

49 Pan, J., Liu, T., Kim, J., Zhu, D., Patel, C., Cui, Z., *et al.* (2009) Enhanced efficacy of recombinant FVIII in noncovalent complex with PEGylated liposome in hemophilia A mice. *Blood*, **114**, 2802–2811.

50 Spira, J., Plyushch, O., Zozulya, N., Yatuv, R., Dayan, I., Bleicher, A., *et al.* (2010) Safety, pharmacokinetics and efficacy of factor VIIa formulated with

PEGylated liposomes in haemophilia
A patients with inhibitors to factor
VIII–an open label, exploratory,
cross-over, phase I/II study.
Haemophilia., **16**, 910–918.

51 Yatuv, R., Carmel-Goren, L., Dayan, I.,
Robinson, M., and Baru, M. (2009)
Binding of proteins to PEGylated
liposomes and improvement of G-CSF
efficacy in mobilization of hematopoietic
stem cells. *J. Control. Release*, **135**,
44–50.

52 Hedner, U., Glazer, S., Pingel, K.,
Alberts, K.A., Blombäck, M.,
Schulman, S., *et al.* (1988) Successful use
of recombinant factor VIIa in patient
with severe haemophilia A during
synovectomy. *Lancet*, **2**, 1193.

53 Yatuv, R., Dayan, I., Carmel-Goren, L.,
Robinson, M., Aviv, I., Goldenberg
Furmanov, M., *et al.* (2008) Enhancement
of factor VIIa haemostatic efficacy by
formulation with PEGylated liposomes.
Haemophilia, **14**, 476–483.

17

Half-Life Extension with Pharmaceutical Formulations: Nanoparticles by the Miniemulsion Process

Katharina Landfester, Anna Musyanovych, and Volker Mailänder

17.1
Introduction

Delivery of bioactive compounds such as therapeutic proteins, peptides and drugs to the place of interest within a living organism is of high interest for the treatment of damaged or diseased tissues. To improve the stability and immunogenicity of the therapeutics, and to extend the half-lives *in vivo*, the biomolecule can either be modified through covalent attachment to the polyethylene glycol chain or be delivered by colloidal carrier (e.g., liposome, micelle, dendrimer, polymeric particles, or capsules) [1–4].

Proteins have several disadvantages when compared with small drug molecules: when brought into the human body they are degraded rapidly. This is especially true for oral applications as they the whole gastrointestinal tract is designed to denature proteins (by the low pH in the stomach) and to enzymatically cleave the polymer chain. But also in other body fluids like blood enzymes actively cleave these therapeutic proteins and thereby destroy their function. If they are small enough and below the molecular weight cutoff of the kidney they may even be filtered in the nephrons of the kidney, and if is they are not actively recovered before they reach the renal calix they are lost. Furthermore they cannot cross barriers like the gastrointestinal barrier, the endothelial [and here especially the blood–brain barrier (BBB) for example] nor the cell membrane as a barrier to intracellular targets as easily as small drugs can. Furthermore it would be desirable to have a slow release of drugs.

To this end several approaches have been investigated with polyethylene glycol (PEG) being the most popular [5]. Here covalent coupling of PEG shields the protein from enzymes and makes the compound increase in molecular weight thereby prolonging the blood half-life dramatically. Other molecules have been used with the same concept in mind (e.g., hyperglycosylation) (for a review see Pisal *et al.* [2]).

Therapeutic Proteins: Strategies to Modulate Their Plasma Half-Lives, First Edition. Edited by
Roland Kontermann.
© 2012 Wiley-VCH Verlag GmbH & Co. KGaA. Published 2012 by Wiley-VCH Verlag GmbH & Co. KGaA.

Liposomal formulations have advanced this field a step further as here the molecule of interest does not need to be chemically altered as it is not covalently attached, but the liposome encloses the protein into a compartment, and thus resembles the most simple cell possible [6]. This gives a drug delivery device so that the release of the drugs can be prolonged, and shields the proteins from the influence of the surrounding enzymes at least temporarily. The drawback of this approach is that the liposomal formulations are usually not stable in blood because of their low critical micelle concentration. This can be altered by crosslinking of the lipids in the liposomes [7]. Also the chemical functionalization has not been trivial so that targeting of the liposomes is hard to achieve. Copolymers are an attractive alternative as they can be designed specifically for the needs of the drug delivery device and have shown some interesting results [4]. But even for these systems the crossing of the gastrointestinal barrier or the BBB has been hard to overcome.

Polymeric particles and nanocapsules in the sub-micrometer size range are of great interest for many biomedical applications [8–15] since they can be produced in a wide size range, have a good colloidal stability, and are chemically resistant. Currently, nanoparticles as specific carriers for therapeutic, contrasting or imaging agents are being developed, generally focusing on their permeability through different kinds of tissue. It is a great advantage that drugs and markers which are encapsulated inside nanoparticles or nanocapsules, can be efficiently protected against enzymatic and hydrolytic degradation and that the half-life extension of pharmaceutical formulations can be increased. Understanding the nanoparticle/cell and nanocapsule/cell interactions, and establishing the effects that influence the intracellular uptake of the nanoparticulates, are important tasks for using such systems as efficient markers and improving the delivery of bioactive agents.

In recent years, the cellular uptake of various polymeric micro- and nanoparticles, prepared from natural or synthetic polymers, has been extensively described in the literature. However, most studies were carried out in different cell systems and with entirely different nanoparticles, that were not always well-characterized, either colloidally or polymer-chemically ;therefore it is difficult to compare these studies. Recent studies demonstrate that the rate and extent of particle uptake can be influenced by many factors, for example, the concentration of nanoparticles in the medium [16, 17], the incubation time and temperature [16, 18], the cell type and density [19, 20], the type of the encapsulated drug, the polymer material [21], the size/shape and surface characteristics of the particles [18, 20, 22–25]. All these parameters determine the particle adsorption/adhesion to and the interaction with the cells. In order to synthesize complex nanoparticles and nanocapsules with a well-defined size, (polymer) composition, morphology, and surface functionality, a synthetic method is required which permits the design of nanoparticles in a wide range to achieve specific application needs.

In this chapter we will mainly focus on the nanoparticles and nanocapsules that are obtained by the miniemulsion technique, and the different possibilities for the application of such nanoparticles and nanocapsules in a biomedical field, as a concept for half-life extension with pharmaceutical formulations.

17.2
Polymeric Nanoparticles

A versatile method for the formation of nanoparticles and nanocapsules is the miniemulsion process where small, stable, and homogeneously distributed nanodroplets are created in a continuous phase by applying high shear [26, 27]. Under high shear, for example, ultrasonication or high pressure homogenization, defined small nanodroplets in the size range between 50 and 500 nm can be obtained. The stabilization of the nanodroplets is guaranteed by a combination of a surfactant with an osmotic pressure agent which is dissolved in the dispersed phase, but also possesses a lower solubility in the continuous phase than the dispersed phase itself. In the case of a direct (oil-in-water) miniemulsion, this agent is a (ultra) hydrophobe, in the case of an inverse (water-in-oil) miniemulsion it represents an (ultra)lipophobe. If the dispersed phase is a monomer, it can be polymerized subsequently inside the droplets without changing the identity of the droplets. The process of miniemulsion polymerization is schematically shown in Figure 17.1 [28, 29].

17.3
Particles Obtained by Radical Polymerization and Their Functionalization

Radical polymerization can be performed with many different vinylic monomers such as styrene, acrylates, methacrylates, fluoroacrylates, acrylamides, etc. Furthermore, a copolymerization between two hydrophobic monomers is well suited to obtain homogeneous copolymer materials. The copolymerization of hydrophobic

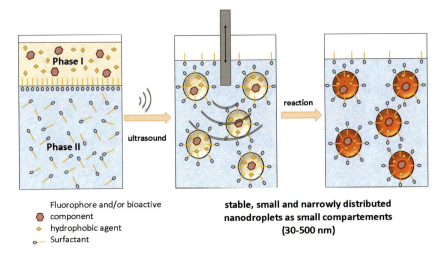

Figure 17.1 Process of miniemulsion polymerization.

and hydrophilic monomers leads to the formation of amphiphilic copolymer particles [30]. A copolymerization with functional monomers permits functionalization of the nanoparticles' surface with different groups in a controlled manner. In this way, the use of acrylic acid (AA) leads to the carboxylic nanoparticles [29]. By introduction of 2-aminoethyl methacrylate hydrochloride (AEMH) or vinyl phosphonic acid as a comonomer of styrene (or methyl methacrylate), particles with surface amino [29], or phosphonic acid [31] groups can be obtained. The density of functional groups on the particle surface can be adjusted by the initial concentration of the comonomer and the type/amount of the surfactant (see Figure 17.2) [28]. The different possibilities for radical polymerizations in miniemulsion systems are also summarized in several reviews [26, 27, 32, 33].

Surface-functionalized polymeric nanoparticles, as obtained by the copolymerization miniemulsion process with styrene and a functional monomer as AA [34], or styrene and vinyl phosphonic acid [35], can also be used as templates for the biomimetic mineralization of hydroxyapatite (HAP) in the aqueous phase [36]. Different densities of the functional groups on the surface were tested (see Figure 17.3). It was found that for a fixed concentration of Ca^{2+} and PO_4^{3-} ions added, the amount of HAP formed on the surface of the particles increased with increasing AA amount. The absence of HAP on latexes prepared with 0 wt% AA for both ionic

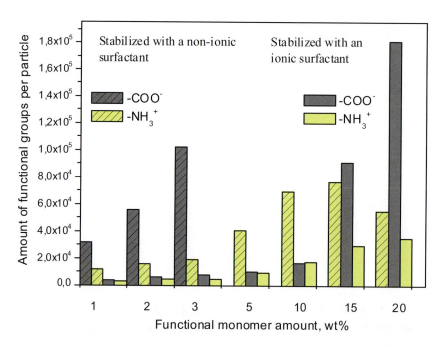

Figure 17.2 Surface functionalization of nanoparticles obtained by the miniemulsion polymerization of styrene with AA or AEMH as functional comonomer using ionic or non-ionic surfactants (non-ionic surfactant: Lutensol AT50 (poly(ethylene oxide)-hexadecyl ether with an EO block length of about 50 units); ionic surfactant in the case of AA: SDS [sodium dodecyl sulfate], in the case of AEMH: CTMA-Cl [cetyltrimethyl ammonium chloride]) [28].

a)

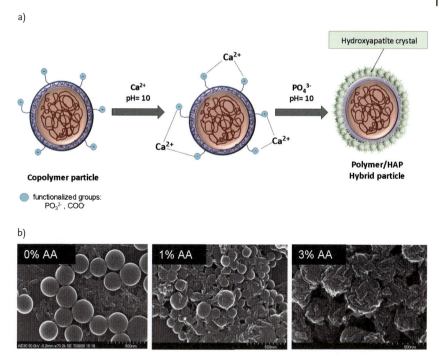

Copolymer particle

● functionalized groups:
   PO₃²⁻ , COO⁻

b)

Figure 17.3 (a) Schematic illustration of HAP formation on a carboxyl-functionalized polymeric nanoparticle by the addition of Ca²⁺ and (PO₄)³⁻ ions; (b) HRSEM images illustrating the HAP formation for samples prepared using non-ionic surfactant (Lutensol AT50) with different amounts of AA. Reprinted with permission from [34]. Copyright (2007) American Chemical Society.

as well as non-ionic surfactant types confirms that the amount of HAP formed depends only on the number of carboxyl groups covalently bonded onto the surface. Such hydroxyapatite/polymer hybrid nanoparticles have a great potential to be used as filler or scaffold for nucleation and growth of new bone material. They offer the feasibility of being injected directly into the damaged part in addition to be applied as coatings on implants. Polymeric nanoparticles coated with HAP for bone repair applications can be used as carriers of drugs and growth factors for better treatment of bone defects and to promote wound healing.

It was shown that the uptake of nanoparticles by the cell is directly influenced not only by the density of surface groups and the total surface charge (cationic versus anionic), but also by several other parameters, including the polymer and surfactant type as well as the cell type. Experiments were performed using carboxyl- or amino-functionalized polystyrene (PS) nanoparticles. HeLa were chosen as an adherent cell line which is well established in cell culture labs, mesenchymal stem cells (MSCs) for their potential in regenerative medicine, KG1a as model for CD34+ hematopoetic stem cells and Jurkat as model for T cells. Furthermore, in order to determine whether the uptake mechanism depends on the nanoparticles surface characteristics, negatively and positively charged PS particles of the same size were used for uptake experiments in HeLa cells in the presence of different inhibitors [37].

From the literature it is known that for particles in the range of a few to several hundred nanometers, the uptake into cells involves several possible mechanisms like pinocytosis, nonspecific endocytosis, or receptor-mediated endocytosis [38]. For the latter case, some receptors like opsonin, lectin and scavenger receptors are known [39]. Endocytosis of polymeric particles in a size-range of several hundred nanometers was studied in various cell types. The mechanisms of endocytosis involved in the uptake of well-defined positively and negatively charged 100 nm fluorescent PS nanoparticles into HeLa cells [37] were evaluated. In order to study the uptake mechanism, nanoparticles were incubated with HeLa cells in the presence or absence of drugs known to inhibit various factors in endocytosis, the intracellular localization was confirmed by confocal laser scanning and transmission electron microscopy (TEM). It was found that endocytosis is highly dependent on dynamin and F-actin, and independent from the particle charge, as shown by inhibition of uptake in the presence of dynasore and cytochalasin D, respectively. Since genistein inhibited the particle uptake, tyrosine specific protein kinases, located in lipid rafts, must be involved in the uptake. These factors suggest an uptake mechanism dependent on dynamin and lipid raft. However, cholesterol sequestration or depletion by filipin or methyl-β-cyclodextrine, respectively, did not hinder the particle uptake.

The particle surface charge has a significant impact on the uptake mechanism: during the endocytosis of positively charged particles, macropinocytosis is an important uptake mechanism as was shown by the strong inhibition of positively charged nanoparticles by 5-(N-ethyl-N-isopropyl)amiloride [37]. The microtubule network and cyclo-oxygenases are also involved in uptake of positively charged particles as particle uptake was hindered in the presence of nocodazole and indomethacin, respectively. Since about 20% of the endocytosis of positively charged particles is inhibited by chlorpromazine it can be concluded that the clathrin-dependent pathway plays only a minor role.

In order to increase the rate of intracellular uptake of nanoparticles some transfection agents can be used [40]. These agents are mostly cationic, positively charged molecules [41]. They are toxic and not approved for clinical use. Hence applications in human trials and therapeutic interventions have so far been prohibited.

Cell surface antigens provide an opportunity for specific targeting by antibody-linked nanoparticles [42, 43]. The first attempts towards a covalent coupling of antibodies to nanoparticles were undertaken by Rolland *et al.* and Akasaka *et al.* [42, 44]. A defined receptor/ligand or antigen/antibody interaction enables the binding of modified nanoparticles on the surface of the target cells followed by a potential cellular internalization [45]. Functionalized PS nanoparticles provide a higher carrier capacity, for, for example, fluorochromes and contrast agents, than the limited amount of binding sides of antibodies alone. This feature makes them especially suited for conjugation with antibodies. Thus antibody-conjugated PS nanoparticles represent a link between a multifunctional carrier and a specific tool.

In order to develop nanoparticles as safe and efficacious delivery systems, surface modifications have been considered as prerequisite to improve stability and pharmacokinetic properties [46].

Polymeric nanoparticles with tumor necrosis factor (TNF) covalently attached to their surface (TNF nanocytes) were shown to be useful carrier systems capable of mimicking the bioactivity of membrane-bound TNF (see Figure 17.4) [47]. Targeted lipid-coated particles (TLP) have been developed in which TNF activity is shielded. The TLPs are composed of an inner single-chain TNF (scTNF)-functionalized, polymeric nanoparticle core surrounded by a lipid coat endowed with PEG for sterical stabilization and a single-chain Fv (scFv) fragment for targeting. Using a scFv directed against the tumor stroma marker fibroblast activation protein (FAP) it was shown that TLP and scTNF–TLP specifically bind to FAP-expressing, but not to FAP-negative cells. Lipid coating strongly reduced nonspecific binding of particles and scTNF-mediated cytotoxicity towards FAP-negative cells. Besides safe and targeted delivery of death ligands such as TNF, TLP should be suitable for various diagnostic and therapeutic applications, which benefit from a targeted delivery of reagents embedded into the particle core or displayed on the core particle surface.

17.4
Other Polyreactions in Miniemulsion

Since the reaction is performed in the small droplets, no diffusion as for example, in emulsion polymerization is required. Therefore, other types of polymerizations can also be carried out in miniemulsion: the anionic polymerization in nonaqueous miniemulsions leads to the polyamide nanoparticles [48], and in aqueous phase, poly(*n*-butylcyanoacrylate) (PBCA) nanoparticles can be synthesized [49]. The miniemulsion technique provides a possibility of obtaining nanoparticles at high solid content. Furthermore, the functionalization of PBCA nanoparticles is possible by using different nucleophiles as initiator for the anionic polymerization of the monomer containing miniemulsion droplets. Additionally encapsulation of various materials is easily achieved [49]. The application of the anionic surfactant SDS permits the preparation of long-term stable dispersions with solid contents of 10% or more and leads to PBCA particles without any covalently bounded molecules on the surface, such as dextran or other steric stabilizers.

The cationic polymerization permits the formation of poly-*p*-methoxystyrene particles [50, 51], the catalytic polymerization enables the synthesis of polyolefin [52, 53] or polyketone particles [54], the ring opening metathesis polymerization leads to polynorbonene nanoparticles [55, 56], and the step-growth acyclic diene metathesis (ADMET) polymerization to oligo(phenylene vinylenes) particles [57]. The polyaddition in miniemulsion permits the formation of polyepoxides [58] or polyurethane particles [59, 60]. Polycondensation processes yield in stable polyester [61] nanoparticles. Polyester particles with a high molecular weight of about $100\,000\,\mathrm{g\,mol^{-1}}$ could also be obtained by enzymatic polymerization in

a)

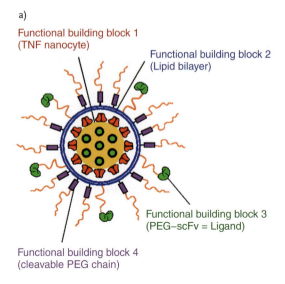

b)

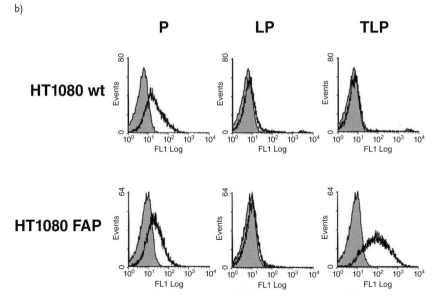

Figure 17.4 (a) Scheme of the composition of bioactive multifunctional composite nanoparticles for the induced drug release of the tumor necrosis factor TNF; (b) Flow cytometry analysis of particles (P), of lipid-coated particles (LP), and targeted lipid-coated particles (TLP) prepared from amino-functionalized (NSL-PMI) P for binding to FAP⁻ HT1080 wild-type (wt) cells or FAP⁺ HT1080 FAP_{hu} cells. (Gray, cells alone; solid line, cells incubated with particle preparations at 3 μg/250 000 cells). Reprinted from [47]. Copyright (2009), with permission from Elsevier.

miniemulsion [62]. The oxidative polymerization of aniline leads to polyaniline nanoparticles [63, 64].

Biodegradable nanoparticles with the size of 80-120 nm were formed by combination of the emulsion/solvent evaporation method and miniemulsion technique using different biocompatible and biodegradable polymers such as poly(L-lactide), poly(D,L-lactide-coglycolide), and poly(ε-caprolactone). Differences between the results of various polymers are found in terms of the particle size and size distribution as well as in the degradation time [65].

17.5
Formation of Nanocapsules and Their Functionalization

The formation of nanocapsules using different synthetic and natural monomers or polymers in order to encapsulate hydrophobic and hydrophilic liquids is of special interest for biomedical application. Here, pharmaceutical formulations can be obtained.

By using the double emulsion technique [66–68], the emulsification/solvent evaporation, the diffusion method [69–71], the salting-out procedure [72], or the layer-by-layer technique [73–76], peptides and proteins could successfully be encapsulated. All these methods usually lead to capsules in a size range over 500 nm. The miniemulsion process is excellently suited for the formation of capsules smaller than 500 nm. For the formation of nanocapsules with a hydrophilic or hydrophobic core, different encapsulation processes, all based on the miniemulsion technique, facilitate the production of capsules with desired properties for a broad range of biomedical applications. A summary of different miniemulsion-based methods for the encapsulation of liquids are summarized in Figure 17.5.

For the encapsulation of hydrophobic liquids, a one-step synthesis in miniemulsion with the phase separation throughout the polymerization process can be used, resulting in polymer nanocapsules with a liquid core. The nanocapsules are formed by a variety of monomers in the presence of a hydrophobic nonpolymerizable oil. Before polymerization the hydrophobic oil and monomer form a common miniemulsion with a homogenous dispersed phase. During the polymerization, the polymer becomes immiscible with the hydrophobic oil and phase separation takes place. The challenge now is to tune the different interfacial tensions of the interfaces polymer/water, polymer/encapsulated oil and encapsulated oil/water in such a way that the morphology consisting of a polymer shell structure surrounding the hydrophobic oil is the thermodynamically preferred one [77]. Amphiphilic oligomers of styrene and maleic acid anhydride were used for a nanoencapsulation via an interfacially confined controlled/living radical miniemulsion polymerization. The oligomers can adsorb at the water/droplets interface facilitating a RAFT living polymerization at the interface [78].

Interfacial reactions are a very elegant way to prepare nanocapsules. The nanometer-sized capsules in direct emulsion could be synthesized employing interfacial crosslinking reactions as polyaddition and polycondensation [79–81], radical [82,

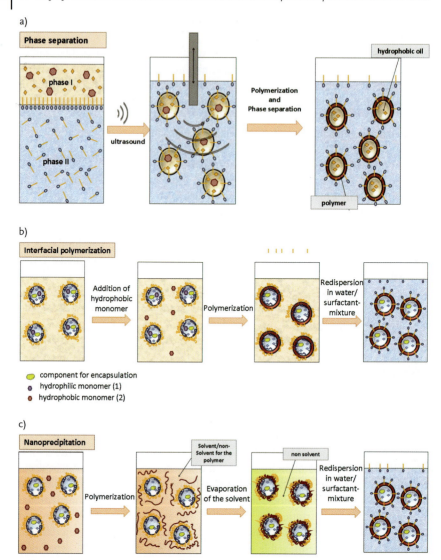

Figure 17.5 Formation of nanocapsules: (a) Phase separation during polymerization: before polymerization, the monomer and the hydropbobic oil form a homogeneous droplet phase, while throughout the polymerization process phase separation occurs leading to the formation of nanocapsules; (b) interfacial polymer reaction on miniemulsion droplets: droplets in an inverse miniemulsion contain the hydrophilic component for encapsulation and monomer 1,the addition of the second monomer leads to a polymerization at the interface forming the nanocapsules which can be transferred to the water phase; and (c) nanoprecipitation: an inverse miniemulsion is formed containing the hydrophilic component for encapsulation inside the aqueous droplets, the continuous phase consists of a solvent/ non solvent mixture for the polymer; during evaporation of the solvent, the polymer precipitates onto the nanodroplets forming the nanocapsules which can subsequently be transferred to the water phase.

83] or anionic polymerization [9, 84]. The polyaddition between the chitosan stabilizer and two biocompatible costabilizers, that is, Jeffamine®D2000 (polyoxypropylenediamine) or Gluadin® (protein based compound), and a linking diepoxide in presence of an inert oil was performed at the interface of direct miniemulsion droplets. Polymeric nanocapsules with thin but rather stable shells were formed (see Figure 17.6b) [85]. Since both water and oil soluble aminic costabilizers can be used, these experiments show the way to a great variety of capsules with different chemical structures.

For many biomedical applications, hydrophilic liquids including dissolved substances, for example, drugs, or therapeutics, are of great interest. Hydrophilic liquids can be encapsulated by using an interfacial polyaddition, radical or anionic polymerization or by nanoprecipitation processes in inverse miniemulsion. The aqueous phase is dispersed in a continuous hydrophobic medium, after formation

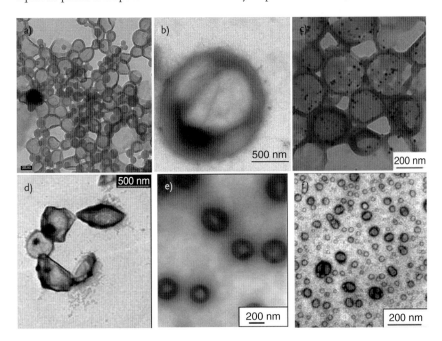

Figure 17.6 (a) Hydrophobic components surrounded by a poly(styrene-co-acrylic acid) shell as obtained by free radical polymerization in direct miniemulsion. Reprinted with permission from [77]. Copyright (2001) American Chemical Society; (b) hydrophobic components inside the crosslinked chitosan capsules as obtained by interfacial polyaddition in direct miniemulsion. Reprinted with permission from [85]. Copyright (2002) American Chemical Society; (c) silver nanoparticles formed by reduction of silver nitrate inside the aqueous core surrounded by a polyurethane shell as obtained by interfacial polyaddition in inverse miniemulsion. Reprinted with permission from [86]. Copyright (2007) American Chemical Society; (d) water-soluble gadolinium complex in polyurethane shell as obtained by interfacial polyaddition in inverse miniemulsion [87]; (e) dsDNA molecules protected by the PBCA shell as obtained by anionic interfacial polymerization in inverse miniemulsion. Reprinted with permission from [88]. Copyright (2008) Springer; (f) aqueous antiseptic agent surrounded by the polymeric shell via interfacial nanoprecipitation in inverse miniemulsion [89].

of the nanocapsules the continuous hydrophobic phase can be replaced by an aqueous environment.

In the case of polyaddition, polyurethanes, polyurea, or polythiourea could be obtained [86, 90]. Starch or other water-soluble natural polymers can be crosslinked with diisocyanate resulting in capsules with stable biodegradable shell (see Figure 17.6b) [91]. Different components can be encapsulated in the aqueous nanocapsules which act as independent nanoreactor, for example, silver nitrate was encapsulated and subsequently reduced to silver nanoparticles inside the nanocapsules (see Figure 17.6c) [86]. For magnetic resonance imaging (MRI) measurements, the capsules can also contain a contrast agent (e.g., Magnevist®, Gadovist®) surrounded by a shell (Figure 17.6d), which is highly permeable for water allowing almost free exchange with the bulk water [87].

Capsules consisting of an aqueous core and a uniform polymeric shell could also be prepared by using the alternating radical copolymerization of hydrophobic maleate esters and hydrophilic polyhydroxy vinyl ethers via the free-radical polymerization process at the interface of water-in-oil droplets [92].

Homogeneous biodegradable PBCA nanocapsules in a size range between 300 and 700 nm, which contain DNA molecules inside, could be obtained by the anionic polymerization of butylcyanoacrylate at the interface of inverse miniemulsion droplets (see Figure 17.6e) [88]. The shell thickness of the capsules can be adjusted between 5 and 40 nm by varying the amount of butylcyanoacrylate used for the synthesis. After polymerization, the capsules could be easily transferred to an aqueous phase in order to be useful for the biologically related experiments. The encapsulation efficiency of DNA in PBCA capsules was about 100% and at least 15% of the total DNA amount was found to be in the form of free, undisturbed double helix chains as determined by resolved agarose gel electrophoresis (see Figure 17.7).

The nanoprecipitation of preformed polymers onto miniemulsion droplets is a further possibility to fabricate nanocapsules. By interfacial deposition of the pre-

a)   b)

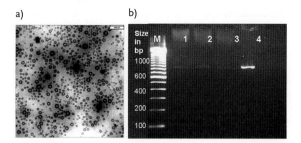

Figure 17.7 (a) TEM micrograph of PBCA nanocapsules obtained by interfacial polymerization; (b) agarose gel electrophoresis of dsDNA (790 bp) recovered from the PBCA capsules synthesized with different types/amounts of surfactant; Lane M: 100-bp ladder; Lane 1: 5 wt% of Span 80; Lane 2: 5 wt% of Tween 80; Lane 3: 5 wt% of Span 80 and Tween 80; Lane 4: unpolymerized miniemulsion prepared with 5 wt% of Span 80 and Tween 80. Reprinted from [88]. Copyright (2008) Springer.

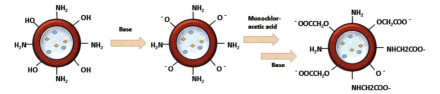

Figure 17.8 Functionalization of nanocapsules consisting of the polyurethane shell by a carboxymethylation reaction.

formed polymer, for example poly(D,L-lactide)-based nanocapsules containing an antitumoral agent [93] or poly(methyl methacrylate) (PMMA) capsules with an entrapped antiseptic agent [89] were successfully prepared (see Figure 17.6f).

In order to trigger cell uptake (e.g., *ex vivo* for marking cells) or to have binding sides for specific molecules (e.g., peptides, antibodies, etc. for active targeting), the nanocapsules need to be functionalized. Recently, the functionalization of well-defined polyurethane nanocapsules with an aqueous core prepared via a polyaddition at the interface of inverse miniemulsion droplets has been demonstrated [94]. The carboxy- and amino-functionalized surface of the nanocapsules can be tailored by an *in situ* carboxymethylation reaction and by physical adsorption of a cationic polyelectrolyte, that is, poly(aminoethyl methacrylate hydrochloride) or poly(ethylene imine) (see Figure 17.8). The increased uptake of amino-functionalized fluorescent nanocapsules by HeLa cells clearly demonstrates the potential of the functionalized nanocapsules to be successfully exploited as biocarriers.

17.6
Encapsulation of Markers and Detection of Nanoparticles in Biological Systems

Nanoparticles without a tracer molecule for example, a fluorescent dye, a radioactive marker or a contrast agent that is used for MRI can be detected in biological systems (cells or organisms) by TEM. This has been done for a variety of nanoparticles. In most cases the nanomaterials were detected inside of membrane surrounded intracellular compartments, most likely endosomes or lysosomes (Figure 17.9). Some authors report that they have found the studied nanoparticles in the cell nucleus [95] or in the cytoplasm. Here special attention should be paid to the fact that cellular membranes can be hard to visualize if no special sample preparation (e.g., high pressure freezing) is applied and therefore nanoparticles may seem to be located in the cytoplasm. As demonstrated in Figure 17.9 the nanoparticle material itself may determine if the nanoparticles are clustered as in the case of the amino-functionalized nanoparticles (see Figure 17.9e) or if they intracellular compartments as single nanoparticles (see Figure 17.9c,d). Particles can also be found extracellularly (see Figure 17.9b).

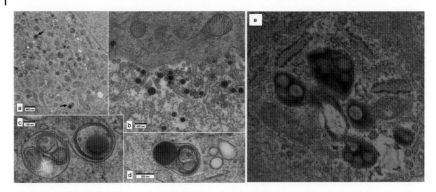

Figure 17.9 Single isoprene nanoparticles in HeLa cells were analyzed after 24 h incubation without added poly(L-lysine) (PLL). Particles were found as single nanoparticles enclosed in intracellular endosomes. (a) shows single particles (arrows) near the nucleus; in (b) particles are located outside of the cell in the microvilli area; (c) and (d) show cellular vesicles with single particles in more detail (preparation method: high-pressure freezing and cryosubstitution), (e) shows a TEM analysis of mesenchymal stem cells (MSC s) incubated for 24 h with amino-functionalized PS nanoparticles.

TEM analysis is excellent technique for the subcelluar studies and is unrivaled when it comes to determine where exactly inside of the cell the nanoparticles are located. The drawback here is that only a few cells are typically studied and that it is not suited for larger areas and whole organs.

However, for biomedical applications from cell labeling to drug delivery, it is important to follow the fate of the nanoparticles in the cells or whole organisms [96]. In some experiments, nanoparticles were labeled with radioactive isotopes (e.g., ^{14}C, ^{125}I). This was achieved through conventional emulsion polymerization [97] as the radio-labeled monomer is chemically identical with the nonradioactive compound, and therefore diffusion through the water phase is not changed.

17.6.1
Fluorescent Component as Marker

Fluorescence labeling of bioactive molecules or particles is more efficient because several dyes with different fluorescent colors can be applied. Such a labeling offers the possibility of tracking cells by fluorescent spectroscopic measurements or fluorescent microscopy in cell cultures and individual cells or even in whole (small) animals [98]. By using a fluorescent molecule as a reporter, the uptake of differentially designed nanoparticles, for example, can be investigated.

In principle, fluorescence labels can be attached to the nanoparticles via reactions on the surface or formation of the particles composed of a fluorescent polymer (e.g., FITC-dextran [99]) is possible. Physically adsorbed dye molecules on the particle's surface can easily desorb or alter the surface of the nanoparticle, which might also affect the response of biological systems to the nanoparticles.

Therefore, the integration of markers inside the particles without changing the particle's surface is desired. Here again the miniemulsion polymerization technique offers the unique possibility for direct introduction of the hydrophobic fluorescent dyes in a polymeric matrix.

The ability of different cell lines to internalize PS nanoparticles with varied densities of amino or carboxy surface-functionalization as obtained by the miniemulsion process was, for example, investigated quantitatively with fluorescence activated cell sorting (FACS) and qualitatively with confocal laser scanning microscopy [100]. Both techniques base on the reliable and uniform distribution of the fluorescent dye inside the nanoparticles [29]. Here, the hydrophobic fluorescent dye *N*-(2,6-diisopropylphenyl)perylene-3,4-dicarbonacidimide (PMI) was used. It could be shown that especially the highly amino-functionalized nanoparticles are favorably internalized in all of the investigated cell lines. As an example, amino-functionalized particles taken up by HeLa cells are presented in Figure 17.10b. Surface functionalization with carboxylic groups (see Figure 17.10a) also slightly enhances the particle uptake compared with plain, nonfunctionalized PS (PS) nanoparticles.

While FACS is not able to distinguish between nanoparticles that adhere to the cell's membrane and nanoparticles that are taken up into the cells, further characterization of the nanoparticle-cell interaction was done by confocal laser scanning microscopy. Here the cell membrane can be stained by a specific fluorescent dye RH414 (see Figure 17.11 pseudo-colored in red), while the nanoparticles are detected in another channel (pseudo-colored in green). As the optical section in a confocal microscope is roughly 500 nm, the intracellular localization can be proven hereby, when the nanoparticles are clearly located inside the plasma membrane.

Besides the above mentioned experiments using PS-based nanoparticles, PMI could be successfully incorporated in phosphonate-functionalized PMMA and PS

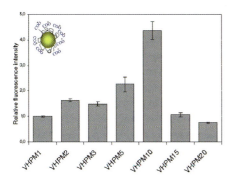
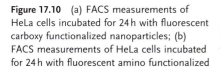
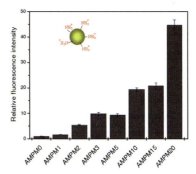

Figure 17.10 (a) FACS measurements of HeLa cells incubated for 24 h with fluorescent carboxy functionalized nanoparticles; (b) FACS measurements of HeLa cells incubated for 24 h with fluorescent amino functionalized nanoparticles [29]. The numbers in the particle name give the amount of functional comonomer (AA or AEMH) that was used for the synthesis. Each bar shows the average normalized fluorescence intensity (nFL).

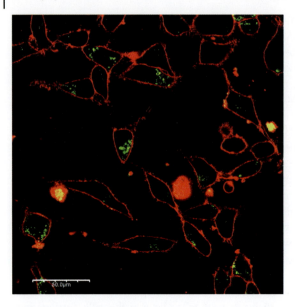

Figure 17.11 Confocal laser scanning microscopy with RH414 as membrane dye (pseudo- red) and the PMI containing nanoparticles (pseudo- green).

[31], polyisoprene (PI), PS-*co*-PI [101], polyester [65] and PBCA [102] particles for the investigation of cellular response to these polymeric nanoparticles. It could be shown, that the internalization in different cell lines depends on the cell line, the polymer and their surface functionalization.

PBCA nanoparticles, prepared with a miniemulsion process, were labeled *in situ* with a fluorescent dye and could also be coated with polysorbate 80. Using the fluorescent polysorbate 80-coated PBCA nanoparticles in an *in vivo* study, direct evidence was found for the presence of nanoparticles entering the brain and the retina of rats [102]. Bone marrow microvascular endothelial cells (BMEC) were used as *in vitro* model for the BBB. The cells showed significant uptake of the particles but no transcytosis was observed *in vitro*. After the application of the particles to the animals in two concentrations, cryosections of the brains and retinas were prepared. In the sections of the rat receiving the lower dose, co-localization of the applied fluorescent particles and the stained endothelial cells was detected in the brain and the retina (see Figure 17.12), indicating particle internalization in the endothelial cells. When applied in higher dosage, the particles could be detected within the brain and the retina, with few co-localized signals, suggesting a passage through the blood–brain and blood–retina barriers.

An alternative to the widely used organic dyes are fluorescent lanthanide-based nanocrystals [103] and semiconductor quantum dots (QDs). Encapsulation in polymeric matrices provides an excellent way to prevent the toxic QDs from direct interaction with the cells. CdS or CdSe quantum dots hydrophobized by trioctylphosphine oxide permits the direct dispersion of the QDs in the monomer and a

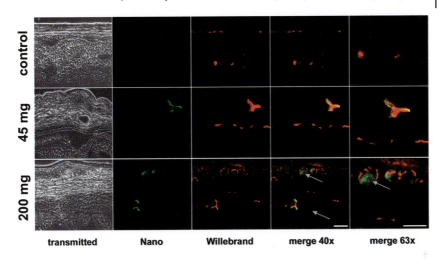

Figure 17.12 Cryosections of the rats' retinas. The left image shows the optical transmission image of the section (40×). Nano: fluorescence created by the PMI-labeled nanoparticles. Willebrand: endothelial cells stained with fluorescent antibody (v. Willebrand-factor primary and anti IgG secondary antibody with fluorescent label). Merge 40× and 63×: image merged from the green and red channel. The arrows indicate particles not co-localized with endothelial cells. A 40× magnification lens was used. The scale bars represent 100 μm [102].

subsequent miniemulsion polymerization procedure. Hybrid particles with PS [104–106] or PBA particles [105, 107] could be prepared, which show good photoluminescent properties. Also CdTe, stabilized by 3-mercaptopropionic acid could be homogenously incorporated in PS nanoparticles using octadecyl-*p*-vinylbenzyldimethylammonium chloride or didecyl-*p*-vinylbenzylmethylammonium chloride as the phase transfer agent [108].

17.6.2
Magnetite as Marker

Polymeric magnetite containing nanoparticles are of great interest for, for example, MRI, hypothermia, and magnetically controlled delivery of anticancer drugs. The advantages of coating magnetic particles with a polymeric layer are the prevention of inorganic magnetic materials from direct contact with the living system, therefore reducing their toxicity, and the ability to functionalize the surface of polymer particles with the biomolecules, which are responsible for possible specific targeting of nanoparticles.

The uniform incorporation of larger amounts of magnetite into polymeric particles is the most crucial point in order to obtain a strong and uniform response from the magnetic hybrid nanoparticles. Based on a co-sonication process, a three-step process was developed for the generation of hydrophobized aggregates of primary magnetite nanoparticles and subsequent encapsulation in monomer droplets [109, 110]. After polymerization, magnetic PS nanoparticles could be obtained. Using this process more than 40 wt% of magnetite could efficiently be encapsulated in PS. By adding comonomers to styrene, surface functionalized magnetic PS nanoparticles could be obtained [110].

For *in vivo* applications, besides being noncarcinogenic, nonmutagenic, and nonallergenic, the nanoparticles should also show no or minimum immunological response, as well as possess the ability to carry numerous therapeutics at sufficient loading concentration. Additionally it is very important that the polymeric mate-rial, which remains in the living organism for an extended period of time after introduction, should be biodegradable and the products of degradation have also to meet the above-mentioned criteria. Here polyester-based particles are of interest which can be obtained by different techniques. For example, magnetic poly(D,L-lactide-coglycolide) (PLGA) [111–113] as well as poly(ε-caprolactone) (PCL) [114] nanoparticles in the size range between 100 and 280 nm were obtained by emul-sification/diffusion technique. PLGA particles loaded up to 13.5 wt.% with mag-netite/maghemite [115] and PCL nanoparticles containing 25-30 wt.% of magnetite [116] as well as an anticancer drug [117] were successfully prepared using this approach. Composite particles were produced from preformed poly(D,L-lactide) polymer [118]. However, usually the particle size obtained by the double-emulsion technique ranges from 100 nm to 2 μm with a broad size distribution. Moreover, the polymeric shell is usually heterogeneous or the magnetite is not completely covered by the polymer. Here the miniemulsion process offers many advantages: narrowly distributed hybrid nanoparticles with homogeneously distributed iron oxide nanoparticles can be obtained. Recently, biodegradable and fluorescent poly-lactide nanospheres loaded with iron oxide particles of different sizes (either 10 or 25 nm) were successfully prepared by the combination of miniemulsion and emulsion/solvent evaporation techniques (see Figure 17.13) [119]. The additional incorporation of a hydrophobic fluorescent dye gives the possibility of employing

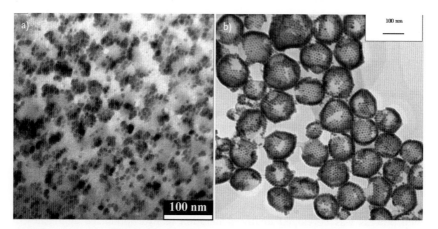

Figure 17.13 (a) PS and (b) polylactide nanoparticles with encapsulated (10 nm) magnetite particles: (a) 40% and (b) 50%.

the obtained composite particles as model markers in order to study, for example, cellular uptake mechanism. So-called dual marker particles are functionalized with fluorescent dye and magnetite simultaneously. The fluorescent dye allows the investigation of cells under the fluorescence or confocal laser scanning microscopy [110, 120].

17.7
Release of Materials

Much effort has been focused on the development of well-defined carriers loaded with catalysts, enzymes or drugs, which have tailored physico-chemical and mechanical properties to release the material selectively at the desired target. The regulation of the release is sufficient and useful to catalyze chemical or biological reactions in certain areas. Therefore, it is necessary to activate its content or release the encapsulated material remotely.

Generally, the release of the loaded therapeutics, drugs, etc. from the biodegradable particles is achieved by degradation of the material. The release profiles can be adjusted through the changes in (co)polymer composition, (co)polymer molecular weight and particle size as the amount of delivered material depends on the total surface area of the carrier [121]. The influence of different processing methods, material properties used as a polymeric matrix and "payload" on the encapsulation efficiency and release kinetics are summarized in the review of Park *et al.* [122]. Rothstein and Little [123] present empirical and mathematical methods to control the release behavior of a broad range of active agents from polyester and polyanhydride matrices.

Typical release mechanisms that are based on a triggered systems usually involve change in pH [124, 125], laser light [126–129], ultrasound [130–132], temperature [133], or enzymatic degradation [134].

For some applications it is important that the nanoparticles/nanocapsules keep the encapsulated materials forever, as for example, in the case of pigments for coatings. However, in many cases, release of the encapsulated material is of interest, either for a long time or rapidly at an instant. Here, different approaches must be applied to fulfill the requirements of release kinetics.

It was shown that for diffusion processes, the release behavior of a copolymeric acrylic matrix with different amounts of encapsulated volatile fragrance is significantly decelerated compared with the pure fragrance miniemulsion. For temperatures significantly lower than the glass transition temperature of the polymer $(T < T_g)$, almost no release can be observed [135]. The increase of the temperature to temperatures around and well above the polymer glas transition temperature $(T > T_g)$ results in an slightly or significantly accelerated release of the fragrance out of the nanoparticles, respectively. Thereby, the release behavior of the volatile compound can be easily tuned by adjusting the release temperature in comparison with the T_g of the polymeric shell. For nanocapsules with photoinitiator, encapsulated in different polymeric shells, it was shown that the capsules are permeable and thus release the initiator in a sufficiently high concentration in the surrounding (monomer) phase to start the polymerization even after a longer time [136]. The release rates directly depend on the type of the polymeric shell and the employed redispersing agent. The studies of the kinetics show the great potential of these capsules for application in polymeric dental filling materials as initiator depots in order to guarantee the required storage time.

A controlled, but sudden, release of the encapsulated material could be achieved by using "switches", which are sensitive to light, temperature, etc. and are encapsulated inside the capsule or (molecularly) incorporated in the shell material. One possibility of obtaining such "switchable" capsules is to embed intact azo-components via radical polymerization in miniemulsion droplets at low reaction temperatures [137]. In spite of the thermal initiation of the polymerization process with the first azo-component decomposing at low temperatures and therefore acting as initiator for the polymerization, the embedded second azo-component maintains its properties so that it can be detonated at a later time at higher temperatures. The nanoexplosion temperature has to be chosen below the T_g of the polymer so that the nitrogen gas, which is developed during the thermal treatment of the capsules (caused by the decomposition of the encapsulated azo-component), builds up an overpressure inside the capsules and effects a blow-out through which the polymeric shell is damaged (see Figure 17.14). This concept encourages a sudden release of materials which are also encapsulated in the nanocapsules. The concept can easily transferred to other "explosives", for example, redox initiators that function at lower temperatures, which liberates a gas inside polymeric nanocapsules during decomposition.

Polyurethane nanocapsules consisting of an aqueous core and a polymeric shell with included azo bonds as obtained via interfacial polyaddition of the monomers toluene-2,4-diisocyanate (TDI) and an azo-containing diol (2,2′-azobis{2-[1-(2-hydroxyethyl)-2-imidazolin-2-yl]propane}dihydrochloride, VA-060) in inverse miniemulsion permit the selective release of encapsulated material by stimuli such as temperature, UV light, or pH change [138]. The capsule degradation, after exposure to the different stimuli, was detected by time-dependent measurement of the fluorescence intensities of the dye sulforhodamine SR101, which is dissolved inside the aqueous capsule's core. The results present proof-of-principle studies of different controlled releases. Depending on the type of stimulus, the release of the fluorescent dye occurs in minutes (mercury arc lamp), in hours (temperature treatment), or in days (pH-induced treatment).

17.8
Conclusion

The miniemulsion process permits the formation of nanosized droplets, which can be functionalized by copolymerization with functional comonomers. Different hydrophobic and hydrophilic liquids and solids can be encapsulated inside the particles or surrounded by a polymeric shell in order to create the nanocapsules as markers for cells and as carrier of drugs, therapeutics with a half-life extension,

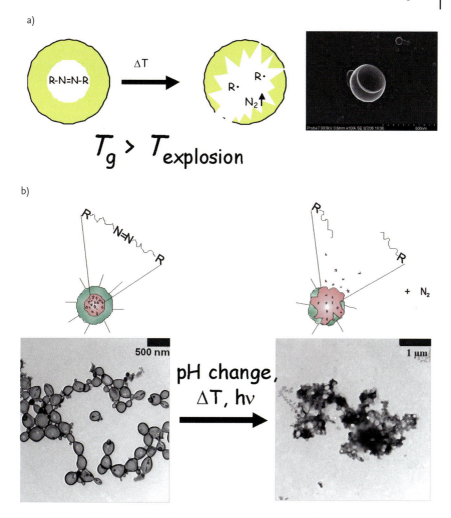

Figure 17.14 (a) Nanocapsules with encapsulated azo-compound after nanoexplosion at elevated temperature [137]; (b) nanocapsules with integrated azo-compound in the shell which can be cleaved by pH change, ΔT or $h\nu$. Reprinted with permission from [138]. Copyright (2010) American Chemical Society.

and for defined delivery. The release of the substances can be tuned in a defined way from the interior of the nanocapsules. Such complex nanoparticles and nanocapsules are ideal candidates for biomedical applications.

Acknowledgment

The Deutsche Forschungsgemeinschaft (DFG, SPP1313) and the Baden-Württemberg Stiftung is greatfully acknowledged for financial support.

References

1 Gilmore, J.L., Yi, X., Quan, L., and Kabanov, A.V. (2008) *J. Neuroimmune Pharmacol.*, **3**, 83.

2 Pisal, D.S., Kosloski, M.P., and Balu-Iyer, S.V. (2010) *J. Pharm. Sci.*, **99**, 2557–2575.

3 Musacchio, T., and Torchilin, V.P. (2011) *Front. Biosci.*, **1**, 1388–1412.

4 Zhang, S., and Uludag, H. (2009) *Pharm. Res.*, **26**, 1561–1580.

5 Top, A., and Kiick, K.L. (2010) *Adv. Drug Deliv. Rev.*, **62**, 1530–1540.

6 Tan, M.L., Choong, P.F., and Dass, C.R. (2010) *Peptides*, **31**, 184–193.

7 Moon, J.J., Suh, H., Bershteyn, A., Stephan, M.T., Liu, H., Huang, B., Sohail, M., Luo, S., Um, S.H., Khant, H., Goodwin, J.T., Ramos, J., Chiu, W., and Irvine, D.J. (2011) *Nat. Mater.*, **10**, 243–251.

8 Moghimi, S.M., Hunter, A.C., and Murray, J.C. (2005) *FASEB J.*, **19**, 311.

9 Allemann, E., Leroux, J.C., and Gurny, R. (1998) *Adv. Drug Deliv. Rev.*, **34**, 171.

10 Panyam, J., and Labhasetwar, V. (2003) *Adv. Drug Deliv. Rev.*, **55**, 329.

11 Matuszewski, L., Persigehl, T., Wall, A., Sehwindt, W., Tombach, B., Fobker, M., Poremba, C., Ebert, W., Heindel, W., and Bremer, C. (2005) *Radiology*, **235**, 155.

12 Gupta, A.K., and Curtis, A.S.G. (2004) *J. Mater. Sci. Mater. Med.*, **15**, 493.

13 Shenoy, D.B., Chawla, J.S., and Amiji, M.M. (2005) *Mater. Res. Soc. Symp. Proc.*, **845**, 369–373.

14 Cegnar, M., Premzl, A., Zavašnik-Bergant, V., Kristl, J., and Kos, J. (2004) *Exp. Cell Res.*, **301**, 223.

15 Delie, F., and Blanco-Príeto, M.J. (2005) *Molecules*, **10**, 65.

16 Davda, J., and Labhasetwar, V. (2002) *Int. J. Pharm.*, **233**, 51.

17 Win, K.Y., and Feng, S.-S. (2005) *Biomaterials*, **26**, 2713.

18 Desai, M.P., Labhasetwar, V., Walter, E., Levy, R.J., and Amidon, G.L. (1997) *Pharm. Res.*, **14**, 1568.

19 Zauner, W., Farrow, N.A., and Haines, A.M.R. (2001) *J. Control. Release*, **71**, 39.

20 Cortez, C., Tomaskovic-Crook, E., Johnston, A.P.R., Scott, A.M., Nice, E.C., Heath, J.K., and Caruso, F. (2007) *ACS Nano*, **1**, 93–102.

21 Pietzonka, P., Rothen-Rutishauser, B., Langguth, P., Wunderli-Allenspach, H., Walter, E., and Merkle, H. (2002) *Pharm. Res.*, **19**, 595.

22 Eldridge, J.H., Hammond, C.J., Meulbroek, J.A., Staas, J.K., Gilley, R.M., and Tice, T.R. (1990) *J Control Release*, **11**, 205.

23 Champion, J.A., and Mitragotri, S. (2006) *Proc. Natl. Acad. Sci. U. S. A.*, **103**, 4930–4934.

24 Decuzzi, P., Godin, B., Tanaka, T., Lee, S.-Y., Chiappini, C., Liu, X., and Ferrari, M. (2010) *J. Control. Release*, **141**, 320–327.

25 Gratton, S.E.A., Ropp, P.A., Pohlhaus, P.D., Luft, J.C., Madden, V.J., Napier, M.E., and DeSimone, J.M. (2008) *Proc. Natl. Acad. Sci. U.S.A.*, **105**, 11613–11618.

26 Landfester, K. (2009) *Angew. Chem. Int. Ed.*, **48**, 4488–4507.

27 Landfester, K. (2006) *Annu. Rev. Mater. Res.*, **36**, 231.

28 Musyanovych, A., Rossmanith, R., Tontsch, C., and Landfester, K. (2007) *Langmuir*, **23**, 5367–5376.

29 Holzapfel, V., Musyanovych, A., Landfester, K., Lorenz, M.R., and Mailänder, V. (2005) *Macromol. Chem. Phys.*, **206**, 2440.

30 Willert, M., and Landfester, K. (2002) *Macromol. Chem. Phys.*, **203**, 825–836.

31 Ziegler, A., Landfester, K., and Musyanovych, A. (2009) *Colloid Polym. Sci.*, **287**, 1261–1271.

32 Asua, J.M. (2002) *Prog. Polym. Sci.*, **27**, 1283.

33 Schork, F.J., Luo, Y.W., Smulders, W., Russum, J.P., Butte, A., and Fontenot, K. (2005) *Adv. Polym. Sci.*, **175**, 129.

34 Ethirajan, A., Ziener, U., and Landfester, K. (2009) *Chem. Mater.*, **21**, 2218–2225.

35 Zeller, A., Musyanovych, A., Kappl, M., Ethirajan, A., Dass, M., Marova, D.,

Klapper, M., and Landfester, K. (2010) *ACS Appl. Mater. Interfaces*, **2**, 2421–2428.

36 Ethirajan, A., and Landfester, K. (2010) *Chemistry*, **16**, 9398 – 9412.

37 Dausend, J., Musyanovych, A., Dass, M., Walther, P., Schrezenmeier, H., Landfester, K., and Mailänder, V. (2008) *Macromol. Biosci.*, **8** (12), 1135–1143.

38 Pratten, M.K., and Lloyd, J.B. (1986) *Biochim. Biophys. Acta*, **881**, 307–313.

39 Schulze, E., Ferrucci, J.T. Jr, Poss, K., Lapointe, L., Bogdanova, A., and Weissleder, R. (1995) *Invest. Radiol.*, **30**, 604–610.

40 Arbab, A.S., Bashaw, L.A., Miller, B.R., Jordan, E.K., Lewis, B.K., Kalish, H., et al. (2003) *Radiology*, **229**, 838–846.

41 Gershon, H., Ghirlando, R., Guttman, S.B., and Minsky, A. (1993) *Biochemistry*, **32**, 7143–7151.

42 Rolland, A., Bourel, D., Genetet, B., and Le Verge, R. (1987) *Int. J. Pharm.*, **39**, 173–180.

43 Bourel, D., Rolland, A., Le Verge, R., and Genetet, B. (1988) *J. Immunol. Methods*, **106**, 161–167.

44 Akasaka, Y., Ueda, H., Takayama, K., Machida, Y., and Nagai, T. (1988) *Drug Des. Deliv.*, **3**, 85–97.

45 Nobs, L., Buchegger, F., Gurny, R., and Allémann, E. (2006) *Bioconjug. Chem.*, **17**, 139–145.

46 Owens, D.E. 3rd, and Peppas, N.A. (2006) *Int. J. Pharm.*, **307**, 93–102.

47 Messerschmidt, S.K.E., Musyanovych, A., Altvater, M., Scheurich, P., Pfizenmaier, K., Landfester, K., and Kontermann, R.E. (2009) *J. Control. Release*, **137**, 69–77.

48 Crespy, D., and Landfester, K. (2005) *Macromolecules*, **38**, 6882.

49 Weiss, C.K., Ziener, U., and Landfester, K. (2007) *Macromolecules*, **40**, 928–938.

50 Cauvin, S., and Ganachaud, F. (2004) *Macromol. Symp.*, **215**, 179.

51 Cauvin, S., Ganachaud, F., Moreau, M., and Hemery, P. (2005) *Chem Commun*, **21**, 2713.

52 Soula, R., Saillard, B., Spitz, R., Claverie, J., Llaurro, M.F., and Monnet, C. (2002) *Macromolecules*, **35**, 1513.

53 Wehrmann, P., Zuideveld, M., Thomann, R., and Mecking, S. (2006) *Macromolecules*, **39**, 5995.

54 Held, A., Kolb, I., Zuideveld, M.A., Thomann, R., Mecking, S., Schmid, M., Pietruschka, R., Lindner, E., Khanfar, M., and Sunjuk, M. (2002) *Macromolecules*, **35**, 3342.

55 Quémener, D., Héroguez, V., and Gnanou, Y. (2005) *Macromolecules*, **38**, 7977.

56 Quémener, D., Héroguez, V., and Gnanou, Y. (2006) *J. Polym. Sci. A*, **44**, 2784.

57 Pecher, J., and Mecking, S. (2007) *Macromolecules*, **40**, 7733.

58 Landfester, K., Tiarks, F., Hentze, H.-P., and Antonietti, M. (2000) *Macromol. Chem. Phys.*, **201**, 1.

59 Tiarks, F., Landfester, K., and Antonietti, M. (2001) *J Polym Sci [A1]*, **39**, 2520.

60 Barrere, M., and Landfester, K. (2003) *Macromolecules*, **36**, 5119.

61 Barrere, M., and Landfester, K. (2003) *Polymer*, **44**, 2833.

62 Taden, A., Antonietti, M., and Landfester, K. (2003) *Macromol. Rapid Comm.*, **24**, 512.

63 Marie, E., Rothe, R., Antonietti, M., and Landfester, K. (2003) *Macromolecules*, **36**, 3967.

64 Bhadra, S., Singha, N.K., and Khastgir, D. (2006) *Synthetic Metals*, **156**, 1148.

65 Musyanovych, A., Schmitz-Wienke, J., Mailaender, V., Walther, P., and Landfester, K. (2008) *Macromol. Biosci.*, **8**, 127–139.

66 Hildebrand, G.E., and Tack, J.W. (2000) *Int. J. Pharm.*, **196**, 173.

67 Bilati, U., Allémann, E., and Doelker, E. (2005) *AAPS Pharm. Sci. Technol.*, **6**, 594–604.

68 Conway, B.R., and Oya Alpar, H. (1996) *Eur. J. Pharm. Biopharm.*, **42**, 42.

69 Quintanar-Guerrero, D., Allémann, E., Doelker, E., and Fessi, H. (1998) *Pharm. Res.*, **15**, 1056.

70 Song, C.X., Labhasetwar, V., Murphy, H., Qu, X., Humphrey, W.R., Shebuski, R.J., and Levy, R.J. (1997) *J. Control. Release*, **43**, 197.

71 Hariharan, S., Bhardwaj, V., Bala, I., Sitterberg, J., Bakowsky, U., and Ravi

Kumar, M.N.V. (2006) *Pharm. Res.*, **23**, 184.

72 Allemann, E., Leroux, J.C., Gurny, R., and Doelker, E. (1993) *Pharm. Res.*, **10**, 1732.

73 Johnston, A.P.R., Cortez, C., Angelatos, A.S., and Caruso, F. (2006) *Curr. Opin. Colloid Interfacce Sci.*, **11**, 203.

74 Caruso, F. (2001) *Adv. Mater.*, **13**, 11.

75 Caruso, F., Spasova, M., Saigueirino-Maceira, V., and Liz-Marzan, L.M. (2001) *Adv. Mater.*, **13**, 1090.

76 Kato, N., Schuetz, P., and Caruso, F. (2002) *Macromolecules*, **35**, 9780.

77 Tiarks, F., Landfester, K., and Antonietti, M. (2001) *Langmuir*, **17**, 908.

78 Luo, Y.W., and Gu, H.Y. (2007) *Polymer*, **48**, 3262.

79 Arshady, R. (1989) *J. Microencapsul.*, **6**, 13.

80 Danicher, L., Frere, Y., and Le Calve, A. (2000) *Macromol. Symp.*, **151**, 387.

81 Torini, L., Argillier, J.F., and Zydowicz, N. (2005) *Macromolecules*, **38**, 3225.

82 Scott, C., Wu, D., Ho, C.-C., and Co, C.C. (2005) *J. Am. Chem. Soc.*, **127**, 4160.

83 Sarkar, D., El-Khoury, J., Lopina, S.T., and Hu, J. (2005) *Macromolecules*, **38**, 8603.

84 Vauthier, C., Dubernet, C., Fattal, E., Pinto-Alphandary, H., and Couvreur, P. (2003) *Adv. Drug Deliv. Rev.*, **55**, 519.

85 Marie, E., Landfester, K., and Antonietti, M. (2002) *Biomacromolecules*, **3**, 475.

86 Crespy, D., Stark, M., Hoffmann-Richter, C., Ziener, U., and Landfester, K. (2007) *Macromolecules*, **40**, 3122.

87 Jagielski, N., Sharma, S., Hombach, V., Mailänder, V., Rasche, V., and Landfester, K. (2007) *Macromol. Chem. Phys.*, **208**, 2229.

88 Musyanovych, A., and Landfester, K. (2008) *Prog. Colloid Polym. Sci.*, **134**, 120.

89 Paiphansiri, U., Tangboriboonrat, P., and Landfester, K. (2006) *Macromol. Biosci.*, **6**, 33.

90 Rosenbauer, E.-M., Landfester, K., and Musyanovych, A. (2009) *Langmuir*, **25**, 12084–12091.

91 Baier, B., Musyanovych, A., Dass, M., Theisinger, S., and Landfester, K. (2010) *Biomacromolecules*, **11**, 960–968.

92 Wu, D., Scott, C., Ho, C.C., and Co, C.C. (2006) *Macromolecules*, **39**, 5848.

93 De Faria, T.J., De Campos, A.M., and Senna, E.L. (2005) *Macromol. Symp.*, **229**, 228.

94 Paiphansiri, U., Dausend, J., Musyanovych, A., Mailänder, V., and Landfester, K. (2009) *Macromol. Biosci.*, **9**, 575–584.

95 Akita, H., Kudo, A., Minoura, A., Yamaguti, M., Khalil, I.A., Moriguchi, R., Masuda, T., Danev, R., Nagayama, K., Kogure, K., and Harashima, H. (2009) *Biomaterials*, **30**, 2940–2949.

96 Mailänder, V., and Landfester, K. (2009) *Biomacromolecules*, **10**, 2379–2400.

97 Singer, J.M., Adlersberg, L., Hoenig, E.M., Ende, E., and Tchorsch, Y. (1969) *J. Reticuloendothel. Soc.*, **6**, 561–589.

98 Stelter, L., Pinkernelle, J.G., Michel, R., Schwartländer, R., Raschzok, N., Morgul, M.H., Koch, M., Denecke, T., Ruf, J., Bäumler, H., Jordan, A., Hamm, B., Sauer, I.M., and Teichgräber, U. (2010) *Mol. Imaging Biol.* **12**, 25–34.

99 Alyautdin, R.N., Reichel, A., Löbenberg, R., Ramge, P., Kreuter, J., and Begley, D.J. (2001) *J. Drug Target.*, **9**, 209–221.

100 Musyanovych, A., Landfester, K., and Mailänder, V. (2009) Polymeric nanoparticles as carrier systems: how does the material and surface charge affect cellular uptake?, in *Clinical Chemistry Research* (eds B.H. Mitchem and C.L. Sharnham), Nova Science Publishers, Inc., pp. 195–220. Chapter VI.

101 Lorenz, M.R., Kohnle, M.V., Dass, M., Walther, P., Höcherl, A., Ziener, U., Landfester, K., and Mailander, V. (2008) *Macromol. Biosci.*, **8**, 711–727.

102 Weiss, C.K., Kohnle, M.-V., Landfester, K., Hauk, T., Fischer, D., Schmitz-Wienke, J., and Mailaender, V. (2008) *ChemMedChem*, **3**, 1395–1403.

103 Toepfer, O., and Schmidt-Naake, G. (2007) *Macromol. Symp.*, **248**, 239–248.

104 Fleischhaker, F., and Zentel, R. (2005) *Chem. Mater.*, **17**, 1346–1351.

105 Esteves, A.C.C., Barros-Timmons, A., Monteiro, T., and Trindade, T. (2005) *J. Nanosci. Nanotechnol.*, **5**, 766–771.

106 Joumaa, N., Lansalot, M., Theretz, A., and Elaissari, A. (2006) *Langmuir*, **22**, 1810–1816.

107 Peres, M., Costa, L.C., Neves, A., Soares, M.J., Monteiro, T., Esteves, A.C., Barros-Timmons, A., Trindade, T., Kholkin, A., and Alves, E. (2005) *Nanotechnology*, **16** (9), 1969–1973.

108 Yang, Y., Wen, Z., Dong, Y., and Gao, M. (2006) *Small*, **2**, 898–901.

109 Ramirez, L.P., and Landfester, K. (2003) *Macromol. Chem. Phys.*, **204**, 22–31.

110 Holzapfel, V., Lorenz, M.R., Weiss, C.K., Schrezenmeier, H., Landfester, K., and Mailänder, V. (2006) *J. Phys. Condens. Matter*, **18**, 2581–2594.

111 Lee, S.J., Jeong, J.R., Shin, S.C., Kim, J.C., Chang, Y.H., Chang, Y.M., and Kim, J.D. (2004) *J. Magn. Magn. Mater.*, **272–276**, 2432.

112 Lee, S.J., Jeong, J.R., Shin, S.C., Kim, J.C., Chang, Y.H., Lee, K.H., and Kim, J.D. (2005) *Colloids Surf. A*, **255**, 19.

113 Wassel, R.A., Grady, B., Kopke, R.D., and Dormer, K.J. (2007) *Colloids Surf. A*, **292**, 125.

114 Gang, J., Park, S.B., Hyung, W., Choi, E.H., Wen, J., Kim, H.S., Shul, Y.G., Haam, S., and Song, S.Y. (2007) *J. Drug Target.*, **15**, 445.

115 Ngaboni Okassa, L., Marchais, H., Douziech-Eyrolles, L., Herve, K., Cohen-Jonathan, S., Munnier, E., Souce, M., Linassier, C., Dubois, P., and Chourpa, I. (2007) *Eur. J. Pharm. Biopharm.*, **67**, 31.

116 Hamoudeh, M., and Fessi, H.J. (2006) *Colloid Interface Sci*, **300**, 584.

117 Yang, J., Park, S.B., Yoon, H.G., Huh, Y.M., and Haam, S. (2006) *Int. J. Pharm.*, **324**, 185.

118 Gomez-Lopera, S.A., Plaza, R.C., and Delgado, A.V. (2001) *J. Colloid Interface Sci.*, **240**, 40.

119 Urban, M., Musyanovych, A., and Landfester, K. (2009) *Macromol Chem Phys*, **210**, 961–970.

120 Mailander, V., Lorenz, M.R., Holzapfel, V., Musyanovych, A., Fuchs, K., Wiesneth, M., Walther, P., Landfester, K., and Schrezenmeier, H. (2008) *Mol. Imaging Biol*, **10**, 138–146.

121 Berkland, C., Kim, K.K., and Pack, D.W. (2001) *J. Control. Release*, **73** (1), 59–74.

122 Ye, M., Kim, S., and Park, K. (2010) *J. Control. Release*, **146** (2), 241–260.

123 Rothstein, S.N., and Little, S.R. (2010) *J. Mater. Chem.*, **21**, 29–39.

124 Dejugnat, C., and Sukhorukov, G.B. (2004) *Langmuir*, **20**, 7265–7269.

125 Mauser, T., Dejugnat, C., and Sukhorukov, G.B. (2004) *Macromol. Rapid Commun.*, **25**, 1781–1785.

126 Skirtach, A.G., Javier, A.M., Kreft, O., Kohler, K., Alberola, A.P., Möhwald, H., Parak, W.J., and Sukhorukov, G.B. (2006) *Angew. Chem. Int. Ed.*, **45**, 4612–4617.

127 Skirtach, A.G., Antipov, A.A., Shchukin, D.G., and Sukhorukov, G.B. (2004) *Langmuir*, **20**, 6988–6992.

128 Radt, B., Smith, T.A., and Caruso, F. (2004) *Adv. Mater.*, **16**, 2184.

129 Angelatos, A.S., Radt, B., and Caruso, F. (2005) *J. Phys. Chem. B*, **109**, 3071–3076.

130 Skirtach, A.G., De Geest, B.G., Mamedov, A., Antipov, A.A., Kotov, N.A., and Sukhorukov, G.B. (2007) *J. Mater. Chem.*, **17**, 1050–1054.

131 Postema, M., Bouakaz, A., and de Jong, N. (2005) *IEEE Trans. Ultrason. Ferroelectr. Freq. Control*, **52**, 1035–1041.

132 Lentacker, I., De Geest, B.G., Vandenbroucke, R.E., Peeters, L., Demeester, J., De Smedt, S.C., and Sanders, N.N. (2006) *Langmuir*, **22**, 7273–7278.

133 Wang, A., Tao, C., Cui, Y., Duan, L., Yang, Y., and Li, J. (2009) *J. Colloid Interface Sci.*, **332** (2), 271–279.

134 De Geest, B.G., Vandenbroucke, R.E., Guenther, A.M., Sukhorukov, G.B., Hennink, W.E., Sanders, N.N., Demeester, J., and De Smedt, S.C. (2006) *Adv. Mater.*, **18**, 1005.

135 Theisinger, S., Schöller, K., Osborn, B., Sarkar, M., and Landfester, K. (2009) *Macromol. Chem. Phys.*, **210**, 411–420.

136 Volz, M., Ziener, U., Salz, U., Zimmermann, J., and Landfester, K. (2007) *Colloid Polym. Sci.*, **285**, 687–692.

137 Volz, M., Walther, P., Ziener, U., and Landfester, K. (2007) *Macromol. Mater. Eng.*, **292**, 1237–1244.

138 Rosenbauer, E.-M., Wagner, M., Musyanovych, A., and Landfester, K. (2010) *Macromolecules*, **43**, 5083–5093.

Index

Page numbers in *italics* refer to information in figures or tables.

Therapeutic Proteins: Strategies to Modulate Their Plasma Half-Lives, First Edition. Edited by
Roland Kontermann.
© 2012 Wiley-VCH Verlag GmbH & Co. KGaA. Published 2012 by Wiley-VCH Verlag GmbH & Co. KGaA.